CAMBRIDGE MONOGRAPHS
ON MECHANICS AND APPLIED MATHEMATICS

GENERAL EDITORS

G. K. BATCHELOR, PH.D., F.R.S.

Professor of Applied Mathematics in the University of Cambridge

J. W. MILES, PH.D.

Professor of Applied Mechanics and Geophysics, University of California, La Jolla

THE FLUID MECHANICS OF LARGE BLOOD VESSELS

T0269341

THE FLUID MECHANICS
OF LARGE
BLOOD VESSELS

T. J. PEDLEY, M.A., PH.D.

Lecturer in Department of Applied Mathematics and Theoretical Physics
University of Cambridge and Fellow of Gonville and Caius College

CAMBRIDGE UNIVERSITY PRESS

CAMBRIDGE

LONDON   NEW YORK   NEW ROCHELLE

MELBOURNE   SYDNEY

CAMBRIDGE UNIVERSITY PRESS
Cambridge, New York, Melbourne, Madrid, Cape Town, Singapore, São Paulo, Delhi

Cambridge University Press
The Edinburgh Building, Cambridge CB2 8RU, UK

Published in the United States of America by Cambridge University Press, New York

www.cambridge.org
Information on this title: www.cambridge.org/9780521226264

First published 1980
This digitally printed version 2008

A catalogue record for this publication is available from the British Library

Library of Congress Cataloguing in Publication data
Pedley, T. J.
The fluid mechanics of large blood vessels
(Cambridge monographs on mechanics and applied mathematics)
Includes index.
1. Hemodynamics – Mathematical models. 2. Fluid mechanics.
3. Blood-vessels. I. Title.
QP105.P4   591.1'1'601531   78-73814

ISBN 978-0-521-22626-4 hardback
ISBN 978-0-521-08956-2 paperback

FOR AVRIL

CONTENTS

PREFACE

Some knowledge of fluid mechanics is required before the circulation of the blood can be understood. Indeed, the single fact that above all others convinced William Harvey (1578–1657) that the blood does circulate was the presence in the veins of valves, whose function is a passive, fluid mechanical process. He saw that these could be effective only if the blood in the veins flowed towards the heart, not away from it as proposed by Galen (129–199) and believed by the European medical establishment until Harvey's time. Harvey was also the first to make a quantitative estimate of the output of blood from the human heart and this, although a gross underestimate (36 oz, i.e. about 1 litre, per minute instead of about 5 litres per minute) was largely responsible for convincing the sceptics that the arterial blood could not be continuously created in the liver and, hence, that it must circulate.

The earliest quantitative measurements of mechanical phenomena in the circulation were made by Stephen Hales (1677–1761) who measured arterial and venous blood pressure, the volume of individual chambers of the heart and the rate of outflow of blood from severed veins and arteries, thereby demonstrating that most of the resistance to blood flow arises in the microcirculation. He also realised that the elasticity of the arteries was responsible for blood flow in veins being more or less steady, not pulsatile as in arteries.

Later in the eighteenth and in the nineteenth centuries, a number of well-known fluid dynamicists interested themselves in the circulation of the blood and made significant contributions to our understanding of it. These included Euler, Daniel Bernoulli (actually a professor of anatomy) and Poiseuille (also a physician; this last name in particular makes it clear how an attempt to solve real, applied problems may often lead to important developments in

fundamental science). One of the greatest polymaths was Thomas
Young (1773–1829), another physician whose research in optics led
to both the acceptance of the wave theory of light and an understanding of the perception of colour. The other important area of his
research concerned the nature of elasticity, in particular the
properties and function of elastic arteries; his theory of wave
propagation in elastic tubes is still recognised as a fundamentally
correct description of the pressure pulse in arteries. It was in his
lecture to the Royal Society of London on this subject (Young,
1809) that he explicitly recognised that 'the inquiry, in what
manner, and in what degree, the circulation of the blood depends on
the muscular and elastic powers of the heart and of the arteries,
supposing the nature of those powers to be known, must become
simply a question belonging to the most refined departments of the
theory of hydraulics'.

It is with some of the more refined departments of the theory of
hydraulics, as applied to the circulatory systems of mammals, that
this book is primarily concerned. The object of the research is the
same as Young's, to understand the physical events that take place
in the normal animal, and thereby make a contribution to physiology. A secondary aim is to make a contribution to medicine, by
analysing particular abnormal or diseased states in the hope of
improving their diagnosis or treatment. In either case it is the
biological (or medical) end that is important, not the mathematical
means that are used to achieve it. The applied mathematics should
be firmly linked to experiment, using, explaining and predicting
experimental results. For this reason considerable space is devoted
to a description of the circulatory system, and, where our fluid
mechanical understanding has come from in-vivo or model
experiments, the experiments are fully described. Furthermore, the
mathematical description of a phenomenon may be complicated,
but if only simple mathematics is needed or available, only simple
mathematics is used. On the other hand, when the phenomenon is
so complex that our understanding is still limited, the only way
forward may be a lengthy analysis of an idealised model of the real
system, with the experimental link as yet absent.

A number of books on the mechanics of the circulation already
exist, but these consist *either* of works aimed principally at a

biological or medical audience *or* of conference proceedings or collections of invited articles. Of the former, some are too simplified for the student of mechanics, and those that are not naturally try to omit the mathematical development of the theory, describing its results in physical terms. (I am a co-author of one such book (Caro *et al.*, 1978); another important one is by McDonald (1960, 1974).) The collections of articles (of which the best is that edited by Bergel (1972*a*)) often do contain mathematical development of the theory, but it is presented in a piecemeal way. In view of the fact that an increasing number of engineers and applied mathematicians are coming into the field of cardiovascular fluid mechanics, the time seems ripe for a research monograph that concentrates on the mathematical analysis, and this book tries to provide it. The contents include some areas of the subject that are already very well understood, such as the propagation of the pressure pulse (chapter 2), but most of the book concerns relatively novel areas where research is still very much in progress. These include the prediction of flow patterns and, in particular, wall shear stresses in arteries (chapters 3–5), which are important both because of their postulated significance in the genesis of arterial disease (§ 1.2) and because they cannot yet be measured. The other major area of work in progress concerns flow in vessels, such as veins, that undergo collapse; several interesting phenomena are observed and their explanation presents an exciting challenge to fluid mechanics that we have as yet only begun to meet.

The subject matter is limited to large blood vessels (the heart, arteries and veins) for two reasons. The first is that the problems presented by the microcirculation are much more difficult because (*a*) there has been little detailed measurement of pressures and velocities there, (*b*) the inhomogeneous and non-Newtonian character of blood cannot be ignored, and (*c*) active changes in the calibre of arterioles are important, and cannot be accounted for in passive mechanical terms. The second reason, probably consequent upon the first, is that my own research has been restricted to vessels in which the Reynolds number, is considerably greater than 1. The topics selected for analysis reflect my research interests over the last 10 years. They include most of the cardiovascular work with which I have been involved in that time, set in its physiological context and broadened by the work of others to form what I hope is a fairly complete account of those topics. The most obvious omission is an analysis of flow through arterial stenoses (constrictions). A lot of experimental and theoretical work has been done on this subject and its neglect can be justified only on the grounds that stenoses are an abnormal geometrical feature, whereas I have elsewhere concentrated on understanding physiologically normal events.

Mention should also be made of the appendix, a separate, self-contained chapter on the behaviour of hot-film anemometers in steady and unsteady flow. This is included partly because it is an area in which I have myself been active, but principally because hot-film anemometry is at the moment the most reliable and widely used method of measuring blood velocity in living arteries. Hot-film measurements form the basic experimental data on which much of the theory of this book is based. Nevertheless, there are a number of inherent difficulties in making accurate measurements in unsteady flow, and I hope that this appendix will serve to warn the theoretical reader to take care in interpreting published data. Perhaps it may also help experimentalists to avoid some of the less obvious pitfalls.

Finally, a note about units. In the text I have endeavoured to quote all data in SI units, but in redrawing diagrams from other works, in particular those containing physiological data, I have usually retained the units employed by the original authors. Pressure units are the most difficult since both millimetres of mercury $(1 \text{ mmHg} = 133 \text{ N m}^{-2})$ and centimetres of water $(1 \text{ cmH}_2\text{O} = 98 \text{ N m}^{-2})$ are commonly used in cardiovascular physiology in place of the newton per square metre (or pascal). I hope that I have managed to avoid confusion by often quoting both SI units and physiological units.

The first draft of this monograph was completed in December 1976 and, together with the review article 'Pulmonary fluid dynamics' (Pedley, 1977), was awarded the Adams Prize of the University of Cambridge for 1975–6.

Thomas Young was part of the great tradition of eighteenth- and nineteenth-century science in paying little regard to the distinction between biological and physical science. By the end of the nineteenth century, however, a gulf was opening up between the two, which has only recently begun to close again (at least in circulatory physiology) by the development of interdisciplinary teams. While it is unrealistic nowadays to expect the applied mathematician to perform all the biological or engineering experiments to which his mathematics is applied, the application will not be useful unless he works closely with those who do perform them. For my introduction to, and continued close contact with, physiological reality I am profoundly grateful to my ex-colleagues at the Physiological Flow Studies Unit, Imperial College, London, especially Colin Caro, Robert Schroter, Anthony Seed and Michael Sudlow. In addition, I have benefited greatly from contact with many senior scientists who have worked at that Unit, especially James Fitz-Gerald, Joseph Milic-Emili, Robert Nerem and Kim Parker. I also owe a lot to my research students Sholaum Springer, Ian Sobey, Simon Farthing and Rosemary Wild, but for whom even fewer of the problems discussed in this book would have been solved, to my associate Christopher Bertram for his collaboration in the computational work of chapter 6, and to everyone, especially Berenice Schreiner, Denise Thomas and Judith Roberts, who co-operated in typing this manuscript with such patience and accuracy. I would finally like to acknowledge a special debt of gratitude to Professor Sir James Lighthill who has taken a keen interest in my work and has strongly supported it since I first turned to physio-logical fluid mechanics in 1968.

Cambridge   T. J. PEDLEY
December, 1978

CHAPTER 1

PHYSIOLOGICAL INTRODUCTION

The overall arrangement of the mammalian cardiovascular system can be summarised briefly as follows. The heart is composed of four chambers arranged in two pairs. The thin-walled atrium on each side is connected through a valved orifice to a thick-walled muscular ventricle; each ventricle connects in turn to a major distributing artery, the mouth of which is again guarded by a valve. The left ventricle is the thicker and leads to the aorta (diameter about 2.5 cm in man), through which oxygenated blood is distributed to the tissues of the body. Large arteries branch off the aorta, smaller ones branch off them, and so on for many subdivisions; the number of branchings along any pathway depends on the particular organ being supplied. The final subdivisions of the arterial tree are the arterioles, which have very muscular walls and internal diameters in the range 30–100 μm. These vessels give rise to the capillaries (diameters down to 4 or 5 μm) across the walls of which the principal exchange of fluids and metabolites between blood and the tissues takes place. The blood passes from the capillaries into the smallest veins (venules) and thence into a converging system of increasingly larger veins, finally merging into the superior and inferior venae cavae which join directly to the right atrium of the heart. (An exception to this pattern is the circulation in the heart muscle itself, which drains directly into the right atrium.) From the right atrium, blood is transferred to the right ventricle, and thence to the pulmonary artery, which leads through a bifurcating system of arteries to the pulmonary capillaries in the walls of the alveoli of the lung, where gas exchange takes place. From there the re-oxygenated blood is returned to the left atrium via the pulmonary veins, and the cycle is repeated. The average time taken for an element of blood to complete one circuit of the system is about one minute (in man).

The aim of this book is to explain the physical processes involved in the circulation of the blood in the large vessels of the circulation. The microcirculation (arterioles, capillaries and venules) is excluded for the reasons given in the preface; it is convenient that the conventional definition of the microcirculation, as consisting of all vessels which cannot be seen except through a microscope (i.e. of diameter less than about 100 μm), conforms closely to a definition more appropriate from the fluid mechanical point of view, namely vessels in which the mean Reynolds number is normally less than 1. Thus we shall be concerned with aspects of the circulation for which fluid inertia is important, and, as it turns out, for which blood can be considered a homogeneous Newtonian fluid (see § 1.2). This chapter is intended to set the scene by describing both the physical properties of the system and the phenomena which it is physiologically important to analyse. We also consider the fluid mechanics of the left ventricle in § 1.3.

Many books and reviews have already been published on the fluid mechanics of the circulation. Among the most important, on which I have leaned heavily for the material of this chapter and the next, are the two editions of *Blood Flow in Arteries* by McDonald (1960, 1974), the two-volume work edited by Bergel (1972a), Bergel & Schultz (1971), chapters 12 and 13 of Lighthill (1975) and the proceedings of various symposia (Attinger, 1964; ASME, 1966; Fung, 1966; Fung, Perrone & Anliker, 1972). Considerable detail on the structure and properties of the whole cardiovascular system, together with much physical (but not mathematical) discussion of the fluid mechanics, is given in a book of which I am a joint author (Caro *et al.*, 1978); most of the information in § 1.1 of this monograph is also contained in that work.

1.1 Anatomy, wall structure and mechanical properties

While the principal aim of the study of cardiovascular mechanics is to understand the circulation in man, most experimental data have been obtained in other mammals, especially dogs. Where the difference is important in what follows, it is mentioned; usually, however, the fluid mechanics is not well enough understood for small inter-species differences to be important except in the matter of scale. Most of the quantitative values to be given will apply to the dog.

1.1.1 *The heart*

The two atria are comparable in structure, and are separated from each other by a common wall, the inter-atrial septum. The veins drain into them without valves. The atrio-ventricular valve on the right side has three cusps (or flaps) while that on the left (the mitral valve) has only two. In each case the edges of the cusps are tethered to the opposite wall of the ventricle by fine cords anchored in small slips of muscle (papillary muscle). These cords have no active function in the opening or closing of the valves, which occur passively (see § 1.3), except to prevent the valves turning inside out and allowing backflow when they have closed. The exit valves from the ventricles (the pulmonary and aortic valves) are similar to each other, each consisting of three cusps that can open to the full cross-section of the artery without coming into contact with the artery wall, because behind each cusp is an outpouching of the artery, or sinus. In the aorta these sinuses are called the sinuses of Valsalva; the coronary arteries branch off two of them. The sinuses have an important function in the operation of the valve, as discussed in § 1.3.

The four valve orifices in the heart are aligned approximately in the same plane (fig. 1.1), and the cusps of each are attached at their bases to a stiff ring of fibrous tissue. The four rings are in turn connected to each other by fibrous tissues, so that the valve plane forms a stiff framework to which the muscles of all the chambers are attached, as are the origins of the pulmonary artery and the aorta. The heart as a whole is slung in a thin but inelastic fibrous bag, the pericardium, which in turn is attached to other structures within the chest, including the spine. The stress–strain relation of the pericardium is, like all fibrous tissues consisting largely of collagen, highly non-linear. Under normal conditions the pericardium is relatively unstretched, and has little effect on pressures and volumes in the heart; in certain diseases, however, fluid accumulates in the pericardium, which becomes taut and constrains the maximum volume of the heart.

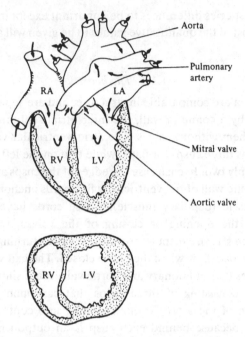

Fig. 1.1. Schematic diagram of a longitudinal and a horizontal cross-section of the heart, showing inflow and outflow tracts and valvular arrangement. Note the differences in the shape of the cross-sections and in the wall thicknesses of the two ventricles. RV, right ventricle; LV, left ventricle; RA, right atrium; LA, left atrium. (After Attinger, 1964.)

The left ventricle is shaped rather like a blunted arrowhead with the valves at its base (fig. 1.1), and with a roughly circular transverse cross-section. When the ventricle contracts, but before the aortic valve opens, the long axis first shortens slightly while the transverse cross-section expands so that the included volume remains constant. However, when ejection begins, the long axis shortens only very little, while the transverse axes shorten by about a third: by the end of systole (the ejection phase) the ratio of long to short axes is about 2.5 to 1, having been about 1.5 to 1 in diastole (the resting phase). A reasonable model of left ventricular shape, for the purposes of fluid mechanical calculations, is that of a prolate spheroid, whose major axis remains constant during ejection, but whose minor axes contract significantly. The two valves would both be at one end of the spheroid. This model involves considerable approximation, but the inaccuracy of measurement of ventricular dimensions *in vivo* is so great that the approximation is not unjustified. The end-diastolic volume of the left ventricle in a 20-kg dog is about 40 cm^3, and about 20 cm^3 is ejected each beat (the corresponding figures in man are 140 cm^3 and 70 cm^3).

The right ventricle has been studied far less thoroughly than the left, but it is known to behave very differently. One wall is functionally part of the left ventricular wall, while the other, free wall is much thinner and has a larger area, so that the cavity of the right ventricle is wrapped round one side of the left ventricle like a pocket, opening at the top into the pulmonary artery and right atrium (fig. 1.1). The operation of the right ventricle is like that of a check-valve pump (Carlsson, 1969): during systole the free wall moves downwards, carrying the open pulmonary valve past the blood in the ventricle; during diastole it moves up again, the closed valve pushing the blood up with it. It is clear from continuity considerations that the amount of blood ejected each beat by the right ventricle must, on average, be the same as that ejected by the left.

The walls of the chambers of the heart consist almost entirely of muscle fibres, interspersed with collagen. Cardiac muscle is a form of striated muscle, as is skeletal muscle, but it differs in its electrical and mechanical properties. For instance, continuous stimulation of skeletal muscle held at a fixed length produces a sustained contraction (or tetanus) with a highly reproducible tension; this is the maximum tension that can be generated by that muscle at that length. Continuous stimulation of cardiac muscle, however, does not produce a tetanus, because the muscle repolarises slowly, so experiments to investigate its intrinsic contractile properties have to be much more elaborate (for example, a series of individual 'twitches' is commonly generated). Further, Hill (1938) showed experimentally that when tetanised skeletal muscle fibres contract against a constant force F, then (a) the rate of heat production is proportional to the speed of shortening V (i.e. equal to aV, say, where a is a constant), and (b) the total rate of energy production is linearly related to F, i.e.

$$(F+a)V = b(F_0 - F), \tag{1.1}$$

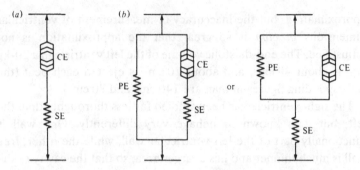

Fig. 1.2. Models of skeletal muscle. (a) A contractile element (CE) and an elastic element in series (SE); there is no tension when CE is relaxed. (b) Two possible arrangements in which a parallel elastic element (PE) supplies a 'resting tension'.

where b and F_0 are also constants. These results were obtained when the length of the muscle fibre at the start of the contraction was short enough for the 'resting tension' to be negligible, in which case a muscle fibre can be modelled as a 'contractile element' in series with a 'series elastic element' (fig. 1.2(a)). At greater lengths there is an initial tension in unstimulated muscle. This has to be accounted for by incorporating a 'parallel elastic element' into the model (fig. 1.2(b)), and by permitting a, b, F_0 in (1.1) to be functions of length, L.

No simple equation like (1.1) is applicable to heart muscle, however. This is partly because heart muscle cannot be tetanised, partly because individual muscle fibres cannot be separated as easily as in skeletal muscle, so the experiments are performed on small strips of (cat) papillary muscle in which some muscle fibres may not be parallel to the sides of the strip, and partly because at all lengths from which contraction is possible there is a significant resting tension. The absence of a tetanised state means that the time from the beginning of a twitch is an important variable in any relation between length, tension and velocity for cardiac muscle fibres, and no single model has been generally agreed. Fung (1970) has proposed the following equation, which can accommodate general non-linear and viscoelastic 'elastic elements':

$$\frac{d\Delta}{dt}(S, \Delta, t) = \frac{\pm b(\Delta)}{a(\Delta) + S} \left| \left[S_0(\Delta)\lambda \sin \frac{\pi(t + t_0)}{2t_m} - S \right] \right|^n. \quad (1.2)$$

In this equation Δ is the length of overlap between the actin and myosin filaments in the muscle sarcomere (a more fundamental measure of contractile element state than fibre length L), S is the stress in the series element, $S_0(\Delta)$ is the maximum attainable isometric tensile stress at overlap length Δ, n is an exponent whose value lies between 0 and 1 (in many cases being in the range 0.5–0.6), t_0 is a phase shift signifying the sudden initiation of active state at the time of stimulation, $2t_m$ is the time to peak activity, λ is an amplitude factor, $a(\Delta)$, $b(\Delta)$ are variable functions corresponding to a and b in (1.1), and the + or − is chosen according to whether the quantity in the square brackets in (1.2) is positive (stretching) or negative (relaxing). Equation (1.2) has the capacity to describe virtually all experimental results, but detailed testing of it, and empirical or theoretical evaluation of the functions and constants involved, are far from complete. Progress is necessarily piecemeal; for example, Brutsaert & Sonnenblick (1969) have shown that S_0 varies approximately linearly with length over a wide range of lengths:

$$S_0 = k_1 L + k_2, \tag{1.3}$$

where k_1 and k_2 are physiological constants.

A major practical difficulty with using (1.2) is the fact that most of the experiments on papillary muscle preparations have ignored two important factors: (a) the ends of the strips of muscle are inevitably damaged when clamped in the apparatus (Krueger & Pollack, 1975), and (b) even if they are not damaged, the strain in the middle section of a strip will differ from that at the constrained ends; this was demonstrated theoretically from finite-deformation elasticity theory by Hunter (1975). All the experiments therefore need to be repeated. Thorough reviews of the state of knowledge of heart muscle mechanics are given by Blinks & Jewell (1972) and by Caro et al. (1978, chapter 11).

Even if a model of the behaviour of heart muscle fibres were generally accepted, applying it to describe the mechanical behaviour of the intact heart would be extremely difficult. This is because of the intricate way in which the muscle fibres are arranged within the heart wall. The left ventricle is the only chamber whose wall structure has been examined systematically. In it there is a continuous distribution of fibre orientation across the wall thickness. The innermost fibres run predominantly longitudinally from the stiff region round the valves (the base) to the other end of the chamber (the apex), which is an elongated cavity with roughly circular cross-section. The next layer of fibres is at a slight angle to the axis of the chamber, so that the fibres form a slight spiral. The angulation of the spiral increases in successively deeper layers, so that approximately half-way through the wall the fibres run circumferentially round the small cross-sections of the chamber. Thereafter the angulation continues, so that the outermost fibres are again longitudinal. Streeter *et al.* (1970), who described this arrangement, also used it as the basis for a prediction of normal and tangential stresses throughout the wall, on the assumption that each muscle fibre exerts the same tension. Predictions were made both for diastole, when the muscle is relaxed, and for systole, when it is contracted. In each case the tangential (hoop) stress is predicted to be greatest in the middle layers of the wall, with a slight bias to the outer surface in systole, while the normal stress is predicted to be greatest on the surface, especially on the inner surface during systole (when the pressure in the ventricle exceeds that outside by about 16 kN m^{-2}, or 120 mmHg). It would be possible to use an equation like (1.3) in Streeter *et al.*'s model, in order to incorporate more details of muscle behaviour, especially during contraction, but without more information on the resting lengths and tensions of the individual fibres, such predictions would be almost worthless. Without them, however, there is no link between the mechanics of individual muscles and that of the intact ventricle, and the latter must be described empirically.

Note that it is important to know something of the stress distribution in the ventricle wall, because the blood supply to the heart muscle is carried in coronary arteries which are embedded in it, and are therefore squeezed when the muscle contracts. Since coronary artery disease is one of the major causes of death in Western society, all information on coronary artery mechanics has potential clinical importance.

From the point of view of the rest of the circulation, contraction of the left ventricle produces a certain flow-rate into the aorta, Q_a, and a pressure, p_a, at its entrance. Thus an index of the effectiveness of the ventricle can, in principle, be obtained by measuring these quantities. However, no satisfactory index of 'cardiac contractility' has yet been proposed, because the chain of events linking muscle performance to aortic pressure and flow-rate has not been fully described. In § 1.3 we analyse a small link in that chain by examining the relation between aortic pressure and flow-rate (relatively easy to measure *in vivo*) and the mean pressure exerted by the ventricular muscle, p_v, together with ventricular volume, V. The former can be related to the average tension in the ventricle wall, and hence to Streeter *et al.*'s model, as long as both V and the shape of the ventricle, as it contracts, are known. However, no accurate way of measuring V is available, and only qualitative observations of the shape have been made, as described above. Quantitative details of the time variation of ventricular and aortic pressures, and of aortic flow-rate, are given in § 1.2.

1.1.2 *The systemic arteries*

The anatomy of the canine aorta and its main branches is illustrated in fig. 1.3, and many of the relevant dimensions are listed in table 1.1 (which was first published in Caro, Pedley & Seed (1974)). The initial part of the aorta, after the sinuses of Valsalva, is relatively straight for about 3 cm and is called the ascending aorta. The aorta then curves, in a complicated three-dimensional way, through about 180° (the 'arch'), giving off two branches to the head and upper limbs (there are normally three branches in man). It then pursues a fairly straight course down through the diaphragm (giving off nine pairs of small intercostal arteries) to the abdomen, where it distributes branches to the abdominal organs. Low down in the abdomen it terminates, forming two iliac arteries and the sacral artery (absent in man). All other large arteries, similarly, are curved and branched in a complicated way; there are relatively few straight segments of artery without branches where the fluid mechanics of long straight tubes can be applied, and more general theories are usually required.

The aorta (like most other arteries) tapers along its length. The rate of taper appears to be quite variable from animal to animal and, presumably, from species to species; however, in the dog, the area

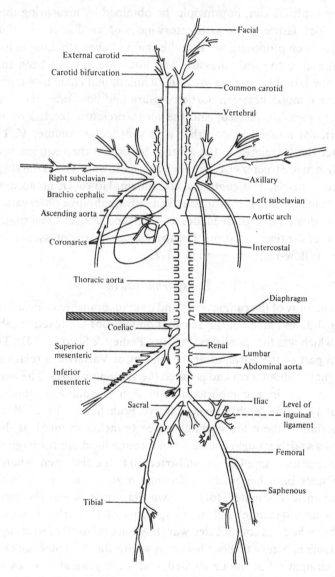

Fig. 1.3. A diagrammatic representation of the major branches of the canine arterial tree. (After McDonald, 1974.)

Table 1.1. *Normal values for canine cardiovascular parameters. An approximate average value, and then the range, is given where possible. All values are for the dog except those for arteriole, capillary, and venule, which have only been measured in smaller mammals*

Site	Ascending aorta	Descending aorta	Abdominal aorta	Femoral artery	Carotid artery	Arteriole	Capillary	Venule	Inferior vena cava	Main pulmonary artery
Internal diameter, d_i (cm)	1.5 (1.0-2.4)	1.3 (0.8-1.8)	0.9 (0.5-1.2)	0.4 (0.2-0.8)	0.5 (0.2-0.8)	0.005 (0.001-0.008)	0.0006 (0.0004-0.0008)	0.004 (0.001-0.0075)	1.0 (0.6-1.5)	1.7 (1.0-2.0)
Wall thickness, h (cm)	0.065 (0.05-0.08)	0.05-0.08	0.05 (0.04-0.06)	0.04 (0.02-0.06)	0.03 (0.02-0.04)	0.002	0.0001	0.0002	0.015 (0.01-0.02)	0.02 (0.01-0.03)
h/d_i	0.07 (0.055-0.084)	0.07 (0.055-0.084)	0.06 (0.04-0.09)	0.07 (0.055-0.11)	0.08 (0.053-0.095)	0.4	0.17	0.05	0.015	0.01
In-vivo length (cm)	5	20	15	10	15 (10-20)	0.15 (0.1-0.2)	0.06 (0.02-0.1)	0.15 (0.1-0.2)	30 (20-40)	3.5 (3-4)
Approximate cross-sectional area (cm^2)	2	1.3	0.6	0.2	0.2	2×10^{-5}	3×10^{-7}	2×10^{-5}	0.8	2.3
Total vascular cross-sectional area at each level (cm^2)	2	2	2	3	3	125	600	570	3.0	2.3
Peak blood velocity (m s^{-1})	1.2 (0.4-2.9)	1.05 (0.25-2.5)	0.55 (0.5-0.6)	1.0 (1.0-1.2)		0.75	0.07	0.35	0.25 (0.15-0.4)	0.7
Mean blood velocity (m s^{-1})	0.2 (0.1-0.4)	0.2 (0.1-0.4)	0.15 (0.08-0.2)	0.1 (0.1-0.15)		0.005-0.01	0.0002-0.0017	0.002-0.005		0.15 (0.06-0.28)
Peak Reynolds number, Re	4500	3400	1250	1000		0.09	0.001	0.035	700	3000
Frequency parameter, α (heart-rate 2 Hz)	13.2	11.5	8	3.5	4.4	0.04	0.005	0.035	8.8	15
Calculated wave speed, c_0 (m s^{-1})	5.8		7.7	8.4	8.5				1.0	3.5
Measured wave speed, c (m s^{-1})	5.0 (4.0-6.0)		7.0 (6.0-7.5)	9.0 (8.0-10.3)	8.0 (6.0-11.0)				4.0 (1.0-7.0)	2.5 (2.0-3.3)
Young's modulus, $E(\times10^2\ \text{kN m}^{-2})$	4.8 (3-7)		10 (9-11)	10 (9-12)	9 (7-11)				0.7 (0.4-1.0)	6 (2-10)

change fits quite well to an equation of the form

$$A = A_0 e^{-\beta x/a_0}, \tag{1.4}$$

where A is aortic area, A_0 and a_0 are the area and radius at a given upstream site, x is the distance from that site, and β is a dimensionless number lying between 0.02 and 0.05. In man the taper is not as smooth as implied by (1.4).

Although arteries taper individually, the total cross-sectional area of the arterial tree increases with distance from the heart, because at most bifurcations the ratio of the summed areas of the two daughter tubes to that of the parent is greater than 1. However, accurate measurements of this area ratio at individual bifurcations *in vivo* are hard to obtain, and measurements in dead animals may not be valid if the arteries are not inflated to physiological pressures, or if they are dissected out of the animal before being measured. Our knowledge is therefore incomplete; the most reliable information concerns the aorta of a dog, as measured by Patel *et al.* (1963*b*) and shown in fig. 1.4. The values of the area ratio, as defined above, range from 0.79 to 1.29. The branches within the chest have ratios close to 1.0, while those in the upper abdomen have slightly higher values, but there is a marked contraction (ratio 0.85–0.90) at the termination of the aorta, which is also found in man (Caro, Fitz-Gerald & Schroter, 1971). It is downstream from the aorta that the area begins to increase sharply.

The angles at which branches come off the aorta and other arteries vary considerably, although most angles off the aorta are closer to 90° than to 0° or 180°; further downstream, smaller branching angles are more common. Most branches can be seen from fig. 1.3 to be asymmetrical; indeed, there are no symmetrical bifurcations in the dog, while in man the only example is the aortic bifurcation.

The coronary arteries (fig. 1.5(*a*)) form a special group both because of their importance in disease and because they are arranged slightly differently. The first branches off the left coronary artery are depicted in fig. 1.5(*b*). The first bifurcation, where the left common coronary artery divides into the left circumflex and left anterior descending arteries, is roughly symmetrical with a branch angle of about 45°, but it is unusual in that the whole junction is set

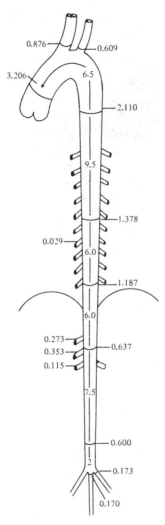

Fig. 1.4. Internal cross-sectional areas (cm^2) at various sites in the canine aorta and its main branches. (After Patel et al., 1963b.)

out on a curved surface, the outside of the heart. Further downstream, both arteries and their branches become increasingly embedded in the heart muscle, with important consequences for their elastic properties.

The walls of all arteries (and veins) have a similar structure and are made up of similar materials, although their proportions vary in

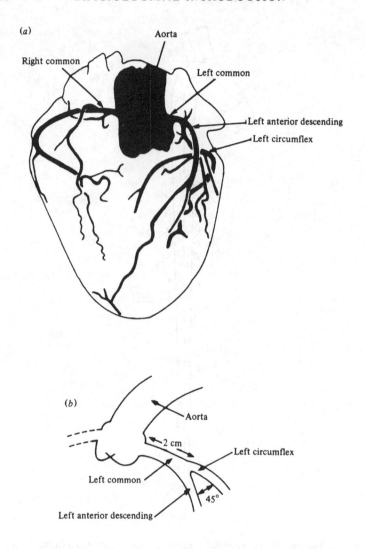

Fig. 1.5. (a) Diagram of the coronary arterial system (after Lusza, 1974), with (b) a schematic diagram of the major arteries on the left side of the heart.

different parts of the circulation. The wall is traditionally divided up into three layers, the innermost (tunica) intima, the (tunica) media, and the outermost (tunica) adventitia. The intima consists of two parts: the endothelium, a single layer of cells, which extends as a continuous lining to all blood vessels, and, surrounding it, a thin subendothelial layer containing collagen fibres. Outside this is the inner boundary of the media, formed mainly by a layer of inter-linked elastin fibres, called the internal elastic lamina. The rest of the media (usually the thickest part of the wall, and that which dominates its elastic behaviour) has a different structure in large central arteries from that in small arteries. In the former it consists of multiple concentric layers of elastic tissue (elastin), separated by thin layers of connective tissue (collagen) and occasional smooth muscle cells. In smaller arteries, the media consists almost entirely of spirally wound smooth muscle cells, arranged in layers, with small amounts of collagen and elastin between them. The outside of the media in each case consists of a thin, external elastic lamina. The adventitia is often as thick as the media, but is less important mechanically because it consists largely of loose connective tissue, containing relatively sparse elastin and collagen fibres.

The elastin, collagen and smooth muscle fibres constitute about 50% of the material of the wall; the rest is largely water, inside or outside cells, which has a negligible effect on the mechanical properties of the wall apart from being effectively incompressible. In large arteries (of dogs) elastin and collagen together constitute about 50% of the dry mass. In the intrathoracic aorta the ratio of elastin to collagen is about 1.5, while in other arteries it is about 0.5. It should be mentioned that the walls of all arteries larger than about 1 mm in diameter have their own blood supply, through small vessels called the *vasa vasorum*. These originate either from the parent artery or from a neighbour, pass into the adventitia, and break up into a capillary network reaching the inner part of the media. The ratio of wall thickness to internal diameter for every large artery is roughly the same, about 0.06 to 0.08, but increases in muscular arteries below about 1 mm in diameter to a value of about 0.4 in arterioles (diameter 0.1 mm).

The elastic properties of the artery wall clearly depend both on the properties of its individual components and on how they are linked together. Elastin is an easily extensible elastic material, individual fibres having non-linear stress–strain relations (fig. 1.6(a)) with a Young's modulus of about 300 kN m^{-2} for strains up to about 40%, but greater stiffness for larger strains (Carton *et al.*,

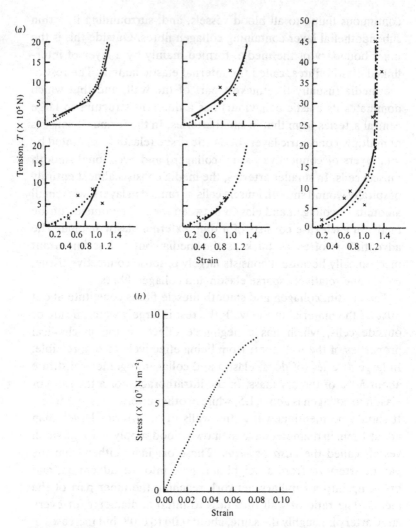

Fig. 1.6. (*a*) Tension–strain graphs for individual elastin fibres; the dotted curve is a mean curve derived from all specimens. (After Carton *et al.*, 1962.) (*b*) Stress–strain curve for human tendon, which is made up largely of longitudinally arranged collagen fibres; the curve is the average for over 50 tendons. (After Benedict *et al.*, 1968.)

1962). Collagen is much stiffer, lengths of human tendon having a Young's modulus of about 10^6 kN m^{-2} (fig. 1.6(*b*)). Smooth muscle has a Young's modulus roughly similar to that of elastin, but its actual value depends on the level of physiological activity, varying from about $100 \, \text{kN m}^{-2}$ in the relaxed state to $1200 \, \text{kN m}^{-2}$ in the active state. Of these three materials, only elastin is purely elastic; both collagen and, especially, smooth muscle show marked visco-elastic properties (creep, stress relaxation and hysteresis), which are reflected in the dynamic properties of artery walls.

As the above description of arterial wall structure shows, the arrangement of the different fibres in the wall is inhomogeneous and very complicated, and to infer the elastic properties of the wall as a whole (regarded, for example, as a thin-walled cylinder) from those of its constituents is virtually impossible. Instead, experiments have been performed on segments of artery, excised from newly dead animals, and with their ends occluded in some way. In these experiments the dimensions which can be accurately measured are length, l, and internal volume, V (or, equivalently if the cross-section is circular, external radius, a_o, linked to internal radius, a_i, by knowledge of the wall thickness, h, which is deduced from a single measured value at particular values of l and a_o, using the assumption of incompressibility; this assumption is accurate for all biological tissue (Patel & Vaishnav, 1972)). In this book the internal cross-sectional area, A, will in general be used as the important transverse dimension; this can be deduced from V and l. The forces tending to deform the artery wall in these experiments are longitudinal tension, T, and transmural pressure, $p_{tm} = p(\text{inside}) - p(\text{outside})$.

Among the first observations to be made are (a) that the length of the artery shrinks by up to 40% of its in-vivo length, l_1, when it is excised, and (b) that the curve relating p_{tm} to A depends strongly on l (Bergel, 1972b). This shows that the artery is normally under considerable tension, and that pressure–area curves measured at lengths other than l_1 have no physiological relevance. All pressure–area relations quoted will be for values of l equal or close to l_1. Another simple observation is that if l is not held constant, but a constant (usually zero) longitudinal tension is applied, then l increases as p_{tm} increases. This immediately reveals that the elastic properties of the artery wall are not isotropic, since a cylinder made of isotropic material with non-zero Poisson's ratio, σ, would shorten in such circumstances (since artery walls are effectively incompressible, σ is approximately 0.5). It is more reasonable to take artery walls to be *orthotropic*, with different elastic constants for stresses applied longitudinally than for stresses applied circumferentially (by means of a distending pressure), but with no extra constants to be defined for stresses applied in other directions.

The static elastic properties of an excised artery wall, then, can be defined by plotting p_{tm} against A for fixed l, and T against l for fixed A. Such curves are shown (for the canine thoracic aorta, and the femoral and saphenous arteries respectively) in fig. 1.7. In each case the *dimension* (A or l) is plotted as the ratio of its actual value to its value at average in-vivo conditions (A_0 is the area at a transmural pressure of 13.3 kN m^{-2}, or 100 mmHg, and l_1 is the in-vivo length). In each case the curve is markedly non-linear, with its slope increasing greatly as A or l exceeds its normal mean value. This would not happen in a linearly elastic material, where $p_{tm} \propto 1 - a_{i0}/a_i$ (a_i, a_{i0} being the actual and the initial internal radii), or in rubber, where $p_{tm} \propto 1 - (a_{i0}/a_i)^2$. Such an increase in slope is required for vessel stability; without it 'blow-outs' would commonly occur. In the most frequently quoted experiments, p_{tm} was not taken below about 2.7 kN m^{-2} (20 mmHg) so that the cross-section of the artery remained circular at all times. It has been proposed that the sharp increase in slope of the curves in fig. 1.7 is associated with the fact that, at low strains, the collagen fibres are not taut, and the relatively extensible elastin bears the load; at higher strains, however, the much stiffer collagen fibres increasingly straighten and support the stress. The analogy of a balloon in a string bag is commonly made (Roach & Burton, 1957). This explanation may well be true in part, but the non-linear properties of elastin itself (fig. 1.6(a)) are probably also important.

In order to predict the speed of propagation of the pressure pulse (chapter 2), the important parameter describing vessel wall properties is the distensibility, D, defined by

$$D = (1/A)(dA/dp_{tm}), \qquad (1.5)$$

which can be computed directly from data such as that of fig. 1.7(a), measured at in-vivo length (Lighthill, 1975, chapter 12). Despite this, it has been conventional to convert such measurements into an effective incremental Young's modulus, E, for circumferential

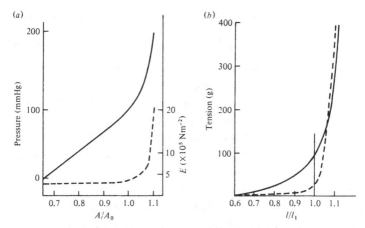

Fig. 1.7. (a) The pressure (continuous curve) and incremental elastic modulus (broken curve) of a segment of dog thoracic aorta, plotted against area divided by in-vivo area (100 mmHg ≈ 13.3 kN m^{-2}). (b) Tension-strain curves for longitudinal extension of specimens of femoral (continuous curve) and saphenous (broken curve) arteries of the dog; l_1 is the length *in vivo*. (After McDonald, 1974.)

extensions of the whole vessel wall, modelled as a uniform cylinder with homogeneous, isotropic (or, exceptionally, orthotropic) walls. The use of an incremental Young's modulus is made necessary by the non-linearity of the relation between hoop stress and circumference, so that the only way to define a single elastic parameter is to consider small departures from a mean, prestressed, in-vivo state, and linearise the stress–strain curves. This may be useful if the amplitude of the pressure pulse is not too large. For an increase in pressure within a thin-walled isotropic arterial segment whose length is held constant, E is related to D by

$$D^{-1} = [E/(1-\sigma^2)]\,(h/d), \qquad (1.6a)$$

where h is the wall thickness, d is vessel diameter and σ is Poisson's ratio, equal to 0.5 since the material is incompressible. This result follows from the classical elasticity of thin shells, and is quoted in a slightly more general form for thick-walled tubes by Bergel (1972b). The corresponding result for an orthotropic tube, with different Young's moduli, E_θ, E_x, and Poisson's ratios, $\sigma_{x\theta}$, $\sigma_{\theta x}$, for applied circumferential or longitudinal stresses, respectively, is

$$D^{-1} = [E_\theta/(1 - \sigma_{x\theta}\, \sigma_{\theta x})](h/d) \qquad (1.6b)$$

(Atabek, 1968; Lighthill, 1975). Using formulae such as these, values of E_θ have been calculated for many canine arteries, and are given in table 1.1, where it can be seen that arteries become stiffer with increasing distance from the heart; the thoracic aorta has a distensibility of about $0.021\ \text{m}^2\ \text{kN}^{-1}$, while D for the femoral artery is about half this. Values of E_x, the incremental Young's modulus for static longitudinal loading, have also been measured in some vessels; for example, Patel & Vaishnav (1972) quote a value of $670\ \text{kN m}^{-2}$ for E_x in the thoracic aorta, while their results give a slightly higher value ($740\ \text{kN m}^{-2}$) for E_θ, at the upper limit of the range quoted in table 1.1.

The experiments discussed above concern static loading, while the actual transverse loading of an artery as the pressure pulse passes is dynamic, and the viscoelastic wall properties come into play. These have been examined by measuring the changes in cross-sectional area when a sinusoidal pressure oscillation of angular frequency ω is superimposed on the mean level (of $13.3\ \text{kN m}^{-2}$). A linear approximation to the results is given by quoting a *complex* incremental Young's modulus (or distensibility) whose imaginary part represents the viscous element of wall behaviour, causing the oscillations in area to lag behind those in transmural pressure. Thus E or E_θ is replaced by $E_{\text{dyn}} + i\eta\omega$ (to use the notation of Bergel (1961b) and McDonald (1974)). Experiments show that, for frequencies of 2 Hz or above, both E_{dyn} and $\eta\omega$ are effectively constant, the former at a value that is greater than the static value of E_θ by a factor of between 1.1 and 1.7, depending on the artery studied (fig. 1.8), and the latter at a value equal to $0.1–0.2 E_{\text{dyn}}$. For lower frequencies, however, both quantities are frequency-dependent, tending to their static values (E_θ and 0 respectively) as $\omega \to 0$. In the dog the heart-rate is about 2 Hz, so the higher-frequency constant values can be used. In man the heart-rate is about 1.2 Hz; it is not known, however, what the lower end of the frequency-independent range is in man. Patel & Vaishnav (1972) report that the longitudinal dynamic modulus is greater than the transverse modulus by nearly 10% (although the static modulus

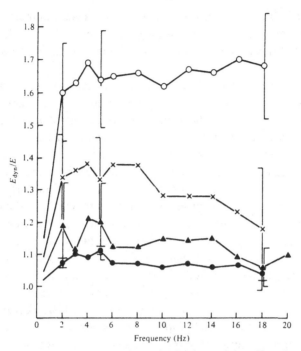

Fig. 1.8. Average values of the dynamic elastic modulus, E_{dyn}, for several arteries expressed as a ratio to the static value, E, plotted against frequency: filled circles, thoracic aorta; filled triangles, abdominal aorta; crosses, femoral artery; open circles, carotid artery. (After Bergel, 1961b.)

is smaller by the same amount). The imaginary part, for isolated arterial segments kept *in situ*, is greater by a factor of 3–4 for longitudinal wall motion than for transverse.

There are two main uncertainties involved in using the mechanical properties of isolated arteries for predictions of *in vivo* mechanics. One is the fact that arteries pass through, and are firmly tethered to, the surrounding tissue, the outer boundary of the adventitia being rather ill defined. This tethering has the effect of greatly inhibiting longitudinal wall movements, which actually means that static experiments on segments of artery whose length is held fixed are more likely to be relevant *in vivo* than if considerable longitudinal wall motions were possible. There must also be some constraint on the radial motion, in that the effective inertia, stiffness and viscosity of the wall will all be increased somewhat, especially the inertia (Patel & Fry, 1966). It is only because wall inertia does not play an important part in pulse-wave propagation (Atabek, 1968) that this radial tethering can be ignored without gross inaccuracy.

The other uncertainty is much more important, and concerns the effect of smooth muscle, which in excised arterial segments is relaxed, but *in vivo* will have an unknown degree of activity; Gow (1972) has devoted a whole chapter to the subject. Three separate effects of smooth muscle contraction can be distinguished. (*a*) If the artery's diameter is held fixed, contraction of the smooth muscle tends to decrease its distensibility. (*b*) However, if the distending pressure is held fixed, muscle contraction reduces the diameter, putting the artery onto a less steep part of the pressure–area curve, and the distensibility is increased. (*c*) The situation *in vivo* is further complicated by the fact that, if all vascular smooth muscle is contracted, including that in the arterioles, the mean arterial blood pressure rises (in order to overcome the increased peripheral resistance), so the net effect on distensibility is unclear. Fortunately, there is a sufficiently small proportion of smooth muscle in the walls of the largest arteries that the effects of its contraction, whether known or not, are small. Otherwise the predictions of pulse-wave velocity made in chapter 2, for example, could not be made. The main influence of smooth muscle is on the viscous behaviour of the wall (Bergel, 1972*b*), and is not known to depend significantly on the state of contraction.

1.1.3 *The pulmonary arteries*

The pulmonary circulation differs from the systemic circulation in several important respects. For example, the mean transmural pressure of large pulmonary arteries is only about 2 kN m^{-2} (as opposed to 13 kN m^{-2} in systemic arteries); the walls of pulmonary arteries are much thinner than those of systemic arteries; and the branching pattern is quite different, many more bifurcations being approximately symmetric, and most of them occurring after only a few (1.5 to 5) diameters of the parent tube (Cumming *et al.*, 1969). The diameter and length of the main pulmonary artery (corresponding to the aorta) in the dog are about 1.7 cm and 3.5 cm respectively; in man the corresponding values are 2.5 cm and

5.0 cm. The main pulmonary artery divides into two arteries supplying the left and right lung, which split into several branches supplying the lobes. These arteries enter the lung itself, and all subsequent arteries are surrounded by a thin sheath of connective tissue and separated from it by fluid. The state of distension of the lung has an important effect on the transmural pressure experienced by pulmonary arteries, and therefore knowledge of it is an essential prerequisite to predictions of the dynamics of the pulmonary circulation. The area ratio of pulmonary branches is less than 1 for two generations, but increases thereafter.

Pulmonary arteries have much thinner walls than systemic arteries of comparable size, h/d taking a value closer to 0.01 than to 0.1. The walls are again usually divided into intima, media and adventitia; the media of all pulmonary arteries of diameter greater than 1 mm consists largely of elastin, with little collagen or smooth muscle. Smaller pulmonary arteries are more muscular, and h/d increases, but only up to a maximum of about 0.1, not 0.4 as in very small systemic arteries. The elastic properties of excised pulmonary arteries have not been as thoroughly studied as those of systemic arteries, but because of their physiological circumstances they have been studied, at least qualitatively, over a wider range of transmural pressures. Fig. 1.9 shows that the main pulmonary artery has a larger Young's modulus than the aorta at transmural pressures typical of the aorta, although the relatively thin wall means that they are not less distensible (see (1.6a) and also table 1.2). At their own normal transmural pressures, however, pulmonary arteries are definitely more distensible than any systemic artery: graphs of pulmonary artery diameter against transmural pressure presented by Maloney, Rooholamini & Wexler (1970) indicate values of D ranging from 0.35 m^2 kN^{-1} for large vessels to 0.88 m^2 kN^{-1} for the smallest. These are over 10 times greater than systemic values. The cross-section of pulmonary arteries remains circular for transmural pressures down to about 1.5 kN m^{-2}, but as p_{tm} is reduced to 0.5 kN m^{-2}, the cross-section becomes markedly non-circular. This can be important *in vivo* because p_{tm} at the top of the lung normally does fall as low as this, and sometimes even becomes negative (see § 1.2). This type of behaviour is discussed more quantitatively in the following subsection, on veins, in which it has been studied in more detail.

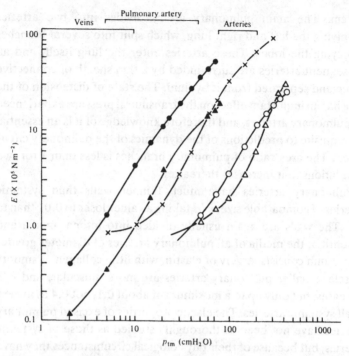

Fig. 1.9. Dependence of effective incremental Young's modulus E on transmural pressure, p_{tm}, for various vessels: filled circles, inferior and superior venae cavae; filled triangles, jugular vein; crosses, pulmonary artery; open circles, descending aorta; open triangles, carotid artery. The normal physiological ranges of transmural pressures for systemic arteries, pulmonary arteries and veins at the level of the heart are also indicated. (After Attinger, 1969.)

Table 1.2. *Distensibility of different vessels at (A) $p_{tm} = 4$ kN m^{-2} and (B) $p_{tm} = 1$ kN m^{-2}*

Vessel	E(kN m^{-2})		h/d		D(m^2 kN^{-1})	
	A	B	A	B	A	B
Venae cavae	3000	420	0.006	0.008	0.04	0.22
Jugular vein	1000	110	0.011	0.015	0.07	0.45
Pulmonary artery	600	106	0.015	0.026	0.08	0.27
Aorta	100	Not given	0.092	0.12	0.08	—

1.1.4 *The veins*

The systemic venous system is not a replica of the arterial system in which the only difference lies in the direction of blood flow. For example, there are many more veins of a given diameter than there are arteries, so that the veins normally contain almost 80% of the systemic blood, although this figure can vary enormously. Furthermore, the transmural pressures are normally negative in some veins, while they are large and positive in others (see § 1.2), and the walls of veins are much thinner than those of arteries. Many veins in the limbs contain valves, which prevent backflow and probably have an effect on the propagation of pulsatile pressures along the venous system. These valves each have two cusps, with sinuses behind them on the downstream side. No survey has been made, as far as I know, of area ratios and branching angles in the veins.

Three layers can be described in the walls of medium and large veins, as in arteries; they are all much thinner, however. The intima consists of the endothelium and only a small amount of elastin. The media consists essentially of a few layers of spirally wound smooth muscle, interspersed with a little elastin and collagen; it is the absence of large quantities of elastin in the media that constitutes the primary structural difference between veins and arteries. The smallest veins do not contain significant smooth muscle. Veins in parts of the body normally below the heart have relatively thick media, though whether this is because they experience a relatively high transmural pressure is not known. In veins the major wall compartment is the adventitia, which contains a network of elastin and collagen fibres, but principally the latter (the ratio of elastin to collagen is about 0.3 in the largest veins, compared with 0.5–1.5 in arteries). The thickness-to-diameter ratio of veins is about 0.01, like that of pulmonary arteries.

The elastic properties of veins have not been studied systematically. The most thorough investigation of a single type of vein is that by Moreno *et al.* (1970) on the inferior vena cava of a dog. Fairly detailed measurements on both venae cavae and the jugular vein (also of a dog) are reported by Attinger (1969). Moreno *et al.* realised that many veins commonly experience both positive and negative transmural pressures, and therefore studied the pressure–

area relations of excised segments of vein over a wide range of such pressures. They also compared their results with those obtained from rubber tubes of comparable diameter and wall thickness; this is important because many model experiments on flow in collapsible tubes, intended as an indication of how veins might behave, are performed with rubber tubes (see chapter 6). Fig. 1.10(a) shows the results of this experiment, while fig. 1.10(b) shows the relation between area and perimeter. From these graphs and from observations of the shape of the cross-section of the tube, we can describe the elastic behaviour of the tubes as p_{tm} is reduced from $4 \, kN \, m^{-2}$ (30 mmHg). When p_{tm} exceeds about $1.5 \, kN \, m^{-2}$ the cross-section of the vein is circular, and the wall is very stiff. In fact $D \approx 0.04 \, m^2 \, kN^{-2}$ when $p_{tm} = 4 \, kN \, m^{-2}$ which is less than that of arteries at the same p_{tm}, although the wall is much thinner, because the collagen (in the adventitia) becomes fully extended at a lower p_{tm} than in arteries. This is best deduced from the results of Attinger (1969), who plots values of E (see fig. 1.9) and h/d for veins, pulmonary arteries and systemic arteries at various pressures. His values for $p_{tm} = 4 \, kN \, m^{-2}$ and $1 \, kN \, m^{-2}$, and the corresponding values of D calculated using (1.6a), are given in table 1.2. Changes in cross-sectional area are accompanied by appropriate changes in perimeter (proportional to $A^{1/2}$). As p_{tm} is reduced from 1.5 to $1.0 \, kN \, m^{-2}$, the vessel still remains circular, but becomes somewhat more distensible, presumably because the elastin or smooth muscle in the wall takes over from the collagen. When p_{tm} falls below about $1.0 \, kN \, m^{-2}$, the vessel becomes elliptical, the area falls more rapidly, and the distensibility rises. The perimeter also continues to fall, although no longer as $A^{1/2}$, and it is this reduction in perimeter, not the change in cross-sectional shape, that makes the larger contribution to the reduction in area (and hence to the distensibility) at least while p_{tm} exceeds about $0.5 \, kN \, m^{-2}$. As p_{tm} falls below this value, however, the change in shape becomes more marked, and makes an increasing contribution to D, which has a maximum of about $3.5 \, m^2 \, kN^{-1}$ when p_{tm} is slightly below $0.5 \, kN \, m^{-2}$. From then on, the slope of the graph falls again. When p_{tm} falls below about $-0.1 \, kN \, m^{-2}$, the vessel cross-section becomes increasingly distorted from its elliptical shape, until at a p_{tm} of $-1.0 \, kN \, m^{-2}$ it is almost completely collapsed, in the dumb-

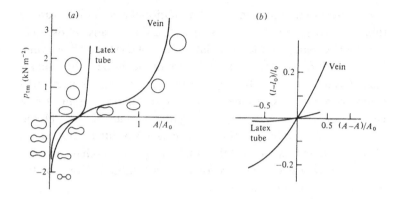

Fig. 1.10. Comparison of the elastic properties of a latex tube with those of an excised segment of a canine vena cava. (a) Transmural pressure as a function of area, A, scaled with respect to the area at zero transmural pressure, A_0. Transverse cross-sections are shown for various areas. (b) Perimeter as a function of cross-sectional area. (Data from Moreno et al., 1970.)

bell configuration shown. The two side-channels may be more nearly closed than shown, because in this state the endothelium becomes corrugated and the cells tend to protrude into the vessel. In this configuration D is again very low, tending to 0 as $A \rightarrow 0$, because A is constrained to be greater than, or equal to 0.

The graphs presented by Moreno et al. (1970) indicate that there is a second peak in D at a negative value of p_{tm}, because A (see (1.5)) has become small more rapidly than dA/dp_{tm}. However, this may not be a real phenomenon, because of the difficulty of measuring very small areas and pressures accurately. Note that a number of workers use the term distensibility to describe the quantity

$$D' = (1/A_0)(dA/dp_{tm}), \qquad (1.7)$$

where A_0 is a reference area, for example the value of A when p_{tm} is 13.3 kN m^{-2}($= 100$ mmHg). This 'pseudo-distensibility' is close to the true distensibility if all areas considered are close to A_0, as for systemic arteries under physiological conditions, but is quite different when a wide range of areas is considered, as here.

It is interesting to note the corresponding behaviour of the rubber tube. The cross-section remains circular until p_{tm} falls almost to 0, and then becomes elliptical and, later, dumb-bell shaped like a vein. However, during collapse, the perimeter of the rubber tube remains almost constant (fig. 1.10(b)), and the area change is associated solely with change of shape. Furthermore, the maximum rate of change of area occurs at a negative value of p_{tm} (about -0.25 kN m^{-2}), while that of a vein occurs at a positive value. It can also be seen that the slope of the graph in fig. 1.10(a) changes much less abruptly for a vein than for a rubber tube, because of the continuing change in perimeter.

The reason for this difference in elastic properties (fully analysed by Moreno et al. (1970)) lies in the fact that the effective Young's modulus of the rubber tube for deformations to an unstressed state, about 2100 kN m^{-2}, is about 40 times that of the vein. Therefore the wall resists both stretching (hence the constant perimeter once collapse has begun) and bending more firmly than in veins. It is the resistance to bending at the points of maximum curvature that provides the stiffness of the collapsed state: the bending moment required to maintain radius of curvature R in a slender beam of breadth b and thickness h (i.e. part of a segment of the vessel wall, sliced longitudinally and of axial length b) is proportional to Ebh^3/R, increasing with E for constant values of b, h and R. This provides a qualitative explanation for the results. A more quantitative explanation was also provided by Moreno et al. (1970), using the analysis described by Love (1927, pp. 423–4) for a bent, linearly elastic rod (or longitudinally uniform cylinder), which leads to the following differential equation for the shape of the cross-section $y(x)$, using the coordinate system of fig. 1.11:

$$-y''/(1+y'^2)^{3/2} = [\tfrac{1}{2}p_{tm}(x^2+y^2)+k]/EI, \qquad (1.8)$$

where $I = h^3/12$ is the moment of inertia of the cross-section per unit length of tube, and k is a constant. Equation (1.8) is solved numerically, subject to the boundary conditions that $y(x)$ and $y'(x)$ should be periodic, and k is determined by knowing either the perimeter or the cross-sectional area at a particular value of p_{tm}. Moreno et al. showed how their numerical solutions gave good agreement between theory and experiment for the relations

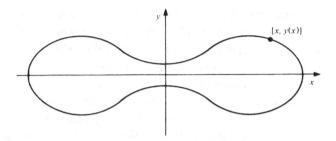

Fig. 1.11. Coordinate system for the buckled cross-section of an elastic tube.

between p_{tm}, A, perimeter and shape, for both rubber tubes and veins. Other analyses (e.g. Flaherty, Keller & Rubinow, 1972a) have analysed the buckling of elastic cylinders in more detail, and have demonstrated that if the wall is very thin, it does not begin to buckle elliptically, but in a periodic shape with circumferential wave number greater than 2, this number increasing as h/d decreases. This accords with experiment (J. M. Fitz-Gerald, personal communication).

The fluid mechanical consequences of the great distensibilities and shape changes of venous cross-sections have not been fully worked out. Model experiments designed to elucidate them, and some preliminary theoretical analysis of these experiments, are described in chapter 6.

Finally, mention should be made of pulmonary veins, which, like pulmonary arteries, were examined by Maloney et al. (1970). They have a structure similar to that of systemic veins, and distensibilities of between 0.17 and 0.35 m^2 kN^{-1}, which are slightly less than those of similarly sized pulmonary arteries at similar transmural pressures (1–2 kN m^{-2}), and comparable with those of systemic veins (see table 1.2).

1.1.5 *The blood*

We conclude this section with a very brief description of the mechanical properties of blood. Much more detail can be found in Whitmore (1968), Charm & Kurland (1972), Cokelet (1972) and Caro et al. (1978, chapter 10). Blood is a suspension of formed elements in plasma. The formed elements, which normally occupy about 45% by volume of the blood (a proportion which is slightly species- and sex-dependent), are the red cells, white cells and platelets. Red cells are by far the most numerous, and completely dominate the mechanical properties. They are flexible, biconcave discs, of diameter about 8 μm (in most mammals) and thickness about 1 μm at the centre and 2–3 μm at the edge; there are about 5 million red cells per cubic millimetre of whole blood. The white cells are on the whole slightly larger, and equally deformable, but are far less numerous (1–2 white cells per 1000 red cells) and therefore dynamically unimportant. The platelets are more numerous than white cells (80–100 per 1000 red cells), but are very small, rounded cells of diameter 2–4 μm. They too are normally unimportant dynamically, but they react very strongly with foreign substances, and therefore play a significant role in the process of blood clotting, as well as being involved in the later stages of the development of atherosclerosis. The plasma is a solution of large molecules, but on the scales of motion and at the rates of shear normally encountered in the blood vessels, it can be regarded as a homogeneous Newtonian fluid of viscosity 0.0012–0.0016 kg m^{-1} s^{-1} at body temperature (37 °C).

Whole blood cannot be regarded as a homogeneous fluid in the smallest blood vessels, because the diameters and spacing of red cells are comparable with capillary diameters. However, in vessels whose diameters exceed 100 μm, blood can be regarded as effectively homogeneous, because the scale of the microstructure is so much smaller than that of the flow. The remaining question to be answered is whether blood can also be assumed to be Newtonian in the large vessels. The standard test of this is to place a quantity of blood in a constant-shear viscometer and to measure its effective viscosity as a function of shear-rate, S. It is found that the measured viscosity is independent of S for $S > 100$ s^{-1} (Whitmore, 1968). We shall see in chapter 3 that the average shear-rate at the walls of arteries is significantly greater than this, and on that basis we shall assume blood to be Newtonian, although it is clear that near the centre of a straight vessel, or in separated regions of recirculating flow, the average value of S will be small. Furthermore, even the wall shear passes through zero twice per cycle (as it reverses), so it is not true that $S > 100\,\text{s}^{-1}$ everywhere at all times. It is not clear how to assess the importance of temporarily non-Newtonian blood on unsteady arterial fluid dynamics; this is a problem that should be investigated, but that will here be ignored. We take blood to be a homogeneous Newtonian fluid of density $1.05 \times 10^3\,\text{kg}\,\text{m}^{-3}$, viscosity $0.004\,\text{kg}\,\text{m}^{-1}\,\text{s}^{-1}$ (at 37 °C) and hence kinematic viscosity about $4 \times 10^{-6}\,\text{m}^2\,\text{s}^{-1}$.

1.2 The physiological relevance of fluid mechanics

In this section, I first outline those physiological phenomena that a fluid mechanical analysis might hope to explain, and then quote the experimental results which suggest that the shear stress exerted by the flowing blood on the artery wall is an important factor in the development of arterial disease.

1.2.1 Blood pressure and transmural pressure

The pressure in the blood at any location is made up of three components: (i) atmospheric pressure, p_0, which is normally taken to be the pressure in the right atrium when its muscles are relaxed; (ii) the hydrostatic pressure, $-\rho g h$, where h is the vertical distance above the right atrium and ρ is the density of the blood; and (iii) the pressure generated by the heart, which we shall denote by p (Lighthill (1975) calls this component the 'excess pressure'). This last component is alone responsible for the motion of the blood, and is commonly called 'blood pressure' (blood pressure is always measured clinically 'at the level of the heart'); p has a mean value of $13.3\,\text{kN}\,\text{m}^{-2}$ in the large arteries, but falls dramatically in the microcirculation, so that the mean pressure in the venae cavae is less than $0.6\,\text{kN}\,\text{m}^{-2}$, and in the right atrium it is even closer to zero.

The pressure p_0 relative to which blood pressures are measured should not be exactly atmospheric, but should be taken to be the pressure inside the chest, which varies with respiration and is normally subatmospheric (by as much as $2\,\text{kN}\,\text{m}^{-2}$) when the lung is expanded. This is because contraction of the ventricular muscle generates a *transmural pressure* between the inside and the outside of the ventricle, the outside being within the chest.

The hydrostatic component of pressure, $-\rho g h$, is important because it determines the transmural pressure of the blood vessels and hence, through their elastic properties (§ 1.1), their calibre. The pressure outside most blood vessels (with the exception of those in the chest and in the skull; see below) is close to atmospheric (in fact, about -0.25 kN m^{-2}; see Wiederhielm, 1972), so p_{tm} is in most cases approximately equal to $p - \rho g h$. Consider a standing man: the mean transmural pressure is about 13 kN m^{-2} at the entrance to the aorta, and about 0.5 kN m^{-2} at the exit from the venae cavae. In the large vessels of the foot (1.1 m below) these values will be increased by about 11 kN m^{-2}, so that both types of vessel will be distended, will have circular cross-section, and will be very stiff (the transmural pressures in some veins can be reduced by voluntary contraction of the skeletal muscles outside them, and can be further alleviated by the action of the valves). In vessels 30 cm above the heart, however (for example in a raised arm), p_{tm} is reduced by 3 kN m^{-2}. This has little effect on arteries, which remain fully open; veins, on the other hand, experience a negative transmural pressure, and collapse therefore occurs (§ 1.1). Venous return is maintained, either continuously, through the small side-channels, which may not completely close (fig. 1.10), or intermittently, as the build-up of upstream pressure forces the veins open. (I am aware of no measurements that unequivocally demonstrate how this venous return is accomplished.) Veins in the skull do not collapse, because the skull acts as a rigid box, any fall in venous volume being accompanied by a corresponding fall in extravascular pressure.

The transmural pressures experienced by pulmonary blood vessels are more difficult to evaluate, because they depend in a complicated way on the state of lung inflation. We consider only arteries and veins; pulmonary capillaries experience different transmural pressures, but are outside the scope of this monograph. The vessels are surrounded by, and probably also linked (by fibres) to, a continuous perivascular sheath; the space between sheath and vessel wall is filled with fluid. The pressure in this fluid, p_{pv}, differs from air pressure in the alveoli of the lung, p_{alv}, for a number of reasons. First, if the neighbouring alveoli are inflated, the alveolar membranes pull outwards on the perivascular sheath, which effectively reduces p_{pv} by an 'elastic stretching pressure', p_{es}.

However, this effect is itself reduced by the elastic recoil pressure of the perivascular sheath itself, less the effective outwards pressure exerted on the vessel wall by the elastic linkages (if any) that span the perivascular space, a net reduction which we may call p_{er}. Thus

$$p_{pv} = p_{alv} - p_{es} + p_{er}.$$

In a static, uniform lung, in which all alveoli are equally inflated, both p_{alv} and p_{es} will be uniform, and p_{pv} will differ from the pressure in the pleural space which surrounds the whole lung, p_{pl}, only insofar as the elastic recoil pressure of the pervascular sheath, p_{er}, differs from that of the pleural membranes. It is common to assume that p_{pv} is equal to p_{pl}, which can be approximately measured by recording the pressure in a balloon inflated in the aesophagus. In a normal man, the average value of p_{pl} is about -0.5 kN m^{-2}, i.e. 0.5 kN m^{-2} below atmospheric, when his respiratory muscles are relaxed. This quantity can be decreased to about -3 kN m^{-2} in a very deep inspiration, or increased to more than 6 kN m^{-2} during a forced expiration. In the living animal the lungs are normally not uniformly inflated, because of the effect of gravity: p_{pl} (and hence p_{pv}) decreases with height by approximately 25 N m^{-2} per cm (in man). In any discussion of the state of distension of pulmonary blood vessels it is important to record (or at least estimate) the perivascular pressure as well as the internal pressure.

1.2.2 Unsteady pressures

The (excess) pressures measured simultaneously in the left ventricle and in the ascending aorta immediately downstream of the aortic valve are shown in fig. 1.12, where the corresponding graphs for the right ventricle and main pulmonary artery are also shown. The left ventricular pressure rises rapidly at the beginning of systole, and soon exceeds that in the aorta so that the valve opens and ejection begins. About half-way through ejection, the two traces cross, so that an adverse pressure gradient is established across the valve, which decelerates the outflow and is maintained as the two pressures fall. The kink in the aortic pressure curve (the 'dicrotic notch') marks the closure of the valve, and thereafter the ventricular pressure falls rapidly as the heart muscle relaxes, while aortic pressure falls more gradually as blood flows out peripherally.

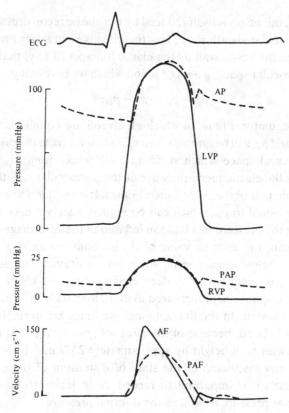

Fig. 1.12. Semi-diagrammatic illustration of pressure and flow-rate occurring simultaneously on the left and right sides of the heart during the cardiac cycle. ECG, electrocardiogram; AP, aortic pressure; AF, aortic flow; LVP, left ventricular pressure; PAP, pulmonary artery pressure; PAF, pulmonary artery flow; RVP, right ventricular pressure. (After Caro *et al.*, 1978.)

Fig. 1.13 shows measurements of pressure made instantaneously at a number of sites along the aorta. The pulse is seen to be increasingly delayed with increasing distance down the vessel, indicating that it is propagated along the aorta as a *wave*. We also note that the shape of the pressure pulse changes dramatically as it propagates: the amplitude increases, the front becomes steeper, and the dicrotic notch disappears. The mean pressure falls, but only very gradually (by about 0.5 kN m^{-2} along the whole aorta). These changes continue as the pulse wave passes into other large arteries

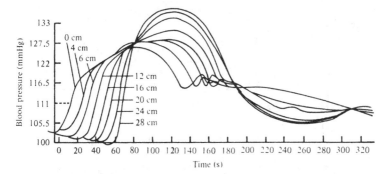

Fig. 1.13. Instantaneous blood pressure records made at a series of sites along the aorta in the dog. 0 cm is at the start of the descending aorta. (After Olson, 1968.)

(fig. 1.14(a)), and it is only in small arteries, with a diameter below about 1 mm, that the amplitude (and the mean) of the pressure pulse begins to fall rapidly. The propagation of the pulse wave is analysed mathematically in chapter 2; the object of any such analysis must be to explain the observed changes.

The pressure waveforms in systemic veins are shown in fig. 1.15; these traces were measured successively at different sites along the same pathway of the same animal. The oscillations in the vena cava are similar to those commonly recorded in the right atrium, and the phase differences between the successive traces clearly reflect the propagation of a wave along the veins from the heart, against the direction of blood flow. The oscillations are almost completely absent in the brachial vein (fig. 1.15(d)), which is almost certainly because there is a valve between that measuring site and those closer to the heart. The role of valves in eliminating pressure oscillations in veins has not been systematically explored. Incidentally, small-amplitude oscillations are observed in very small veins, and are thought to be the attenuated remnants of the arterial pulse. Their amplitude is greater in conditions of vasodilatation, when the microcirculatory attenuation is less severe (see chapter 2; and Zweifach, 1974).

Typical pressure waveforms recorded in the main pulmonary artery and the left atrium (representative of large pulmonary veins) of a dog are shown in fig. 1.16. The former has a similar shape to the

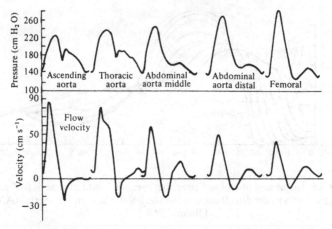

Fig. 1.14. Matched records of (*a*) pressure and (*b*) average velocity at different sites in the arteries of a dog. (From McDonald, 1974.)

aortic waveform (albeit with much smaller amplitude), but there is a marked difference between the waveforms in smaller pulmonary arteries (not shown) and those in smaller systemic arteries (fig. 1.13), because there is no obvious change in shape as the wave propagates along the pulmonary arteries.

1.2.3 *Velocity waveforms*

At any location in the cardiovascular system, the motion of the blood is driven by the local pressure gradient, which in turn is determined by the propagation of the pressure pulse. Thus the fluid mechanical problems of the circulation can be divided into two parts: first, an explanation of the pressure waveforms already described, which is embarked on in chapter 2; secondly, an analysis of the blood motions driven by the pressure gradients, which is introduced in chapter 2, but forms the main subject of chapters 3 to 5. Here, however, we describe measurements of blood velocities.

It was only in the 1960s that it became possible to measure blood velocities or flow-rates in arteries with any degree of confidence. The most common device used on animals was (and still is) the cuff electromagnetic flowmeter. This is placed round a vessel, a uniform transverse magnetic field is applied across the vessel, and the e.m.f. generated in a perpendicular direction when the conducting blood

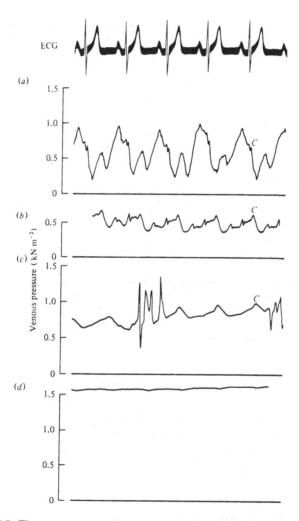

Fig. 1.15. The pressure waveforms measured sequentially with a single catheter at four successive sites in the human venous system. These sites are, from (a) to (d), the superior vena cava, the subclavian vein, the axillary vein and the brachial vein. The four waveforms are aligned with the ECG record (top trace) so that the phase difference between the vena cava and the subclavian vein can be seen; C represents the same event in successive records. (After Caro et al., 1978.)

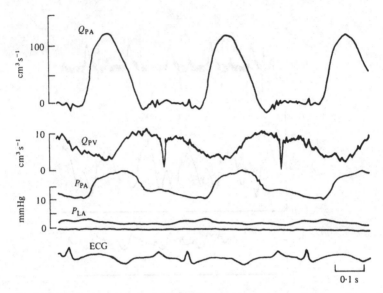

Fig. 1.16. Experimental records of pulmonary pressures and flow-rates in a
conscious dog, at rest. *Q*, blood flow-rate; *P*, pressure; PA, pulmonary
artery; PV, pulmonary vein, within 2 cm of left atrium; LA, left atrium;
ECG, electrocardiogram. The sharp spikes in the flow-rate traces during
diastole are artefacts. (From Milnor, 1972.)

flows is measured. From it can be inferred the volume flow-rate in
the vessel or, equivalently, the average velocity (averaged across the
cross-section) as long as the internal diameter of the vessel is
known. The cuff flowmeter is reliable as long as the velocity profile
in the vessel is axially symmetric (although it does impose some
restriction on the radial motion of the vessel wall and therefore
might interfere with pulse propagation), and can, in principle, be
made insensitive to asymmetries in rectilinear flow (Bevir, 1970).
Mills (1972) discusses the operation of these flowmeters in detail.
Mills & Shillingford (1967) also designed a catheter-tip electromagnetic flowmeter, which effectively averages the velocity over
a small region near the probe and can therefore measure velocities
at points within the vessel, as long as the velocity profile is locally
approximately flat. A more satisfactory device for the measurement
of local velocities is the hot-film anemometer, because it effectively
averages over a *very* small region and does not require a flat profile.

This was originally developed and used for blood measurements by three different groups (Ling *et al.*, 1968; Schultz *et al.*, 1969; Seed & Wood, 1970*a*). The principles of operation of such an anemometer are analysed in the appendix, where particular attention is paid to the behaviour of the probe in unsteady flow. Another method for measuring local velocities is the pulsed ultrasonic Doppler flowmeter (Peronneau *et al.*, 1969); this is, in principle, better than any because it does not require invasion of the blood vessel by a probe, but it has not yet been widely used. The measurements to be presented below are either of average velocity/flow-rate (recorded with a cuff electromagnetic flowmeter) or of local velocity in the centre of the vessel (recorded with a hot-film or a catheter-tip electromagnetic probe). The two will be approximately the same if, and only if, the velocity profile is effectively flat over most of the vessel cross-section, with only thin boundary layers at the wall. In the largest vessels this is indeed the case.

Average velocity waveforms at the entrances of the aorta and the pulmonary artery are given in fig. 1.12, and the variation in the waveform with distance along the systemic arterial tree is depicted in fig. 1.14(*b*). Blood is ejected from the heart for about one-third of the period of the beat (this fraction increases to about one-half at higher heart-rates), and the ejection phase is followed by a brief period of backflow as the aortic valve closes. The amount of backflow increases with distance from the heart, since it can be accommodated by expansion of that part of the aorta between the measurement site and the heart. The small backflow at the entrance to the aorta can be accounted for by the closed aortic valve bulging back (there is normally negligible leakage) and by diastolic flow into the coronary arteries. The shape of the velocity waveform and its variation with distance must also be explained by an analysis of the pulse wave (chapter 2). Mean and peak velocities in different canine vessels are given in table 1.1; the maximum velocity in the aorta normally exceeds the mean by a factor of 5 or 6. Note that the word 'mean' is used to refer to a time mean, and the word 'average' to denote a cross-sectional average. A sequence of point-velocity waveforms, measured in the centres of the arteries in question with a catheter-tip electromagnetic probe, is shown in fig. 1.17 with the corresponding pressure waveforms. Their general features

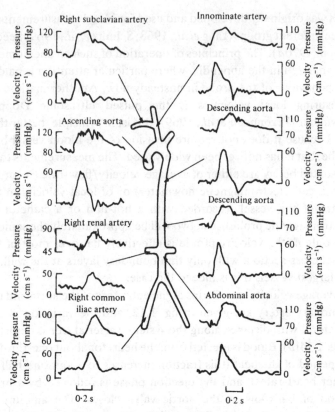

Fig. 1.17. Simultaneous pressure and blood velocity waveforms at numerous points in the human arterial tree. All were taken from one patient with the exception of the right renal artery and the right common iliac artery. (After Mills *et al.*, 1970.)

correspond to those of fig. 1.14(*b*), but show more fluctuation, suggesting that the high-frequency components of the waveforms are averaged out by the cuff flowmeter.

Velocity waveforms in coronary arteries are different from those in other systemic arteries because of their special circumstances. During systole, when the velocity in the aorta becomes very large, the ventricular muscle is contracted, squeezing the coronary blood vessels within it. Their resistance to flow is therefore very large, and flow is sluggish. In diastole, however, the ventricular muscle is relaxed, the coronary vessels open wide, and a vigorous flow

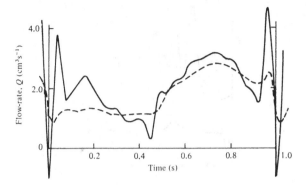

Fig. 1.18. Flow-rate waveform in the left circumflex coronary artery of a dog; continuous curve is predicted from measured pressure gradient, broken curve is measured directly with a cuff electromagnetic flowmeter. (After Atabek *et al.*, 1975.)

through them is established. A measured flow-rate waveform showing this (in a dog) is presented in fig. 1.18. Point-velocity measurements in the same arteries of horses reveal a further interesting phenomenon which is absent in dogs. This is shown in fig. 1.19, indicating relatively high-frequency (10 Hz) and large-amplitude oscillations superimposed on the basic waveform. Their presence in horses has been confirmed by another group of workers (Wells *et al.*, 1974), and the search for a satisfactory explanation is still going on. An intriguing question is whether these oscillations depend on large size, and if so, whether or not they are normally present in man.

Typical flow-rate waveforms in the pulmonary artery and pulmonary vein are shown in fig. 1.16. Data on the average velocity waveforms in the venae cavae (of man) are presented in fig. 1.20, which shows that in each cardiac cycle there are two main oscillations in flow velocity, out of phase with the pressure oscillations: one when the ventricles contract and one when they are relaxed. The contraction of the right *atrium* is associated with low velocities in the venae cavae.

Before turning to velocity profiles, it is worth noting that both pressure and velocity waveforms, being essentially periodic, can be subjected to Fourier analysis. This makes analysis of the relation

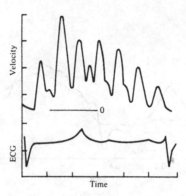

Fig. 1.19. Centre-line velocity waveform in the left circumflex coronary artery of a horse; note the relatively high-frequency oscillations, absent in fig. 1.18. Peak Reynolds number $\hat{R}e = 308$, $\alpha = 3.16$. (After Nerem *et al.*, 1974*b*.)

between the two waveforms particularly simple as long as a linear theory is applicable, and it is shown in chapter 2 that such is normally the case. Patel, de Freitas & Fry (1963*a*) demonstrated that the pressure waveform could be very accurately described with a mean and six periodic components, while 10 periodic components were necessary for comparable accuracy of the flow-rate waveform. This means that velocity probes must be able to respond accurately to frequencies up to 20 Hz (in dogs – only about 12 Hz in man). We may note in passing that a probe designed to measure the wall shear-rate (which involves differentiation of the velocity) must be accurate up to about 100 Hz (50 Fourier components). This is shown in § 3.1, where the difficulties of designing a suitable wall shear probe are briefly discussed.

1.2.4 Velocity profiles

Velocity profiles have been measured in a number of arteries using a hot-film anemometer, with the film mounted on a needle inserted through the artery wall and carefully positioned, but I know of no such measurements in veins. The artery most commonly studied is the canine aorta (Ling *et al.*, 1968; Schultz *et al.*, 1969; Seed & Wood, 1971; Clark & Schultz, 1973), because it is the only one, in a dog, to have an internal diameter greater than about 1 cm; the probes are usually 1–2 mm across, and in any smaller artery they

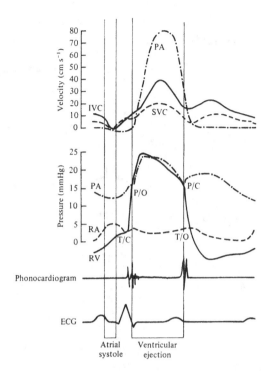

Fig. 1.20. A composite diagram of pressure and velocity of flow into and out of the right side of the heart, constructed from a number of different measurements. The top curves show the velocities in the inferior and superior venae cavae (IVC, SVC) and the pulmonary artery (PA), then come the pressures in the right side of the heart (RA, RV) and pulmonary artery, and then the phonocardiogram and ECG. These are included to serve as timing references. T/C, tricuspid valve closed; P/O, pulmonary valve open; P/C, pulmonary valve closed; T/O, tricuspid valve open. (After Wexler et al., 1968.)

would distort the flow, making accurate measurement of the velocity profile impossible. A few measurements in relatively smaller vessels have been made in the horse (Nerem et al., 1974a). The recording of velocity profiles is complicated, because the profiles vary throughout the cardiac cycle, and care has to be taken in recording velocities at different positions in the vessel cross-section, but at the same stage of the cycle.

The mean velocity profiles at various stations in the canine aorta are presented in fig. 1.21. They can be seen to be roughly flat

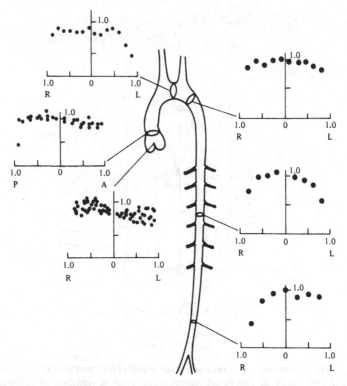

Fig. 1.21. Mean velocity profiles in the aorta of a dog. For each graph, the mean velocity at each site on the vessel diameter is expressed relative to the centre-line mean velocity (ordinate), and positions on the diameter are scaled with respect to the vessel radius (abscissa). R, L: right, left; P, A: posterior, anterior. (After Schultz, 1972.)

throughout, although a relatively thick boundary layer is apparent in the abdominal aorta. Because of the size of the probe and the motion of the artery wall as the wave passes, accurate measurements closer than 1–2 mm from the wall are unobtainable. The same observation, of a fairly flat profile surrounded (one infers) by thin boundary layers, is made throughout the cycle in the ascending aorta (fig. 1.22(a)). Near the aortic valve, the profile remains symmetric, but as the flow enters the arch, a marked skew develops during peak forward flow, with higher velocities towards the posterior wall of the aorta, which is the inside of the curve (fig. 1.22(a) and (b), from measurements made by W. A. Seed and N. B.

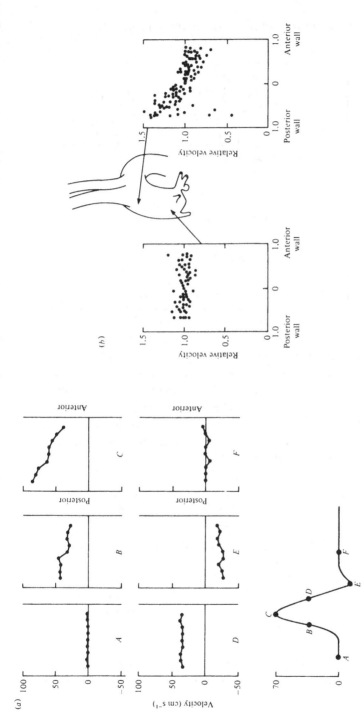

Fig. 1.22. (a) Instantaneous velocity profiles at six different times during the cardiac cycle in the ascending aorta of the dog. The lower figure shows (schematically) the velocity waveform, with the six times marked on it. (b) Velocity profiles at peak systolic velocity across two diameters in the ascending aorta of the dog (velocities scaled with respect to centre-line velocity; radial positions scaled with respect to vessel radius). The profiles at both sites are from several dogs, and were measured in the plane of curvature of the aortic arch. (After Caro et al., 1978.)

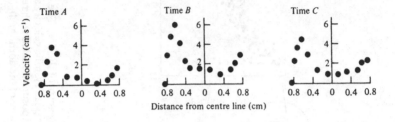

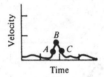

Fig. 1.23. Velocity profiles in the terminal aorta of a horse at various times during the cardiac cycle as indicated; measurements are in the plane of the aortic trifurcation. (After Nerem *et al.*, 1974a.)

Wood). These profiles were measured in the principal plane of curvature of the arch. Profiles have also been measured in a perpendicular plane (Clark & Schultz, 1973) and also show a skewing, towards the left wall. These skews are presumably associated with the geometry of the aorta, i.e. the curvature of the arch, the branches off it and possible misalignment between the ventricular outflow tract and the ascending aorta itself. Furthermore, in diastole, when the average blood velocity has fallen to zero, measurements show a skewed profile immediately downstream of the aortic valve; this may be related to the flow into the coronary arteries, but may also be influenced by disturbances generated during systole. Part of the purpose of chapters 3 to 5 is to explain the observed velocity profiles.

Profiles in the descending aorta remain remarkably flat throughout systole, with a slight skew towards the right wall, but with little apparent influence of the large branches upstream. The profile tends to become rounder with distance from the heart, as indicated by the mean profiles of fig. 1.21.

Schultz *et al.* (1969) reported rather distorted aortic and pulmonary artery velocity profiles in human patients with mitral, aortic or pulmonary valve disease.

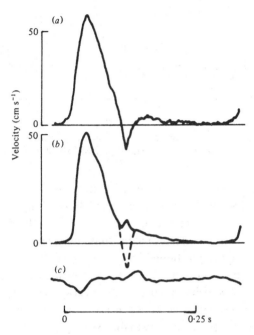

Fig. 1.24. Simultaneous aortic velocity measurements with (a) an electromagnetic flow cuff and (b) a hot-film anemometer located on the vessel
centre line. The signal from a second hot-film, (c), indicates when reversal
occurs. (After Clark & Schultz, 1973.)

Similar aortic velocity profiles were obtained in the horse (Nerem
et al., 1974a). In this animal, however, it is possible to make
measurements in the terminal part of the abdominal aorta, distal to
the mesenteric and renal branches, as well as in the external iliac
and mesenteric arteries themselves. These measurements reflect
the complexity of the local geometry, with marked skews and
asymmetries. One striking feature is that just before the iliac
bifurcation, peak aortic velocities are off the centre of the tube (fig.
1.23), while in the iliac artery the profile is rounded and approximately symmetric.

Finally, some profiles have been measured in horse coronary
arteries (Nerem et al., 1974a). Profiles in the left circumflex artery
are shown in fig. 1.19, measured in a plane perpendicular to the
plane of the common coronary artery bifurcation, which as we have
already seen lies on a curved surface. There is a marked skew

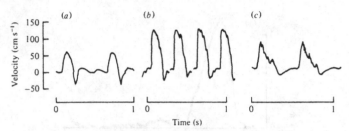

Fig. 1.25. Velocity waveforms recorded in the descending thoracic aorta of dogs: (a) undisturbed, (b) disturbed, (c) highly disturbed. (After Nerem et al., 1972.)

towards the outside of this curve at all stages in the cycle, and this skew is also observed in the left anterior descending artery.

We should note that the presence of relatively flat velocity profiles in most large arteries means that measurements made by a probe only roughly positioned in the centre of the tube (e.g. on a catheter) will be representative of the centre-line velocity, and that the centre-line velocity is close to the average velocity. Fig. 1.24 shows the correspondence between the waveforms of centre-line velocity and of average velocity in the ascending aorta. In regions of complicated geometry, of course, such a correspondence cannot be assumed (cf. fig. 1.23).

1.2.5 *Disturbed or turbulent flow*

Most velocity measurements in the arteries of anaesthetised normal dogs show a smooth variation with time which suggests that random disturbances are absent and that the flow is laminar. Occasionally, however, high-frequency disturbances are noted in the aorta, occurring just after peak forward velocity and either dying out rapidly or persisting until the blood comes virtually to rest in diastole (fig. 1.25). Fourier analysis of the disturbed waveforms shows that the fluctuations cover a wide and continuous frequency spectrum from at least 500 Hz (above which the amplitude is too small for accurate measurement) down to below 25 Hz, where they become indistinguishable from the high-frequency components of the underlying waveform (Nerem & Seed, 1972). The randomness and non-reproducibility from beat to beat of these fluctuations indicate that the disturbances can correctly be referred to as turbulence, although the frequency spectrum is, of course, not the same

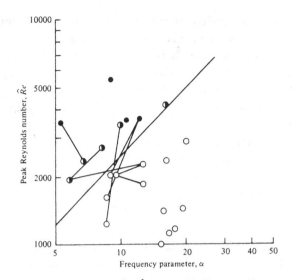

Fig. 1.26. Peak Reynolds number, $\hat{R}e$, and frequency parameter, α, for the descending thoracic aorta of dogs: open circles, undisturbed flow; half-filled circles, disturbed flow; filled circles, highly disturbed flow; continuous line, (1.9). Points joined together are for the same dog in different physiological states. (After Nerem *et al.*, 1972.)

as that of steady, fully developed pipe flow turbulence; Parker (1977) has analysed the spectrum using a sophisticated ensemble-averaging technique.

By using drugs to vary the heart-rate and cardiac output of their experimental animals, Nerem & Seed (1972) showed that an initially laminar flow could be made to become turbulent. Furthermore, the borderline between laminar and turbulent flow occurred for values of frequency and flow-rate that are close to those of unanaesthetised animals. This can be seen from fig. 1.26, in which the axes are peak Reynolds number, $\hat{R}e$, and frequency parameter, α $(=\frac{1}{2}d(2\pi f/\nu)^{1/2}$, where d is the diameter of the aorta, f is the heart-rate and ν is the kinematic viscosity of blood). The line

$$\hat{R}e = 250\alpha \qquad (1.9)$$

roughly separates the turbulent dogs from the laminar. It is clear that normal dogs may be on either side of this line, which no doubt explains the controversy that has grown up between those who assert that turbulence is not normally present and those who assert that it is. A simple scaling analysis of the whole animal kingdom, using the plentiful data on heart-rate and cardiac output quoted by Stahl (1967), suggests that $\hat{R}e$ is proportional to $\alpha^{2.3}$ as body size varies, and that for resting animals this line (on a log-log plot) crosses the line describing (1.9) at a point corresponding to an animal mass of about 25 kg (Pedley, 1978). Roughly speaking, therefore, animals larger than dogs may be expected normally to have turbulent aortic flow and smaller animals may be expected not to. Exercise would make turbulence more likely. Certainly the flow in a horse's aorta *is* normally turbulent (Nerem *et al.*, 1974*a*), and measurements by W. A. Seed (personal communication) indicate that man, too, normally has .turbulent flow in the aorta. A wider range of experiments would be of great interest. As far as this book is concerned, however, the main interest lies in the origin of the turbulence. Is it a local flow instability, or does it represent the convection to the measuring site of disturbances already present in the ventricle? This question is addressed in § 5.3.

Seed's observations in man conflict with the conventional clinical view that (*a*) 'aortic murmurs' heard through a stethoscope reflect the presence of turbulence, and (*b*) murmurs are not heard in normal man, but are primarily heard in conditions of aortic valve stenosis, when the valve cannot open properly and a turbulent jet is ejected each beat. Therefore (*c*) there is normally no turbulence. One cannot doubt the observations (*a*) and (*b*), but only the logic leading to (*c*). It is quite possible that the vigorous disturbances in a turbulent jet are audible at the chest wall while those in the transitorily disturbed normal aortic flow are not: Seed also reports that faint murmurs are heard through a stethoscope placed on the surface of a normal artery in which the flow is turbulent, but not through one placed on the chest. If stethoscopes of greater sensitivity could be designed, then it might be possible to hear murmurs in normal man.

Apart from its possible diagnostic significance, the presence of turbulence in an artery is known to have an important effect on the material of which the artery wall is comprised. It is an old clinical observation that the walls of a blood vessel often show localised dilatation just downstream of a partial obstruction to the flow, called a stenosis. Roach (1972) has shown that this 'post-stenotic dilatation' occurs regardless of the origin of the obstruction, as long as it is severe enough to cause turbulence in the downstream flow (and not so severe that the blood flow-rate is cut down to a laminar trickle). Roach also showed that by exposing segments of artery to small-amplitude oscillatory disturbances, then if the disturbance frequency was in the range 25–500 Hz, the wall became significantly more distensible (the higher frequencies having a relatively greater effect on older arteries). This is the range of frequencies present in the turbulent jet downstream of a stenosis over the physiological range of Reynolds numbers, and the conclusion that post-stenotic dilatation is a reflection of such an increased distensibility is inescapable. One is immediately tempted to ask what might be the effect of intermittent turbulence in the aorta, assuming that it normally does occur. The human aorta tends to become stiffer with age, while more peripheral arteries tend to become less stiff (Learoyd & Taylor, 1966). The former effect could conceivably be a chronic response to the degenerative action of arterial turbulence, which could not become apparent in the relatively short-term studies of Roach and her colleagues. A comparison of the changes in aortic distensibility brought about with age in large (turbulent) and small (laminar) animals would be of interest.

The fact that turbulence has an important effect on arterial wall structure means that it would be expected also to influence any physiological process that depends on wall structure. In particular, it may have an influence on the permeability of artery walls to various chemicals, and hence, as discussed in the next subsection, on the initial stages of arterial disease.

1.2.6 *Wall shear stress and arterial disease*

It is becoming increasingly accepted that haemodynamic factors play a role in the initiation of atherosclerosis. This belief is based on a number of observations. It has been noted that fatty streaks, thought by many to be the precursors of the true atheromatous plaques that constitute the disease, are distributed preferentially at certain locations on the artery walls (Caro *et al.*, 1971). Particular sites include the outer walls of arterial junctions (and not the flow dividers), the inside walls of curves (e.g. the aortic arch), the carotid sinus (an outpouching of the carotid artery, not associated with valves), and the abdominal rather than the ascending aorta. Further experiments in which animals (particularly rabbits and dogs) have been fed high-cholesterol diets have indicated that the distribution of the fatty streaks that develop after a few weeks is roughly the reverse of that just described (see, for example, Newman & Zilversmit, 1962; Cornhill & Roach, 1976). These observations have suggested the hypothesis that the patchy distribution of fatty streaks reflects a non-uniformity in the permeability of artery walls to the lipoproteins on which cholesterol is transported in or out. In the cholesterol-fed animals, the concentration of these substances in the blood is higher than that in the wall, and it is suggested that fatty streaks occur in regions of relatively high permeability, where the rate of influx is greater. In normally fed man, on the other hand, the concentration is thought to be higher in the wall than in the blood, so the lipoproteins can escape more readily through high-permeability regions, which are therefore spared from fatty streaking. This conjectured relation between fatty streaking and wall permeability is still rather tenuous, because the forms in which cholesterol enters and leaves the wall and in which it accumulates in the wall are different, and the links between them are not well understood (Porter & Knight, 1973).

Circumstantial support for the above hypothesis comes from experiments performed *in vivo* with marked large molecules, such as Evans Blue dye (bound to albumin in the plasma) or radioactively labelled fibrinogen or lipoprotein (Fry, 1973; Schwartz *et al.*, 1974; Nerem, Mosberg & Schwerin, 1976). After such substances have been circulating in the blood for some time, some of them will have been transported into the vessel walls, the amount accumulating at any site being roughly proportional to the permeability of the wall there. Post-mortem examination shows that the regions of greatest accumulation, and thus of greatest permeability, are well correlated with the regions in which fatty streaks develop in cholesterol-fed animals.

Although it is possible that some regions of the wall are more permeable than others for genetic reasons, it is more conventional to suppose that the different regions have been subjected to different external influences, in particular, to different mechanical stresses exerted by the flowing blood. These consist of the pressure and the wall shear stress. There is evidence (Fry, 1973) that an elevated mean pressure enhances wall permeability to large molecules if it results in a marked stretching of the endothelial cell layer (but not otherwise, ruling out pressure-driven filtration as the mechanism for mass transfer); indeed, chronic high blood pressure is known to be a predisposing factor in the development of atherosclerosis. However, an elevated mean pressure would be distributed more or less uniformly over the artery walls, and interest has therefore centred on the role of the wall shear stress. This varies both with time throughout the cardiac cycle and with position in the arterial tree, and there is considerable evidence linking it with wall permeability.

Using excised segments of artery, Fry and his colleagues have shown that an elevated steady wall shear stress, τ_w, increases the permeability of the artery wall (Fry, 1973). Indeed, Carew (1971) found a good correlation between permeability and τ_w^2, while Caro & Nerem (1973) fitted their data to $\tau_w^{0.29}$, and Nerem et al. (1976) found $\tau_w^{0.38}$ to be the best fit for steady shears, but reported a stronger dependence on the amplitude of oscillatory shear stresses (with zero mean), especially when this amplitude exceeded about 5 N m^{-2}. They also found an influence of oscillatory *pressure* (at a frequency of 1 Hz), which is consistent with the idea already encountered that fluctuating forces influence artery wall structure (Roach, 1972). Fry (1973) also quotes evidence that turbulence in the flow can enhance wall permeability.

Fry (1968) has shown further that if the shear stress is maintained for a short time at the (unphysiologically) high level of 40 N m^{-2} or more, then the endothelial surface is irreversibly damaged, and its permeability greatly enhanced. Fry has suggested that this high-shear damage might be a factor in atherogenesis. However, Caro et al. (1971) noted that the regions in which fatty streaks develop are those in which the wall shear stress would be expected to be low if the flow in the arteries were steady. This is consistent with the hypothesis that fatty streaks develop in regions of low permeability, and has led to the further hypothesis that fatty streaks and early atheromatous plaques develop preferentially in regions of low wall shear.

Before this hypothesis can be established, several aspects of the process must be more fully understood. These include (i) the link between wall permeability to large molecules and atheroma; (ii) the mechanism by which such large molecules are transported across the endothelium (probably 'pinocytosis' – see Weinbaum & Caro (1976) for a convincing theoretical analysis of the process); (iii) the way in which the wall shear stress influences this mechanism; and (iv) the actual distribution of wall shear stress in the arteries. There is scarcely even a hypothesis to explain (iii), although the wall shear is known to affect the endothelial cell structure, because the cell nuclei are found to align themselves with the direction of the prevailing shear (Flaherty *et al.*, 1972*b*). Caro *et al.* (1971) originally suggested that the rate-controlling step in the transport process was the transfer across a diffusion boundary layer in the blood, and, in a long footnote to that paper, Lighthill rightly argued that the unsteady components of shear stress could have no influence on this. The important haemodynamic factor was therefore thought to be the *mean* wall shear stress. However, Caro & Nerem (1973) have since shown that the diffusion boundary layer cannot be the rate-controlling step, and the reasons for restricting attention to the mean shear stress are no longer valid. The importance of the unsteady components will depend crucially on the time-scale of the events involved in the mass-transfer process, and very little is known of these. Therefore, in pursuing (iv) above, it is important to determine the distribution of both the mean and the fluctuating components of shear stress in large arteries. One must be able to determine the shear-stress variations over a very small length-scale, for, as Cornhill & Roach (1976) have shown, the distribution of fatty streaking round the entrances to the tiny intercostal arteries (fig. 1.3) in cholesterol-fed rabbits is markedly and reproducibly non-uniform, being greatest on the downstream lip. They also noticed a marked non-uniformity from one pair of intercostals to the next, possibly reflecting the development of the flow as it straightens out after the aortic arch, or possibly being a consequence of the distortion caused to the aortic flow by the removal of a little blood into the upstream intercostals (see § 5.2).

It has not yet proved possible to measure unsteady wall shear accurately either *in vivo* or in rigid casts of arteries. This is because

(a) the velocity gradient varies rapidly across a thin boundary layer (less than 2 mm thick) near the arterial wall, so any probe must be accurately embedded in the wall: any protuberance can significantly affect the local distribution of shear stress (Lutz *et al.*, 1977); (b) the frequency response of a shear-stress probe must be much better than that of a velocity probe (see § 3.1); and (c) a hot-film shear-stress probe (the only device that has so far been seriously suggested) rectifies the signal, and therefore cannot be accurate during shear reversal (see § 3.1 and the appendix). It is therefore necessary to study arterial wall shear stress theoretically. Part of chapter 2 and much of chapters 3 to 5 are devoted to this end.

For more biochemical details concerning fatty streaks and atherosclerosis, see Porter & Knight (1973) and Lighthill (1975, chapter 13).

1.2.7 *Korotkoff sounds*

This section should not end without mention of a fascinating problem that has intrigued both clinical physiologists and fluid dynamicists for many years, and for which there is still no completely satisfying explanation. This is the phenomenon of Korotkoff sounds, which are used by clinicians to measure systolic and diastolic arterial blood pressure. A cuff containing an inflatable bag is wrapped round the upper arm, and the bag is inflated to pressures well above the systolic pressure, so that the brachial artery is completely closed off. A stethoscope bell is placed over the artery peripherally, and nothing can be heard through it. Then the cuff pressure is gradually reduced, and when it falls below a certain value a characteristic sharp tapping sound is heard each beat. As the cuff is further deflated, this sound becomes louder and is often accompanied by a brief murmur reminiscent of that associated with aortic valve stenosis. Subsequently the sound becomes noticeably muffled, and finally it disappears altogether. The cuff pressure at which the sounds first appear is usually taken to be systolic pressure, and that at which they are muffled is taken to be diastolic pressure; however, simultaneous, independent, direct measurements suggest that these values can be in error by up to 25% unless the cuff length is at least 1.6 times the arm diameter (see Anliker & Raman, 1966; and Steinfeld *et al.* 1974). Correct interpretation of the pressures

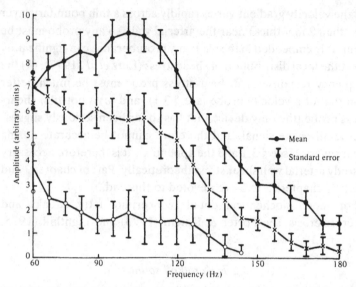

Fig. 1.27. Graph of amplitude against frequency for the high-frequency components of Korotkoff sound: filled circles, sound preceding muffling; crosses, muffled sound; open circles, sound following muffling (background). (After McCutcheon *et al.*, 1967.)

measured in this way will be impossible until there is full understanding of the mechanism by which Korotkoff sounds are produced.

A thorough experimental investigation of the phenomenon was carried out by McCutcheon & Rushmer (1967). They performed a Fourier analysis of the sounds heard before, during and after muffling, showing that it was the frequency components between 60 and 180 Hz that are diminished when muffling occurs (fig. 1.27). By observing the passage through the occluded artery of radio-opaque dyes, these authors were able to show that when the first sound was heard (at a cuff pressure just *below* systolic), a narrow jet could briefly be seen emerging into the stagnant blood downstream of the cuff. The jet soon broke up into turbulence. Observations on casts, made by injecting substances into the arteries of a limb at arterial pressure, while the limb was compressed by a cuff, showed that the cross-sections of the arteries in this situation were non-circular, with smaller area and circumference than usual, which recalls the behaviour of excised veins under compression. Chapter 6 is devoted to the mechanics of flow through flexible tubes under compression, and includes a description of both model experiments and the theories designed to elucidate them. A discussion of some of the mechanisms put forward to explain Korotkoff sounds is included.

1.3 Fluid mechanics of the left ventricle

In this section we examine some of the fluid mechanical phenomena that accompany the filling and emptying of the left ventricle, as a prelude to the analysis of the arterial system, which occupies the next four chapters. We discuss first the mitral valve, then the aortic valve and finally the overall pressure–flow-rate relations of left ventricular ejection.

1.3.1 *The mitral valve*

The mitral valve is set off the axis of the left ventricle (fig. 1.1) and the jet of fluid that enters through it as the ventricle fills is not symmetrical. Model experiments by Bellhouse (1972) showed that when the jet hit the far wall of the ventricle, it curved round and generated what he described as an asymmetrical vortex ring, with a much more extensive vortex in the outflow tract beneath the aortic valve than on the opposite (posterior) side. This confirms the observations of Taylor & Wade (1970) who took cine-angiographic pictures of flow patterns in the left ventricle *in vivo*. Bellhouse (1972) also showed, in experiments with both life-sized and very large model ventricles (in the latter only a very weak vortex developed), that when ventricular contraction begins, the presence of the vortex helps the mitral valve to close before significant backflow develops. He also showed that in both models, even when operated at the rather low frequency of 0.5 Hz, the cross-sectional area of the aperture began to decrease well before the forward velocity into the ventricle reached its maximum, i.e. while the fluid was still being accelerated forwards.

A simple model of the initiation of mitral valve closure can be developed if we ignore the asymmetry of the jet and the vortex, and model the ventricle as an axisymmetric chamber with the mitral valve on the axis of symmetry (fig. 1.28). Suppose the cusps to form

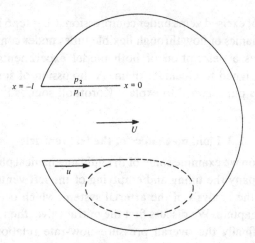

Fig. 1.28. Model for mitral valve closure. The left ventricle is taken to be spherical, and the valve cusps are assumed to form a cylindrical inflow tube. Dotted curve represents the separated vortex sheet. For explanation of symbols, see text.

a parallel-sided orifice when fully open, and suppose that the inflow forms a parallel stream of velocity $U(t)$. If a typical radius of the ventricle is a, then the viscous diffusion time is a^2/ν (ν is the kinematic viscosity of blood) which in man is over half a minute, much greater than the 1 s which is the duration of the cardiac cycle. Therefore the effects of viscosity will be small except in thin boundary layers on the ventricular walls, and in the separation of the stream from the edge of the cusp to form a vortex sheet, which will begin to roll up as indicated in the lower half of fig. 1.28. Apart from this vortex sheet, we assume the flow to be irrotational. The valve will begin to close when the average pressure on the outside of the cusp exceeds that on the inside. These pressures can be estimated from the unsteady form of Bernoulli's equation. On the inside of the cusp, of length l, the longitudinal velocity is U, and the velocity potential ϕ can be written Ux ($x = 0$ at the cusp tip, as shown in fig. 1.28). Thus the pressure, p_1, is given by

$$p_1/\rho + \dot{U}x + \tfrac{1}{2}U^2 = p_0/\rho + \tfrac{1}{2}U^2,$$

where p_0 is the pressure at $x = 0$. The average pressure is then

$$\frac{1}{l}\int_{-l}^{0} p_1 \, dx = p_0 + \tfrac{1}{2}\rho l \dot{U}. \qquad (1.10)$$

On the outside of the cusp, the velocity, u, will vary from zero at the stagnation point at $x = -l$ to a value that may reach U at $x = 0$ when the vortex is strong. As a rough approximation let us take

$$u = U(1 + x/l) \quad \text{and} \quad \phi = U(x + x^2/2l). \tag{1.11}$$

Substitution into Bernoulli's equation then gives the average pressure on the outside to be

$$\frac{1}{l}\int_{-l}^{0} p_2 \, \mathrm{d}x = p_0 + \tfrac{1}{3}\rho U^2 + \tfrac{1}{3}\rho l \dot{U}. \tag{1.12}$$

Thus the difference in the average pressures, from outside to inside, is

$$\tfrac{1}{3}\rho(U^2 - \tfrac{1}{2}l\dot{U}). \tag{1.13}$$

This is negative only when $\dot{U} > 0$ and U^2 is not too great; it changes sign before $\dot{U} = 0$. In one of his examples, Bellhouse took

$$U \approx U_0(1 - \cos 2\pi f t),$$

where $U_0 = 0.6$ m s^{-1}, $f = 0.83$ Hz. The cusp length in his model was 2.8 cm. Change of sign of the quantity (1.13) is then predicted to occur at $t = 0.15$ s. In fact, Bellhouse observed it to occur in his life-sized ventricle at $t \approx 0.33$ s (his fig. 12) and presumably our prediction is an underestimate because the vortex does not induce such a vigorous motion on the outside of both cusps as assumed in (1.11). If we neglect the vortex altogether, and suppose the outside pressure to be uniform at $p = p_0$, then (1.13) would be replaced by $-\tfrac{1}{2}\rho l \dot{U}$, which changes sign at peak forward flow when $\dot{U} = 0$, i.e. at $t = 0.60$ s. In Bellhouse's large-ventricle model, where only a very weak vortex is developed, the valve begins to close at $t = 0.5$ s, which is consistent with the above ideas.

A more accurate theoretical model of mitral valve behaviour would require specification of the flow while the valve was opening and closing as well as while the cusps were parallel. It would also involve calculation of the position of the separated vortex sheet, which could be done by the methods used by aerodynamicists to compute the rolling-up of a vortex sheet off a delta-wing. However, this would all require lengthy numerical computation, and is probably not worth it unless very subtle fluid dynamical factors prove to be important in the design of artificial valves.

1.3.2 The aortic valve

Bellhouse & Talbot (1969) and Bellhouse (1969, 1972) have also performed experiments on models of the tricuspid aortic valve, placed across a tube with a sinus downstream of each cusp. Fig. 1.29 shows a cross-section of one sinus when the valve is open. Flow-visualisation studies showed that (a) the valve cusps opened smoothly; (b) when the tip of a cusp passed the sinus ridge R, a three-dimensional vortex motion was initiated in the sinus in which flow entered each sinus at the middle of the ridge, curled back along the sinus wall, then along the cusp to flow out into the main stream at the points of attachment of the cusp free margins; (c) the vortex motion and the position of the cusp (projecting slightly into the sinus) remained roughly steady until some time after the maximum velocity in the aorta, then (d) the cusp began to close smoothly, being over 72% closed before the aortic flow reversed; (e) only 2% of the net forward flow leaked back through the valve; and (f) the velocity profile in the aorta just downstream of the valve was at all times flat (this is very important for aortic fluid dynamics). In a model without sinuses, the cusps, when open, touched the wall of the aorta and the valve closed suddenly and unevenly *after* the velocity had reversed its direction, with up to 25% leakback.

These observations show that the sinuses are essential for smooth valve closure. The vortex motions are not themselves vital, however, as indicated by experiments in which they were rather weak but smooth closure still occurred (van Steenhofen & van Dongen, 1979). This is consistent with a theory like the second of those described above for the mitral valve, in which no vortex motion is present, the uniform pressure in the sinus being equal to p_A, the pressure in the aorta at the downstream edge of the cusp. This leads to the prediction that the valve will close from a parallel-sided open position when the aortic flow begins to decelerate ($\dot{U} = 0$). The presence of the vortex is expected to aid valve closure, however, as in the case of the mitral valve. Furthermore, for the cusps to protrude stably into the sinuses, in steady flow or at peak flow, the dynamic pressure of the vortex is required to balance the slight pressure excess on the aortic side of the cusp due to the divergence of the flow there.

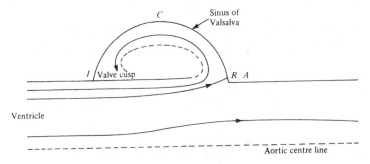

Fig. 1.29. Sketch of the plane of symmetry of a sinus of Valsalva, showing the formation of a vortex in the sinus.

The simplest possible quantitative model of the relations between unsteady pressures and velocities in the aorta and sinuses was given by Bellhouse (1972). Suppose the sinuses to be hemispheres of radius a, and suppose that the upstream and downstream velocity in the aorta (at stations I and A in fig. 1.29) is $U(t)$. Then

$$p_A - p_I = -2\rho a \dot{U}, \tag{1.14}$$

while the pressure on the sinus ridge, R, is stagnation pressure, so

$$p_R - p_A = \tfrac{1}{2}\rho U^2. \tag{1.15}$$

In the experiments of Bellhouse and Talbot (1969), the maximum velocity in the sinus (at point C) was found to be about $0.9U$; if we *assume* it to be equal to U, then we also obtain

$$p_R - p_C = \tfrac{1}{2}\rho U^2, \tag{1.16}$$

so

$$p_C - p_I = -2\rho a \dot{U}, \tag{1.17}$$

while from (1.14) and (1.15),

$$p_R - p_I = \tfrac{1}{2}\rho U^2 - 2\rho a \dot{U}. \tag{1.18}$$

Bellhouse (1972) reported measurements of U, p_R, p_C, p_I as functions of time, and compared (1.16), (1.17), (1.18) with experiment. The results are shown in fig. 1.30, and agreement is seen to be excellent.

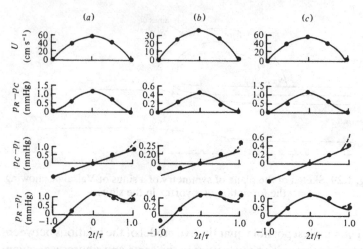

Fig. 1.30. Pressure and velocity measurements for three different ventric-ular ejections through the model valve. (a) Duration of systole $\tau = 0.575$ s, and peak velocity $u_{max} = 56.7$ cm s^{-1} (b) $\tau = 0.91$ s, $u_{max} = 35.0$ cm s^{-1} (c) $\tau = 1.17$ s, $u_{max} = 56.7$ cm s^{-1}. The continuous line represents the simple theory derived in the text; the broken line represents a more complicated theory. (After Bellhouse, 1972.)

In his earlier papers, Bellhouse developed a much more compli-cated theory than this, involving half a Hill's spherical vortex in each sinus, but the theory was still incomplete since it required an empirically based assumption about the strength of the vortex (such as the assumption that the velocity at C is equal to U). Further-more, it resulted in negligible improvement in the predictions of pressures in terms of velocity (fig. 1.30). As with the mitral valve, a more complete theory will require the nettle of the rolling-up vortex sheet to be firmly grasped.

1.3.3 *The dynamics of left ventricular ejection*

In experiments on isolated strips of muscle, the muscle is made to contract against a known load, at a known rate of shortening. In order to apply the results of these experiments to the intact ventri-cle, therefore, it is important to know the load against which the ventricle contracts. A convenient definition of this 'afterload' is in terms of the average pressure in the left ventricle and the rate of change of its volume. However, a more physiologically useful definition would be in terms of the pressure and flow-rate at the entrance to the aorta, which are easier to measure, and which are related through the pressure–flow relations of the vessels downstream, i.e. through the 'input impedance' of the vascular system (see chapter 2); indeed, Milnor (1975) has suggested using this impedance as a *definition* of 'afterload'. In this subsection, therefore, we present a simple model relating aortic pressure and flow-rate to ventricular pressure and volume (this model has been previously presented by Pedley & Seed (1977)). The link, involving the momentum of the blood in the ventricle, is not entirely trivial, and gains added importance from the observation by Noble (1968) that the blood momentum can contribute significantly to ventricular ejection. Thus in the latter stages of systole the ventricular muscle may still be shortening when ventricular pressure, and hence the load, is extremely small, a situation that is not normally encountered in papillary muscle experiments.

Any motion of the blood in the ventricle before the aortic valve opens, including both the vortex behind the mitral valve cusps and the motion resulting from the change in shape of the ventricle as it contracts at constant volume (see § 1.1), is ignored. When ventricular pressure p_v exceeds aortic pressure p_a (fig. 1.12) the aortic valve opens and blood is ejected with a flat velocity profile (as discussed in the last subsection and in § 1.2). Let the velocity of the blood being ejected be $U(t)$. This is related to the rate of change of ventricular volume, $\dot{V}$, by the continuity equation

$$\dot{V} = -AU, \tag{1.19}$$

where A is the cross-sectional area of the aortic valve ring, assumed constant (we ignore the details of aortic valve operation). Hence

$$V(t) = V_0 - A\int_0^t U \, dt, \tag{1.20}$$

where V_0 is the end-diastolic volume of the ventricle, and the valve opens at $t = 0$.

We suppose that the force of contraction of the ventricular muscles can be represented by a single ventricular pressure, p_v, which is the uniform pressure of the fluid in contact with the ventricular wall. This would be strictly true only if the ventricle wall was a spherical shell, of radius R, with a uniform, isotropic tension T in its walls, so that $p_v = 2T/R$. The use of p_v in the following equations provides an effective definition of it for a non-spherical ventricle. We now apply the integral momentum equation to the blood in the ventricle; this states that the net longitudinal force exerted by p_v against the aortic pressure p_a is balanced by the rate at which momentum is ejected from the ventricle plus the rate of change of momentum of the fluid remaining in the ventricle. That is

$$-\int_S \frac{p}{\rho}(\mathbf{i} \cdot \mathbf{n}) \, dS = \rho \int_S (\mathbf{u} \cdot \mathbf{n})(\mathbf{u} \cdot \mathbf{i}) \, dS + \rho \frac{d}{dt}\int_V (\mathbf{u} \cdot \mathbf{i}) \, dV$$

$$(1.21)$$

where S is the surface of the control volume, V, comprising the surface of the ventricle, S', and that of the aortic opening, S_o; $\mathbf{n}$ is the outward normal to S, and $\mathbf{i}$ is a unit vector in the direction of the axis of the aorta (fig. 1.31).

The left-hand side of (1.21) is equal to

$$(p_v - p_a)A,$$

which defines p_v if the ventricle is not a sphere. The first term on the right-hand side of (1.21) consists of a contribution $\rho A U^2$ from the aperture S_o, assuming that, as observed, the valve ring remains stationary as the ventricle wall moves, and another contribution $-\rho B U^2$ from the ventricle surface S'. In the crudest approximation, $B U^2$ would be neglected compared with $A U^2$, but it can be calculated for any given ventricular shape. In general we write

$$(A - B)U^2 = \beta A U^2, \qquad (1.22)$$

and for simplicity take β to be a constant ($= 1$ to first approximation). The case of a spherical ventricle is worked out in detail below.

The second term on the right-hand side of (1.21) is the rate of change of momentum of the fluid in the ventricle. We expect this momentum to be roughly proportional to the volume of the ventricle times the aortic velocity, say

$$\int_V (\mathbf{u} \cdot \mathbf{i}) \, dV = IVU, \qquad (1.23)$$

where I is a number, in general depending on t, but constant if the ventricle and the flow within it remain geometrically similar at all

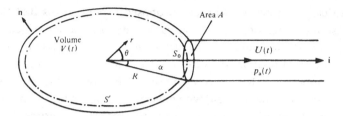

Fig. 1.31. Schematic model of the left ventricle, showing variables and control surface (broken curve) defined in the analysis of ventricular ejection given in the text. (After Pedley & Seed, 1977.)

times. We can calculate I for particular shapes of the ventricle, and again we do so below for the case of the sphere.

Equation (1.21) now reduces to

$$p_v - p_a = \rho \beta U^2 + (\rho/A)(d/dt)(IVU)$$

$$= \rho(\beta - I)U^2 + \rho(d/dt)(IU)\left(\frac{V_0}{A} - \int_0^t U \, dt\right). \quad (1.24)$$

This equation may be used in two ways. (i) If the quantities β, I, V_0/A are known, then (1.24) can be used to infer p_v at all times from measurements of aortic pressure, p_a, and velocity, U. This would enable a comparison to be made between the value of p_v measured by the insertion of a catheter into the ventricle and that defined here; both p_a and U could be measured simultaneously with a single aortic catheter. (ii) Alternatively, if the catheter method of measuring p_v is accepted, one could use two catheters to measure U, p_a and p_v simultaneously (as did Noble, 1968), and from these measurements some or all of the unknown quantities in (1.24) can be inferred. In particular, V_0 and A are very hard to measure accurately *in vivo*, complicated analysis of pairs of X-ray photographs, or the invasive use of yet more instruments, being required (Bergel, 1972c). At the time when $\dot{U} = 0$, (1.24) could be used to estimate $\beta - I$ (assuming β and I to be constants), and then the results at two other times could be used to infer I and IV_0/A. Comparison of the experimental results with (1.24) at all other times would constitute a check on the theory leading to that equation. Also the values inferred for β and I could be used to check predictions of those quantities, as made below for a spherical ventricle. A drawback with this proposed experiment is that the measured values of p_v would already contain some contribution from blood momentum; care would have to be taken to find (empirically) the optimum location of a ventricular catheter.

It is appropriate here to compare the magnitudes of the terms on the right-hand side of (1.24) with measured values of the difference between ventricular and aortic pressure, from the experiments of Noble (1968). According to the theory for a spherical ventricle, presented in the next subsection, the value of $\beta - I$ is predicted to decrease from 0.90 to 0.83 during a contraction in which the ventricular volume falls from 50 cm^3 to 25 cm^3 (representative values for a normal dog), while I is predicted to increase from 0.11 to 0.16. Thus the assumption that these quantities remain constant (at 0.87 and 0.14, say) is a reasonable first approximation. For a peak aortic velocity of 1.2 m s^{-1} (table 1.1), the first term on the right-hand side of (1.24) then takes a maximum value of about 1.3 kN m^{-2} (9.5 mmHg). With a heart-rate of 2 Hz, the maximum positive and negative values of acceleration can be taken to be about ±75 m s^{-2} (Noble, Trenchard & Guz, 1966). Taking $V_0 = 50$ cm^3 and $A = 1.8$ cm^2 (corresponding to an aortic diameter of 1.5 cm), the last term in (1.24) is thus predicted to vary in the range ±2.7 kN m^{-2} (±20 mmHg). Now in conscious dogs, with heart-rates that varied from 1.5 to 2.0 Hz, Noble (1968) recorded a maximum value of $p_v - p_a$ between 1.6 and 2.3 kN m^{-2} (12 and 17 mmHg), and a minimum between -0.5 and -1.9 kN m^{-2} (-4 and -14 mmHg). Although a detailed comparison of the predicted and the observed pressure differences as functions of time cannot be made from Noble's data, and although the exact relation between the value of p_v measured by him and that defined here remains somewhat obscure, the experimental values are fairly close to the predictions, and confirm that the present model is a reasonable one. Predictions from a more accurate, spheroidal model of the ventricle will clearly be useful.

We may note that, although the values quoted above for the difference between ventricular and aortic pressure are small compared with mean aortic pressure (13.3 kN m^{-2}), the larger values amount to about a half of the variation of aortic pressure which takes place each beat (the 'pulse pressure'). Thus the time

variation of canine ventricular pressure is not accurately represented by that of aortic pressure. (Similar results are predicted for
man.) Furthermore, in a subject in which the peak aortic velocity is
as high as $2 \, \mathrm{m \, s^{-1}}$ (as may happen either with a stenosed or an
incompetent aortic valve), the difference between ventricular and
aortic pressure is predicted to be even greater and indeed to be a
significant proportion of mean aortic pressure.

Another way of looking at ventricular dynamics is by way of the
energy equation, which may be written

$$\frac{\partial}{\partial t} \int_V \tfrac{1}{2}\rho q^2 \, \mathrm{d}V + \int_{S'} (\mathbf{u} \cdot \mathbf{n})(p + \tfrac{1}{2}\rho q^2) \, \mathrm{d}S = - UA(p_a + \tfrac{1}{2}\rho U^2),$$

$$(1.25)$$

where $q = |\mathbf{u}|$, the fluid speed. The right-hand side is a surface
integral, like the second term on the left-hand side, evaluated over
the surface S_0 spanning the entrance to the aorta. This integral
represents the rate at which energy is delivered to the aorta. If
(1.25) is integrated with respect to time, from the time of valve
opening to the time of valve closure, then we see that the work done
by the ventricular muscle (the left-hand side) equals

$$-\int (p_a + \tfrac{1}{2}\rho U^2) \, \mathrm{d}V$$

and this is approximately equal to $-\int p_a \, \mathrm{d}V$ if the dynamic pressure
$\tfrac{1}{2}\rho U^2$ is neglected. This 'stroke work' has for many years been taken
as an index of cardiac contractility, despite the fact that it depends
strongly on aortic pressure, which is determined as much by
peripheral conditions as by the direct action of the ventricle (Blinks
& Jewell, 1972). Skalak (1972) and McDonald (1974) give full
accounts of the energetics of the whole circulatory system. It is clear
that for given peripheral conditions the energy equation is a simple
and adequate representation of overall ventricular behaviour (at
least as long as the kinetic energy terms can be neglected), but I
believe that the momentum equation is more suitable for examining
ventricular dynamics throughout ejection.

1.3.4 The case of a spherical ventricle

Here we give more precise calculations of the quantities β and I
(see (1.22) and (1.23)) for the case where the ventricular surface S'

is part of a sphere S of radius R. The exit control surface S_o is taken to be the completion of the sphere S. We take the valve ring to remain fixed while the ventricular wall moves towards it during contraction. We further take the valve ring to be a circle of radius $R \sin \alpha$, which normally remains approximately constant while the ventricle contracts, so α will increase as R decreases. We assume that viscous effects are negligible, since the viscous diffusion time is large compared with the period of the heart beat. Thus the motion will be irrotational as well as axisymmetric, and can be described by a velocity potential ϕ. This is the solution in $r < R$ of

$$\nabla^2 \phi = 0; \qquad \partial\phi/\partial r = f(\cos\theta) \quad \text{on } r = R$$

where

$$f(\cos\theta) = U\cos\theta \quad 0 \leqslant \theta \leqslant \alpha$$
$$= \dot{R}(1 - \cos\theta/\cos\alpha) \quad \alpha < \theta \leqslant \pi. \qquad (1.26)$$

(see fig. 1.31). The function $f(\cos\theta)$ is the normal velocity on the boundary of the sphere; its value in $\alpha < \theta \leqslant \pi$ derives from the facts that the velocity of a point of the ventricle wall is

$$\dot{R}\mathbf{n} - (\mathrm{d}/\mathrm{d}t)(R\cos\alpha)\mathbf{i}$$

and that $R \sin \alpha$ is constant. The solution for ϕ can be expressed in terms of Legendre polynomials as

$$\phi = U \sum_{n=0}^{\infty} \alpha_n \frac{r^n}{R^{n-1}} P_n(\cos\theta);$$

the boundary conditions imply

$$U\alpha_n = \frac{2n+1}{2n} \int_{-1}^{1} P_n(\mu) f(\mu)\, \mathrm{d}\mu. \qquad (1.27)$$

Now the linear momentum per unit mass, from (1.23), is equal to

$$\mathbf{i} \cdot \int_V \nabla\phi\, \mathrm{d}V = \int_S \phi\cos\theta\, \mathrm{d}S$$

$$= U \int_{-1}^{1} \sum_{0}^{\infty} R\alpha_n P_n(\mu) \cdot \mu \cdot 2\pi R^2\, \mathrm{d}\mu$$

$$= \tfrac{4}{3}\pi R^3 U\alpha_1.$$

Hence $I = \alpha_1$, and this can be calculated from (1.27), using (1.26) and the facts that

$$AU = -4\pi R^2 \dot{R} \quad \text{and} \quad A = \pi R^2 \sin^2 \alpha. \qquad (1.28)$$

The result is

$$I = [(1-c)/16c](2+13c+11c^2+7c^3-c^4)$$

where $c = \cos \alpha$; this is tabulated for various values of α in table 1.3. It can be seen that the assumption of constant I as the ventricle contracts will be very approximate if α varies sensibly.

Table 1.3. *Hydrodynamic constants for a spherical left ventricle at different values of the angle subtended by the aortic opening*

α	$\cos \alpha$	$I = \alpha_1$	β_1	β_2	β	$\beta - I$
18.2°	0.95	0.097	0.93	0.098	1.03	0.93
25.8°	0.90	0.19	0.87	0.12	0.99	0.80
31.8°	0.85	0.27	0.81	0.13	0.94	0.67
36.9°	0.80	0.35	0.74	0.14	0.88	0.53
45.6°	0.70	0.50	0.62	0.13	0.75	0.25
60°	0.50	0.75	0.36	0.11	0.47	-0.28
78.5°	0.20	1.27	-0.69	0.028	-0.66	-1.93
Dog						
20.0°	0.94	0.12	0.92	0.10	1.02	0.90
24.5°	0.91	0.17	0.88	0.12	1.00	0.83
Man						
23.1°	0.92	0.15	0.90	0.11	1.01	0.86
29.5°	0.87	0.24	0.83	0.13	0.96	0.72

The quantity β is derived from the first integral on the right-hand side of (1.21), in which $(\mathbf{u} \cdot \mathbf{n})$ is the same as $f(\cos \theta)$ while

$$\mathbf{u} \cdot \mathbf{i} = f(\cos \theta) \cos \theta - (1/R)(\partial\phi/\partial\theta) \sin \theta.$$

Hence

$$\beta AU^2 = 2\pi R^2 \int_{-1}^{1} f^2(\mu)\mu \, d\mu + 2\pi R \int_{-1}^{1} f(\mu)\frac{\partial\phi}{\partial\mu}(1-\mu^2) \, d\mu;$$

we rewrite the first term as $\beta_1 A U^2$ and the second as $\beta_2 A U^2$. The first integral is easy to evaluate, and yields

$$\beta_1 = (1/96c^2)(-c^6 + 55c^4 + 8c^3 + 45c^2 - 8c - 3),$$

which is also tabulated in table 1.3. The second integral, derived from the tangential components of velocity on the spherical surface, must be evaluated from the series solution for ϕ. After some reorganisation, this gives

$$\beta_2 A U^2 = 2\pi R^2 U \sum_{n=1}^{\infty} \alpha_n \int_{-1}^{1} f(\mu)(1-\mu^2)P_n'(\mu)\,d\mu;$$

the integrand can be rewritten using the standard recurrence relations for $P_n(\mu)$, so that

$$\beta_2 A U^2 = 2\pi R^2 U \sum_{n=1}^{\infty} \frac{n(n+1)}{2n+1}\alpha_n \int_{-1}^{1} f(\mu)[P_{n-1}(\mu) - P_{n+1}(\mu)]\,d\mu$$

$$= 4\pi R^2 U^2 \sum_{n=2}^{\infty} \frac{(n-1)n\alpha_{n-1}\alpha_n}{4n^2 - 1} \tag{1.29}$$

from (1.27), after some manipulation. Thus evaluation of β_2 requires evaluation of the α_n and summation of the above series.

The values of $I(=\alpha_1)$, β_1, β_2, β and $\beta - I$ are given for various values of α in table 1.3. It immediately becomes apparent that, if α changes appreciably during ventricular contraction, then so will β (if α is large) and I, and the assumptions of constant β and constant I will be inadmissible. In practice, however, quite considerable changes in ventricular volume can take place without large changes in α. For example, for a dog in which (a) the end-diastolic ventricular volume is 50 cm^3, (b) the volume after ejection is 25 cm^3, and (c) the diameter of the aortic ring is 1.50 cm, the values of α at the beginning and end of the contraction (assuming a spherical ventricle) are 20.0° and 24.5°. The corresponding values for I and β can be seen from table 1.3 to vary by 42% and 2% respectively over this range of α. Similarly, in a man with an end-diastolic volume of 140 cm^3, an end-systolic volume of 70 cm^3, and an aortic ring diameter of 2.5 cm, the values of α before and after contraction are 23.1° and 29.5°. The corresponding variations in I and β are then 60% and 5% respectively. Thus the assumption of constant β (and of constant $\beta - I$) is reasonably good, but that of constant I is not. An improved theory based on a spheroidal model of the ventricle is now under way; the experiments necessary to validate (1.24) and to assess the catheter measurement of ventricular pressure are being performed by Dr. W. A. Seed of the Charing Cross Hospital, London.

CHAPTER 2

PROPAGATION OF THE PRESSURE PULSE

It is the propagation of the pulse that determines the pressure gradient during the flow at every location in the arterial tree, so it is important to begin the mathematical analysis of arterial fluid mechanics with a description of this propagation. The most concise and easily comprehensible outline of the subject is that by Lighthill (1975, chapter 12), and I shall frequently refer to his account in what follows.

2.1 One-dimensional theory, uniform tube, inviscid fluid

2.1.1 Basic theory

It is necessary, as in most branches of applied mathematics, to analyse a simple model before introducing the many modifying features present in reality. We therefore start by considering the propagation of pressure waves in a straight, uniform, elastic tube, whose undisturbed cross-sectional area and elastic properties are independent of the longitudinal coordinate, x. We also take the blood to be inviscid, as well as being homogeneous and incompressible (density ρ); the last two assumptions are made throughout this book. The neglect of viscosity is based on the observation (§§ 1.2, 1.3) that the velocity profiles in large arteries are approximately flat, suggesting that the effect of viscosity is confined to thin boundary layers on the walls; this is confirmed mathematically below. We further suppose that the wavelengths of all disturbances of interest are long compared with the tube diameter, so that the velocity profile will remain flat at all times, and the motion of the blood can be represented by the longitudinal velocity component $u(x, t)$, where t is the time. In any comparison with experiment, u would correspond to the average velocity across the cross-section.

In this one-dimensional model, the (excess) pressure is taken to be $p(x, t)$ and the tube cross-sectional area to be $A(x, t)$. The equation of mass conservation is

$$A_t + (uA)_x = 0, \qquad (2.1)$$

where a suffix denotes partial differentiation, and the momentum equation is

$$u_t + uu_x + (1/\rho)p_x = 0. \qquad (2.2)$$

The elastic properties of the tube wall are taken to be represented by a single-valued relation between p and A:

$$p = P(A). \qquad (2.3)$$

This assumption neglects many features exhibited by a real artery, such as (a) the effect of distension at one value of x on the value of A at another, and (b) the viscoelastic nature of the wall. However, it is the simplest assumption consistent with the one-dimensional model, and $P(A)$ can, in practice, be specified by virtue of the experiments described in § 1.1. Note too that, since p is the *excess* pressure, the *transmural* pressure of an artery, as well as its undisturbed cross-sectional area and elastic properties, will normally depend on x because of the hydrostatic pressure gradient inside the vessel but not outside.

The distensibility of the tube, from (1.5), is

$$D = 1/[AP'(A)];$$

we define a velocity c given by

$$c^2 = (\rho D)^{-1} = (A/\rho)P'(A), \qquad (2.4)$$

and the last term in (2.2) becomes $c^2 A_x/A$. The familiar Riemannian theory for one-dimensional waves (see, for example, Lighthill, 1978, § 2.8) can now be used: adding $\pm c/A$ times (2.1) to (2.2) leads to the conclusion that, on the characteristic curves $C_\pm$ such that

$$dx/dt = u \pm c, \qquad (2.5)$$

the quantities (Riemann invariants)

$$R_\pm = \frac{1}{2}\left[u \pm \int_{A_0}^{A} \frac{c}{A}\,dA\right] \qquad (2.6)$$

are constants, where A_0 is the undisturbed area. Thus non-linear waves are propagated in the $\pm x$-directions with speeds $u \pm c$.

If at all times $c \gg u$, the theory can be linearised, (2.1) and (2.2) can be combined into the wave equation

$$p_{tt} = c^2 p_{xx},$$

and any pressure disturbance is predicted to propagate (in either direction) with constant speed c_0, where the suffix zero means that the quantities in (2.4) are to be evaluated in the undisturbed state. If a mean fluid velocity U_0 were present in the tube, and if $u \ll U_0$, the theory could again be linearised, and the speeds of propagation of the wave $(U_0 \pm c_0)$ would still be constant.

For the case of a circular tube of wall thickness h and undisturbed diameter d, whose response to circumferential stresses can be represented by an effective incremental Young's modulus E, and provided no additional longitudinal stresses are applied (so that the tube will shorten when it is distended), then D is given by (1.6) without the factor $(1 - \sigma^2)$, and

$$c_0 = (Eh/\rho d)^{1/2}. \tag{2.7a}$$

If the longitudinal tethering of artery walls is taken into account, so that radial stretch induces longitudinal tension, D is given by (1.6a) or (1.6b) and c_0 by

$$c_0 = (Eh/\rho d)^{1/2}(1 - \sigma_{\theta x}\sigma_{x\theta})^{-1/2}$$
$$\approx 1.15(Eh/\rho d)^{1/2} \tag{2.7b}$$

when $\sigma_{x\theta} = \sigma_{\theta x} = 0.5$.

The linear theory leading to (2.7a) was first performed by Young (1809), and that value of c_0 is known as the Moens–Korteweg wave speed after two Dutch scientists who rediscovered it in 1878.

2.1.2 *Comparison with experiment*

If values of E measured in static experiments on arteries are inserted into (2.7a), then values of c_0 are predicted that range from about 5.8 m s^{-1} in the aorta (of a dog) to nearly 10 m s^{-1} in peripheral arteries (see table 1.1). These predictions can be compared with measured values of the wave speed, which are also quoted in table 1.1 (see Bergel (1972b) or McDonald (1974, p. 418) for more extensive lists and references to the original experiments). Wave speed is not in fact very easy to measure *in vivo* because of the change in shape of the pressure pulse with distance along the arterial tree (fig. 1.14 or 1.17), which is primarily a consequence of wave reflection (§ 2.3). The only reliable methods are *either* to measure the speed of propagation of the initial part (the 'foot') of the pulse where the pressure begins to rise (McDonald, 1974, p. 394), *or* to introduce short trains of artificial, high-frequency, sinusoidal waves and measure their propagation speed (Anliker, Histand & Ogden, 1968a). These methods allow a pair of pressure measurements to be made (with two manometers) before reflected waves have time to distort the signal. The two methods give similar results, although the frequencies used by Anliker *et al.* (1968a) are higher than those important in the pressure pulse. Note that the increase of wave speed with distance from the heart means that any measured values can be regarded only as averages over the distance between manometers.

Considering the crudeness of the linear theory leading to (2.7a), the agreement between prediction and observation (table 1.1) appears to be remarkably good. Actually the agreement is not quite as good as indicated, because (a) the dynamic, not the static, Young's modulus should be used, which is greater by a factor of about 1.2 (p. 20), and (b) (2.7b) should be used instead of (2.7a). This has the effect of increasing the predictions of c_0 by about 25%, so that they exceed the observations considerably, and we should seek mechanisms that might reduce the predicted wave speed. Nevertheless, the qualitative agreement suggests that the simple linear model, incorporating merely wall elasticity and fluid inertia, contains the most essential features of pulse propagation.

The linear theory indicates that the speed of propagation of small-amplitude waves, c_0, depends on the distensibility of the tube in its undisturbed state. But as we saw in § 1.1, the distensibility of arteries falls as the distending pressure, and hence the undisturbed cross-sectional area, increases. Therefore the wave speed should increase with blood pressure. In particular, the wave speed during systole ($p \approx 16 \text{ kN m}^{-2}$) will be greater than that during diastole ($p \approx 10.7 \text{ kN m}^{-2}$); if the difference between systolic and diastolic wave speed is significant, then the use of linear theory to describe the propagation of the pulse would be inappropriate. This was tested experimentally in the canine aorta by Histand & Anliker

(1973), who superimposed short trains of high-frequency waves on the natural pressure pulse during both systole and diastole, and measured the propagation speed. Their results were complicated by the fact (already noted) that the propagation speed is increased by the local average convection velocity, which is zero in diastole but significant in systole. They therefore measured the speed of propagation upstream as well as downstream, and took the average to indicate the intrinsic wave speed at the pressure concerned. The difference between upstream and downstream wave speeds normally reached about 1 m s^{-1} in systole (indicating a rather low peak flow velocity of 0.5 m s^{-1}), but in some animals rose to 2 m s^{-1}. More important, perhaps, is the fact that the intrinsic wave speed was greater in systole than in diastole by about 1 m s^{-1}, or 20% of its normal measured value (the 'foot-to-foot' wave speed should be roughly the same as the diastolic value, since the pressure and flow velocity are only beginning to increase when this part of the pulse passes). These results were confirmed and extended by experiments in which the aorta was occluded below or above the measuring site, so that the mean pressure could be raised or lowered beyond the normal range (fig. 2.1).

As well as the non-linearity of the pressure–area relation, fluid mechanical non-linearities may also affect the validity of linear theory if u/c is not much less than 1 at all times. In the thoracic aorta of a normal dog, $u_{max}/c \approx 0.20$–0.25 (see table 1.1), and a similar ratio obtains in man. The ratio becomes smaller in smaller arteries, where u_{max} decreases and c increases. Thus the crude linear theory is expected to give a reasonable first approximation to the description of wave propagation, with errors of at most about 25% arising from fluid mechanical and elastic non-linearities in the aorta. The fluid mechanical non-linearities become less important in smaller arteries, but the elastic ones do not because the amplitude of the pressure pulse increases peripherally (fig. 1.14).

Non-linearities assume greater importance in abnormal conditions where either (a) the artery walls become very distensible, so that both c falls (increasing u_{max}/c) and the area change for a given pressure pulse increases, or (b) the output of the ventricle per beat is significantly increased, so that u_{max} is increased. The latter occurs, for example, when the aortic valve does not close properly (i.e. is

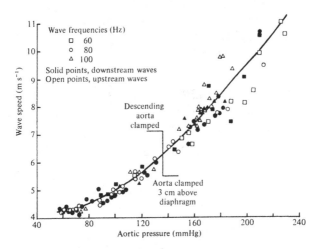

Fig. 2.1. Wave speed plotted against pressure in the canine aorta. Different pressures were obtained by occluding the aorta either above or below the test segment. Each point represents the average of a peak and a successive trough of a small sine wave. (After Histand & Anliker, 1973.)

'incompetent'), so that a large proportion of the blood ejected by the ventricle in systole flows back into it again in diastole. The heart responds to this condition (over a long period) by becoming greatly enlarged and increasing the volume ejected in an attempt to maintain the peripheral circulation. The amplitude of the pulse is then very large; indeed, one way of diagnosing the condition is by a very strong pulse in the limbs, called the arterial 'pistol shot'. It is virtually certain that non-linear theory is required to describe such a pulse; indeed, Anliker, Rockwell & Ogden (1971) propose that it is the manifestation of a hydraulic jump, or shock, in the arterial system (see below).

2.1.3 *Attenuation*

Before developing a non-linear analysis, we should consider one further aspect of the experiments of Anliker *et al.* (1968*a*) and of Histand & Anliker (1973). This concerns the attenuation of waves, which must take place in any real system, but which the peaking of the pressure pulse (fig. 1.14(*a*)) effectively obscures in large arteries. The introduction of short trains of high-frequency waves is an ideal way of studying attenuation, and Anliker *et al.* (1968*a*)

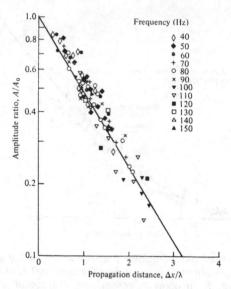

Fig. 2.2. Attenuation of sinusoidal pressure waves in a dog's aorta, with A/A_0 plotted on a logarithmic scale against propagation distance in wavelengths. $\Delta x = 6$ cm; attenuation constant $k = 0.75$. (After Anliker, 1972.)

were able to show convincingly that small-amplitude sinusoidal pressure waves in the aorta had a constant rate of attenuation per wavelength (fig. 2.2). That is, the pressure in the artery takes the form

$$p = p_1 + p_2\, e^{-kx/\lambda}\, e^{i\omega(t-x/c)}, \tag{2.8}$$

where p_1 is the underlying blood pressure, p_2 is a measure of the amplitude at $x = 0$, and $\lambda = 2\pi c/\omega$ is the wavelength of the wave. Anliker et al. (1968a) found that the attenuation constant k normally lies between 0.7 and 1.0 for waves propagating peripherally in the canine aorta (i.e. attenuation of 50–63% per wavelength; this would have a small effect on the natural pulse in the aorta because the wavelength for a fundamental frequency of 2 Hz is at least 2.5 m). Histand & Anliker (1973) found that upstream propagated waves suffer greater attenuation, with k between 1.3 and 1.5.

The source of this wave attenuation lies primarily in the visco-elastic properties of the vessel wall. This follows from calculations made on the assumption that all attenuation comes from blood viscosity, which predict $k \approx 0.04$ for the aorta (§ 2.2). However, if we use a complex Young's modulus (or distensibility) to describe the viscoelastic properties of the artery wall, i.e. $E_\theta = E_{dyn} + i\eta\omega$ (p. 20), and put this into (2.7a or b) for c, we deduce a value of k given by

$$k = 2\pi \tan \tfrac{1}{2}\theta \quad \text{where} \quad \tan \theta = \eta\omega/E_{dyn}. \tag{2.9}$$

This gives a value of k between 0.47 and 0.62 for $\eta\omega/E_{dyn}$ between 0.15 and 0.2, and, combined with the prediction for blood viscosity, approximately accounts for the observed k of 0.7–1.0.[†] It is not obvious why the attenuation of upstream waves in arteries should be significantly greater; it may be related to the taper of the arteries (§ 2.3).

We may note further that Anliker, Wells & Ogden (1969) measured the propagation characteristics of pressure waves travelling towards the heart in the canine vena cava by a method similar to that used in arteries. They found a wave speed rising from about 2.5 m s^{-1} at a (physiological) distending pressure of 1 kN m^{-2} to 5.0 m s^{-1} at a distending pressure of 2.3 kN m^{-2}. They also found approximately constant attenuation per wavelength, but with a value of k that ranged from 0.6 to 3.3 in different animals at different pressures.

2.1.4 Non-linear effects

The theory leading to (2.5) and (2.6) already incorporates non-linearities; their effect in such an inviscid, uniform-tube model is to cause the front of any wave to steepen, because the wave speed increases with both u and p. Such steepening of the pressure pulse is commonly observed (§ 1.2). Ultimately this steepening creates a discontinuity, and a shock is formed. Of course, the non-dissipative equations (2.1)–(2.3), from which the existence of a shock is predicted, will break down before a discontinuity actually appears, because the neglected dissipative terms become important. When a steady shock (or wavefront) is set up, these terms exactly balance the non-linear terms that are responsible for the steepening. The

---

[†] Some very recent experiments suggest that there is still a considerable gap, however (Dr C. D. Bertram, personal communication).

following prediction of when the shock will form will be accurate only if the dissipative terms are very small except for a short time before the steady state is set up. Suppose that blood is ejected from the ventricle ($x = 0$) into a uniform tube with velocity $U(t)$, which is positive for $t \geq 0$ (at least for some time) and zero for $t < 0$. Assume the flow and the tube to be undisturbed for $t < 0$. Then the usual methods (Lighthill, 1978) can be applied to the $x > 0$, $t > 0$ quadrant of the (x, t)-plane. Write

$$\int_{A_0}^{A} \frac{c}{A} \, dA = V(c)$$

where $V(c_0) = 0$ and $V(c)$ is assumed to be a single-valued function with a unique inverse. For $x > c_0 t$, $u = 0$ and $c = c_0$. Now all C_- characteristics originate (at $t = 0$) in the region $x > c_0 t$, so that, from (2.6), $u = V(c)$ everywhere. Hence, on the C_+ characteristics in the region $0 < x < c_0 t$, $u + V(c) = 2u = $ constant, and so $c(=c(u))$ is also constant, and the characteristics are straight, a typical one having the equation

$$x = \{U(\tau) + c[U(\tau)]\}(t - \tau).$$

Two C_+ characteristics first cross, and hence a shock is formed, at a value of t equal to the minimum of

$$\tau + F[U(\tau)]/\dot{U}(\tau)F'[U(\tau)] \qquad (2.10)$$

for values of τ such that $\dot{U}(\tau) > 0$, where $F(U) \equiv U + c(U)$.

A simple example of a function $P(A)$, which has the property that the tube becomes less distensible as A increases, is

$$P = \tfrac{1}{2}\rho c_0^2 A^2 / A_0^2 + \text{constant.} \qquad (2.11)$$

This is very convenient for working out numerical examples since $c = c_0 A / A_0$ and $V(c) = c - c_0$, so that $F(U) = 2U + c_0$. Let us also choose a simple but realistic form for $U(t)$:

$$\begin{aligned} U(t) &= U_0[1 - (1 - t/t_0)^2] & 0 \leq t \leq 2t_0, \\ &= 0 & t > 2t_0; \end{aligned} \qquad (2.12)$$

this is plotted in fig. 2.3; $2t_0$ is the duration of systole, while U_0 is the maximum systolic velocity. The minimum value of (2.10) for $0 \leq \tau \leq t_0$ (values of τ for which $\dot{U} > 0$) occurs at $\tau = 0$, because U is an increasing function and $\dot{U}$ a decreasing function. The time at which

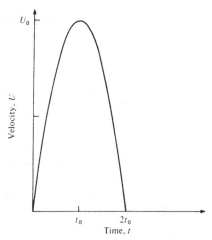

Fig. 2.3. Sketch of the velocity waveform used in the calculation of the time and place at which a hydraulic jump is expected to develop.

a shock forms is therefore $t_s = c_0 t_0 / 4U_0$, and the position at which it forms is $x_s = c_0^2 t_0 / 4U_0$. For a normal dog's aorta, with $c_0 \approx 5$ m s^{-1}, $U_0 \approx 1.2$ m s^{-1}, $t_0 \approx \frac{1}{12}$ s (based on a heart-rate of 2 Hz, and systole lasting one-third of the cycle), t_s is about 0.09 s, and x_s about 43 cm. This is somewhat greater than the length of the aorta, and would in practice be greater still because of the increase of c_0 and decrease of U_0 with distance from the heart; dissipative effects would also increase the prediction. Thus the occurrence of shocks is not normally to be expected. However, if U_0 is increased to, say, 3 m s^{-1} in a diseased system, x_s is predicted to be only about 17 cm, which is quite feasible, and means that there is a *prima facie* case for examining the mechanics of shocks in elastic tubes.

We analyse such a shock, represented as a step jump in A, u and p in the same way as a hydraulic jump on a channel (Lighthill, 1978, § 2.12). Take axes fixed in the shock in order to make the problem steady, and consider the flux of mass, momentum and energy across the boundaries of the control surface S (fig. 2.4). Conservation of mass implies

$$u_1 A_1 = u_2 A_2, \qquad (2.13)$$

where subscripts 1,2 represent upstream and downstream conditions respectively; note that u_1 is actually the speed of propagation

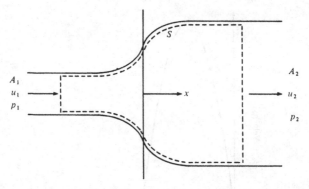

Fig. 2.4. Control surface, S, used to derive the jump conditions for a hydraulic jump in an elastic tube; the axes are fixed in the shock.

of the shock into fluid initially at rest. We take $A_1 < A_2$ as drawn. Conservation of momentum is a bit more complicated than in the case of the hydraulic jump, because of the contribution to the momentum flux of the pressure exerted on the fluid by the non-parallel walls of the shock. The equation is

$$A_1(p_1 + \rho u_1^2) + \int_1^2 p \, dA = A_2(p_2 + \rho u_2^2), \qquad (2.14)$$

where the integral is the extra pressure force term. We must recognise that within the shock our simple model of the elasticity might break down, so that p in the integral may not be equal to $P(A)$ (see (2.3)). Finally, energy conservation requires that energy cannot be gained by fluid as it passes the shock, i.e.

$$p_1 + \tfrac{1}{2}\rho u_1^2 = p_2 + \tfrac{1}{2}\rho u_2^2 + R, \qquad (2.15)$$

where $R \geq 0$.

These equations can be manipulated in various ways. Equations (2.13) and (2.14) lead to

$$\rho u_1^2 (A_1/A_2)(A_2 - A_1) = \bar{A}(p_2 - p_1), \qquad (2.16)$$

where we define

$$\bar{A}(p_2 - p_1) = A_2 p_2 - A_1 p_1 - \int_1^2 p \, dA$$

$$= \int_1^2 A \, dP - \int_1^2 (p - P) \, dA.$$

The first term on the right is equal to $\int_1^2 \rho c^2 \, dA$, which is greater than $\rho c_1^2 (A_2 - A_1)$ as long as c^2 is a monotonically increasing function of A and $A_2 > A_1$ (as we suppose). Thus if $p = P(A)$ throughout the shock, then

$$(u_1^2/c_1^2) > (A_2/A_1) > 1.$$

Similarly we can show $u_2^2 < c_2^2$ in this case. Thus upstream of the shock the flow is supercritical, and downstream it is subcritical, as for hydraulic jumps. However, this result may not be true if $p \neq P(A)$ in the shock and if $\int_1^2 (p - P) \, dA$ is sufficiently large.

Equations (2.13)–(2.15) also yield

$$R\bar{A}/\rho(u_1^2 - u_2^2) = \tfrac{1}{2}\bar{A} - A_1 A_2/(A_1 + A_2). \tag{2.17}$$

Again, if $p = P(A)$ in the shock, $R > 0$ requires $u_1 > u_2$ (as shown in fig. 2.4) as long as

$$\frac{\int_1^2 A(dP/dA)\, dA}{P(A_2) - P(A_1)} > \frac{2A_1 A_2}{A_1 + A_2}. \tag{2.18}$$

This condition imposes a constraint on the function $P(A)$, that the average area, weighted by the slope of the pressure–area curve, should be greater than the harmonic mean area. The condition (2.18) is satisfied when $P(A)$ is a linear function,

$$P = \alpha + \beta A \quad (\beta > 0),$$

for the left-hand side is then equal to $\tfrac{1}{2}(A_1 + A_2)$; clearly, therefore, (2.18) is satisfied for any function $P(A)$ for which dP/dA increases with A, including our simple quadratic case (2.11). It is interesting to note that the two sides of (2.18) are equal for the case

$$P = \tfrac{1}{2}\rho c_0^2 (1 - A_0^2/A^2),$$

chosen by Olsen & Shapiro (1967) to represent a rubber tube. Thus a discontinuity can propagate without any energy loss, which explains Olsen & Shapiro's observation that arbitrary non-linear waves could propagate without change of form.

We have concluded, then, that for realistic pressure–area relations, obtaining throughout the shock, there should be some energy loss in the fluid ($R > 0$). This energy loss, as in the case of the hydraulic jump, may be associated with the significant viscous dissipation that would occur if the flow emerging from the shock formed a separated jet that became turbulent, or (in the case of a weak shock) it could be propagated away in non-linear waves, for which a two-dimensional theory would have to be developed. However, there is another source of dissipation available in the case of blood vessels, and that is the viscoelastic properties of their walls. We have seen already that this is responsible for much more wave attenuation than blood viscosity, and the analysis below indicates that it is the most probable source of dissipation for cardiovascular shocks, since the presence of viscoelasticity would limit the wall slope at the shock, even for a strong shock, to a value below that at which the blood flow would separate. Thus significant turbulent dissipation is unlikely to occur.

If all energy loss takes place in the vessel walls, there is none in the fluid, and R in (2.15) is zero. However $p \neq P(A)$ throughout the shock because energy is dissipated at any location as the vessel area increases. This means that, with axes fixed in the tube wall, we should write

$$p = P(A) + N(A, \partial A/\partial t), \qquad (2.19)$$

where N is a monotonically increasing function of $\partial A/\partial t$ with $N(A, 0) = 0$ (Kivity & Collins, 1974). The simplest form of N, corresponding to the complex Young's modulus used above, is

$$N = g(A)(\partial A/\partial t). \qquad (2.20)$$

For a small-amplitude sinusoidal wave of the form $e^{i\omega(t-x/c)}$, superimposed on a basic state of cross-sectional area A_0, the linearised theory based on (2.1) and (2.2) leads to the prediction that

$$D^{-1} = \rho c^2 = A_0[P'(A_0) + i\omega g(A_0)].$$

Thus we have

$$\eta\omega/E_{\mathrm{dyn}} = \mathrm{Im}\,(D^{-1})/\mathrm{Re}\,(D^{-1}) = \omega g(A_0)/P'(A_0), \qquad (2.21)$$

which is related to the attenuation constant k, (2.9). Note that since $\eta\omega$ is observed to be independent of ω, $g(A_0)$ will be frequency-dependent.

If we now revert to the frame of reference fixed in the shock (fig. 2.4), the problem once more becomes steady, $\partial/\partial t$ becomes

$u_1\, \partial/\partial x$ and we obtain

$$p = P(A) + u_1 g(A)\, dA/dx.$$

Thus

$$\int_1^2 (p - P)\, dA = u_1 \int_1^2 g(A)\left(\frac{dA}{dx}\right)^2 dx > 0,$$

and (2.17) with $R = 0$ leads to

$$u_1 \int_1^2 g(A)\left(\frac{dA}{dx}\right)^2 dx = \int_1^2 A\, \frac{dP}{dA}\, dA - \frac{2A_1 A_2}{A_1 + A_2}[P(A_2) - P(A_1)].$$

$$(2.22)$$

For a shock corresponding to a given area change, (2.22) will yield an estimate for the slope (related to dA/dx) as long as u_1 is known; (2.13), combined with (2.15) with $R = 0$, yields

$$\rho u_1^2 = [2A_2^2/(A_2^2 - A_1^2)][P(A_2) - P(A_1)], \qquad (2.23)$$

which determines u_1.

We consider a particular example in which $P(A)$ is a quadratic function, given by (2.11), and $g(A)$ is a constant, ρg_0, so that (2.23) gives

$$u_1 = A_2 c_0 / A_0$$

and (2.22) gives

$$g_0 \int_1^2 \left(\frac{dA}{dx}\right)^2 dx = \frac{c_0 (A_2 - A_1)^3}{3 A_0 A_2}.$$

To estimate the wall slope, let L be a scale for the distance over which the area change $\Delta A = A_2 - A_1$ takes place; then

$$\Delta A / L = c_0 (\Delta A)^2 / 3 g_0 A_0 A_2. \qquad (2.24)$$

For the aorta, suppose $A_1 = A_0 \approx 1.8$ cm^2, and take $\Delta A = 0.7$ cm^2 normally (corresponding to a pressure pulse of 5.3 kN m^{-2}) and take $\Delta A = 1.8$ cm^2 in a diseased state (when the pressure pulse is 13.3 kN m^{-2}); then $(\Delta A)^2/3A_0 A_2$ varies between 0.04 and 0.17. We take $c_0 = 5$ m s^{-1} and from (2.21) with $\eta \omega / E_{dyn} = \frac{1}{6}$ and $\omega = 20\pi$ (taken to be a typical frequency of shock passage) we take $g_0 = 3.7 \times 10^3$ s^{-1}, and therefore obtain

$$\Delta A / L \approx 0.005 \text{ cm } (\Delta A = 0.7 \text{ cm}^2) \quad \text{or} \quad 0.023 \text{ cm } (\Delta A = 1.8 \text{ cm}^2).$$

If the tube is circular, this corresponds to a wall slope of $0.05°$ ($\Delta A = 0.7$ cm^2) or $0.22°$ ($\Delta A = 1.8$ cm^2), which in either case is too gradual for flow separation to take place. The slope would be greater only if ω were greater and g_0 smaller.

In this particularly simple example, the 'shock structure' can be calculated exactly, as shown by Kivity & Collins (1974). The constancy of the flow-rate, uA, and pressure head, $p + \frac{1}{2}\rho u^2$, means that (2.19) can be written

$$N(A, u_1 \, dA/dx) = P(A_1) - P(A_2) + \frac{1}{2}\rho u_1^2 (1 - A_1^2/A^2).$$

Taking $P(A)$ of the form (2.11), and N of the form (2.20) with $g(A) = \rho g_0$, and recognising that $A_1 = A_0$ and $u_1 = A_2 c_0/A_0$, we obtain

$$(2g_0 A_2/c_0) \, d\alpha/dx = (1/\alpha^2)(\alpha_2^2 - \alpha^2)(\alpha^2 - 1),$$

where $\alpha = A/A_0$ and $\alpha_2 = A_2/A_0$. This can be integrated to give

$$\left(\frac{\alpha - 1}{\alpha + 1}\right)\left(\frac{\alpha_2 + \alpha}{\alpha_2 - \alpha}\right)\alpha_2^3 = \exp\left[\frac{c_0}{g_0 A_0} \frac{(\alpha_2^2 - 1)}{\alpha_2}(x - x_0)\right],$$

where x_0 is the location of the shock centre. The longitudinal length-scale revealed by the solution is $(g_0 A_0/c_0)[\alpha_2/(\alpha_2^2 - 1)]$, which agrees dimensionally with (2.24).

The application of this model to arteries is rather unsatisfactory because of the assumption of constant g_0, since experiments show that g_0 is inversely proportional to ω for sinusoidal disturbances of angular frequency ω greater than 4π s^{-1}. Now a suitable choice for ω in the case of a passing shock is some multiple, β, of $2\pi c_0/L$, where L is the axial length-scale. But then the quantity L drops out of (2.24), which serves merely to determine β, and appears to permit an arbitrary value of L. What this means is that the details of shock thickness and structure in an artery will depend on rather more subtle features of the wall viscoelastic properties than can be described by (2.20) with constant $g(A)$. Presumably the low-frequency contributions, describing the slow departure of the tube from its cylindrical shape far upstream and downstream dominate the expression for shock thickness, since a constant g_0 is appropriate there. The problem clearly merits further study. Collins *et al.* (1976) have applied the analysis of Kivity & Collins (1974) to experimentally produced shocks in Silastic tubes, and used their observations to infer the viscoelastic properties of the tubes. They propose using the same technique in arteries, but the doubts expressed above indicate that their results will have to be examined with caution.

2.2 Two-dimensional theory, uniform tube, viscous fluid

While the above one-dimensional approach is believed to contain most of the features pertinent to the pulse wave in arteries modelled as uniform tubes, certain elastic phenomena cannot be described by that theory. Furthermore, the effects of viscosity can be incorporated in a one-dimensional theory only after an analysis of the velocity profile in a parallel-sided tube has been separately performed (Lighthill, 1975, chapter 12). We therefore present here a *linearised* analysis of the axisymmetric motion of a viscous fluid contained in a circular tube whose wall is a viscoelastic, orthotropic membrane, tethered to a rigid external structure by linear constraints. The analysis closely follows and extends the paper of Atabek (1968), which is an excellent example of this type of theory, based on, but going further than, the original work of Womersley (1957). Other relatively recent examples are the papers of Anliker & Dorfman (1970) and of Jones, Anliker & Chang (1971). Most of these papers have examined the effects of the various parameters at great length, but with a purely numerical approach. The purpose of this section is to bring out the main qualitative features of the results in a simple analytical way.

Equations of motion of the wall

Let the undisturbed state of the tube be that of a uniform, circular thin-walled cylinder, radius a, density ρ_w and wall thickness h, with a uniform longitudinal tension S per unit circumference, and a circumferential tension T per unit length corresponding to transmural pressure P_0 ($T = P_0 a$). We shall use cylindrical polar coordinates (x, r, θ) and restrict attention to axisymmetric disturbances in which no displacements occur in the θ-direction (this last condition rules out axisymmetric torsional waves, which have been analysed by Anliker & Maxwell (1966)). Let the displacement of a point in the wall be $(\xi, \eta, 0)$, functions of only x and t, and let the wall be subjected to *external* stresses $(X, P_0 + Y, 0)$. Let the perturbation tensions in the axial and circumferential directions be s' and t' respectively.

The equations of motion of the wall can be deduced from the linearised version of the standard equations of motion of axisymmetric shells (Flügge, 1973), and are

$$\left.\begin{array}{l} \rho_w h \dfrac{\partial^2 \xi}{\partial t^2} = \dfrac{(S-T)}{a} \dfrac{\partial \eta}{\partial x} + \dfrac{\partial s'}{\partial x} + X, \\[3mm] \rho_w h \dfrac{\partial^2 \eta}{\partial t^2} = \dfrac{T\eta}{a^2} - \dfrac{t'}{a} + T \dfrac{\partial^2 \eta}{\partial x^2} + Y. \end{array}\right\} \tag{2.25}$$

Stress–strain relations
In an orthotropic viscoelastic shell the stress perturbations can be most easily related to the displacements if we restrict attention to sinusoidal disturbances of angular frequency ω, in which all variables are proportional to $e^{i\omega t}$. Then a linear relation between stress, strain and rate of strain reduces to a linear relation between stress and strain, as long as all variables are allowed to be complex. We obtain

$$\left.\begin{array}{l} t' = B_{11}\eta/a + B_{12}\,\partial\xi/\partial x, \\[2mm] s' = B_{21}\eta/a + B_{22}\,\partial\xi/\partial x, \end{array}\right\} \tag{2.26}$$

where in terms of (complex) Young's moduli and Poisson's ratios, different in the x- and θ-directions, we have

$$\left.\begin{array}{l} B_{11} = E_\theta h/(1 - \sigma_\theta\sigma_x),\; B_{22} = E_x h/(1 - \sigma_\theta\sigma_x), \\[2mm] B_{12} = E_\theta h \sigma_x/(1 - \sigma_\theta\sigma_x) = B_{21} = E_x h \sigma_\theta/(1 - \sigma_\theta\sigma_x), \end{array}\right\} \tag{2.27}$$

(see Atabek (1968), and Patel & Vaishnav (1972)).

Stresses exerted on the wall
Two types of stress are exerted on the wall. (i) There are the hydrodynamic stresses exerted by the fluid within the tube, and equal to

$$[p - 2\mu\,\partial v/\partial r]_{r=a} \quad \text{and} \quad -\mu[\partial u/\partial r + \partial v/\partial x]_{r=a}$$

in the r- and x-direction, respectively, where $(u, v, 0)$ is the fluid velocity vector, and p is the fluid pressure perturbation. (ii) There are also tethering stresses exerted by the material outside the tube.

We model these (following Patel & Fry, 1966) by supposing that the external tissue provides added inertia, stiffness and (viscoelastic) damping. Thus we take

$$
\left.
\begin{aligned}
X &= -\mu\left[\frac{\partial u}{\partial r}+\frac{\partial v}{\partial x}\right]_{r=a} - M_1\frac{\partial^2\xi}{\partial t^2} - L_1\frac{\partial\xi}{\partial t} - K_1\xi, \\
Y &= \left[p-2\mu\frac{\partial v}{\partial r}\right]_{r=a} - M_2\frac{\partial^2\eta}{\partial t^2} - L_2\frac{\partial\eta}{\partial t} - K_2\eta.
\end{aligned}
\right\}
\tag{2.28}
$$

Equations of motion and kinematic boundary conditions for the fluid
The linearised axial and radial momentum equations are

$$
\left.
\begin{aligned}
\frac{\partial u}{\partial t} &= -\frac{1}{\rho}\frac{\partial p}{\partial x}+\nu\left(\frac{\partial^2 u}{\partial r^2}+\frac{1}{r}\frac{\partial u}{\partial r}+\frac{\partial^2 u}{\partial x^2}\right), \\
\frac{\partial v}{\partial t} &= -\frac{1}{\rho}\frac{\partial p}{\partial r}+\nu\left(\frac{\partial^2 v}{\partial r^2}+\frac{1}{r}\frac{\partial v}{\partial r}-\frac{v}{r^2}+\frac{\partial^2 v}{\partial x^2}\right),
\end{aligned}
\right\}
\tag{2.29}
$$

while the continuity equation is

$$
\frac{1}{r}\frac{\partial}{\partial r}(rv)+\frac{\partial u}{\partial x}=0.
\tag{2.30}
$$

The linearised kinematic boundary conditions at the wall are

$$
u|_{r=a}=\frac{\partial\xi}{\partial t}, \qquad v|_{r=a}=\frac{\partial\eta}{\partial t},
\tag{2.31}
$$

while on the axis we have

$$
v=\frac{\partial u}{\partial r}=0 \quad\text{on}\quad r=0.
\tag{2.32}
$$

Statics
In static experiments to measure the elastic properties of excised blood vessels, it is usual to define 'incremental' values of the elastic parameters (§ 1.1). An incremental value for B_{11}, for example, would be equal to a times the increment of circumferential tension, t', for unit increment in radial strain, η, longitudinal strain being zero (see the first of equations (2.26)). Consider an experiment, therefore, in which ξ is held to zero and an increase in internal pressure, p, is applied. Then (2.25) and (2.26), with (2.28) in which

radial tethering is absent, lead to

$$pa = [(B_{11} - T)/a]\eta.$$

Thus the incremental value of B_{11} is in fact $(B_{11} - T)$, and it is this value that is measured, and through (2.27) related to incremental Young's modulus, etc.

We find similarly, by considering an experiment in which p remains zero, but a small extra longitudinal stress ΔX is applied to the artery, that

$$[B_{22} - B_{12}(B_{21} + S - T)/(B_{11} - T)] \, \partial^2 \xi/\partial x^2 = -\Delta X,$$

while if η could be held zero we would have

$$B_{22} \, \partial^2 \xi/\partial x^2 = -\Delta X, \quad (B_{12}/a) \, \partial \xi/\partial x = p.$$

Thus it is clear that the incremental value of B_{21} is $B_{21} + S - T$, while B_{12} and B_{22} are unchanged. For a full discussion of the experiments needed to measure the incremental values of B_{11}, B_{21} etc., see Patel & Vaishnav (1972).

Dispersion relation

The problem is now completely specified. We reduce it to one in which only ordinary differential equations have to be solved by supposing every variable to be proportional to $e^{i\omega(t - x/c)}$, where ω is the real angular frequency and c is the complex wave speed. We denote the corresponding amplitude of every quantity by the suffix 1. The linearisation of the fluid mechanical equations is valid as long as

$$|u_1/c| \ll 1 \tag{2.33}$$

always; the linearisation of the elasticity requires

$$|\eta_1/a| \ll 1 \quad \text{and} \quad |\xi_1/a| \ll 1,$$

which are both equivalent to (2.33) from the kinematic boundary conditions (2.31), together with the continuity equation (2.30). We simplify the equations even further by making the long-wavelength approximation, $|\omega a/c| \ll 1$, which is certainly valid *in vivo* (rather more so than the linearity approximation), and which means that (a) the $\partial^2/\partial x^2$ terms on the right-hand sides of (2.29) are negligible, (b) the second of equations (2.29) reduces to $\partial p_1/\partial r = 0$, and (c) the second term in each square bracket in (2.28) is much less than the first. This assumption is not a necessary condition for solving the equations, but it makes the solution less cumbersome.

The first of equations (2.29) now reduces to Bessel's equation, and the solution satisfying the boundary condition at $r = 0$ is

$$u_1(r) = (p_1/\rho c_0)[c_0/c + A J_0(i^{3/2}\alpha r/a)/J_0(i^{3/2}\alpha)], \quad (2.34)$$

where c_0 is a scale for c, which can conveniently be taken equal to the Moens–Korteweg speed (see (2.7a)), A is an as yet undetermined constant, and α is the *Womersley parameter* (Womersley, 1955):

$$\alpha = a(\omega/\nu)^{1/2}, \quad (2.35)$$

Equation (2.30) then yields

$$v_1(r) = \frac{i\omega a p_1}{c_0^2 \rho} \cdot \frac{c_0}{c}\left[\frac{c_0}{c} \cdot \frac{r}{2a} + \frac{A J_1(i^{3/2}\alpha r/a)}{i^{3/2}\alpha J_0(i^{3/2}\alpha)}\right].$$

The four unknowns ξ_1, η_1, p_1 and $p_1 A$ now satisfy four homogeneous equations: the two equations (2.31) and the two equations (2.25), using (2.26) and (2.28), reduce respectively to

$$\left.\begin{array}{l} (k' + A)(a p_1/\rho c_0^2) = i(\omega a/c_0)\xi_1, \\[2mm] (k'^2 + k'FA)(a p_1/\rho c_0^2) = 2\eta_1, \\[2mm] \left(B'_{22}k'^2 + \dfrac{K'_1}{(\omega^2 a^2/c_0^2)}\right)\left(\dfrac{\omega a}{c_0}\xi_1\right) + ik'B'_{21}\eta_1 + \dfrac{iF}{2}\left(\dfrac{a p_1 A}{\rho c_0^2}\right) = 0, \\[2mm] -ik'B_{12}(\omega a/c_0)\xi_1 + (B'_{11} + K'_2)\eta_1 - a p_1/\rho c_0^2 = 0. \end{array}\right\} \quad (2.36)$$

Various complex dimensionless parameters have been used here:

$$k' = \frac{c_0}{c}, \qquad K'_j = \frac{a}{\rho c_0^2}[K_j + i\omega L_j - \omega^2(M_j + \rho_w h)] \quad (j = 1, 2),$$

$$B'_{11} = \frac{B_{11} - T}{\rho a c_0^2}, \qquad B'_{12} = \frac{B_{12}}{\rho a c_0^2},$$

$$B'_{21} = \frac{B_{21} - T + S}{\rho a c_0^2}, \qquad B'_{22} = \frac{B_{22}}{\rho a c_0^2},$$

and

$$F(\alpha) = 2J_1(i^{3/2}\alpha)/i^{3/2}\alpha J_0(i^{3/2}\alpha). \quad (2.37)$$

The four equations (2.36) have a solution only if the quantity k'^2 satisfies the quadratic equation

$$k'^4(1-F)[B'_{22}(B'_{11}+K'_2)-B'_{12}B'_{21}]$$
$$+k'^2\{F(B'_{12}+B'_{21}-\tfrac{1}{2}B'_{11}-\tfrac{1}{2}K'_2)-2B'_{22}$$
$$+[K'_1/(\omega^2a^2/c_0^2)](B'_{11}+K'_2)(1-F)\}+F-2K'_1/(\omega^2a^2/c_0^2)=0.$$
$$(2.38)$$

This dispersion relation has two roots for k'^2, indicating the existence of two types of wave, only one of which will be the pressure wave analysed one-dimensionally in § 2.1. Note that if $k'=k'_r+ik'_i$ (where k'_r is always chosen to be >0), the speed of propagation of the wave is c_0/k'_r and its attenuation constant k from (2.8), is given by

$$k=-2\pi k'_i/k'_r, \qquad (2.39)$$

so the imaginary part of k' must be negative for attenuation. Atabek (1968) calls e^{-k} the 'transmission per wavelength'. To investigate the character of the waves whose dispersion relation is (2.38), we examine various particular cases.

In all cases, the solution will depend on the value of α through the function $F(\alpha)$, from (2.37). The quantity α is defined in (2.35), and can be interpreted physically in various ways. The ratio of tube radius to the thickness of the oscillatory (Stokes) boundary layer on the tube wall is proportional to α. When α is large, the boundary layer is thin and the velocity profile effectively flat across the core (this follows from the expression (2.34) for $u_1(r)$, since, as $y\to\infty$,

$$J_0(i^{3/2}y)\sim\frac{e^{y/\sqrt{2}}}{\sqrt{(2\pi y)}}\exp\left\{i\left(\frac{y}{\sqrt{2}}-\frac{\pi}{8}\right)\left[1+O\left(\frac{1}{y}\right)\right]\right\},$$

so that

$$1-\frac{u_1(r)}{u_1(0)}\sim\frac{A}{k'}\left(\frac{a}{r}\right)^{1/2}\exp\left[-\frac{\alpha}{\sqrt{2}}(1+i)(1-r/a)\right]$$

as $\alpha\to\infty$, which is familiar as the Stokes layer profile). When α is small, the tube is fully occupied by the innermost portion of the boundary layer, and the flow is quasi-steady $(J_0(z)\sim1-\tfrac{1}{4}z^2$ as $|z|\to0$, so

$$u_1(r)\propto k'+A[1-i(\alpha^2/4)(1-r^2/a^2)]$$

as $\alpha \to 0$, which is a constant plus Poiseuille flow). α^2 can also be thought of as the ratio between the viscous diffusion time and the period of the oscillation, or as an unsteady Reynolds number. The asymptotic expansions of $F(\alpha)$ at large and small values of α are

$$F(\alpha) \sim (2/\alpha i^{1/2})[1 + (1/2\alpha) + O(\alpha^{-2})] \quad \text{as } \alpha \to \infty$$
$$\sim 1 - i(\alpha^2/8) - (\alpha^4/48) + O(\alpha^6) \quad \text{as } \alpha \to 0.$$

In large arteries α is large (see § 1.2): in the aorta, $\alpha \approx 13$ (dog) and 17 (man). For a heart-rate of 2 Hz, α would be approximately 1 in a vessel of diameter 0.03 cm; $\alpha \ll 1$ in the microcirculation (vessels <0.01 cm in diameter).

2.2.1 Isotropic, elastic walls; no wall inertia, no initial stresses, no tethering

Here $S = T = K_1' = K_2' = 0$, $B_{11} = B_{22} = B$ (say), which is real, and $B_{12}' = B_{21}' = \sigma B$. Furthermore, the Moens–Korteweg wave speed c_0 is given by

$$c_0^2 = Eh/2\rho a = B(1 - \sigma^2)/2\rho a,$$

so (2.38) reduces to

$$(1 - \sigma^2)F - [4 + F(1 - 4\sigma)]k'^2 + 4(1 - F)k'^4 = 0. \qquad (2.40)$$

We can also set $\sigma = \frac{1}{2}$ for incompressible material. The solutions of (2.40) are complex and depend on α through $F(\alpha)$, from (2.37).

$\alpha \to \infty$

As $\alpha \to \infty$, the two roots of (2.40) are approximately

$$k'^2 = 1 + \frac{9}{8\alpha i^{1/2}} \quad \text{and} \quad k'^2 = \frac{3}{8\alpha i^{1/2}}\left[1 + \frac{1}{2\alpha}\left(1 + \frac{7}{4i^{1/2}}\right)\right];$$

the former represents the Moens–Korteweg wave (speed c_0) with little attenuation ($k \approx 9\pi/8\sqrt{2}\alpha \approx 2.5/\alpha$; see (2.39)), and the latter a much faster wave with speed $c_0(8\alpha/3)^{1/2} \sec \frac{1}{8}\pi$ and attenuation constant $k \approx 2\pi \tan \frac{1}{8}\pi \approx 2.6$ reflecting a constant attenuation of 93% per wavelength. However, since the speed of this wave is so large, the wavelength is also large, and the attenuation per Moens–Korteweg wavelength would be only 34% for $\alpha = 13$.

It is interesting to note that, for the first type of wave, a frequency of 40 Hz leads to a value of α of about 60, and hence to a value of $k \approx 0.04$. This is not adequate to make up the difference between the value predicted from observed viscoelastic wall properties, (2.9), and the observed value of $k(=0.7–1.0)$, although its small magnitude may explain why Anliker et al. (1968a) did not observe the frequency-dependence of attenuation predicted here, while McDonald & Gessner (1968) did observe it at somewhat lower frequencies (up to 20 Hz). Note that for $\alpha = 13$ (dog's aorta) $k \approx 0.19$, which is not negligibly small. Note too that the wave speed is predicted to be slightly frequency-dependent, since $c_0/k_r' \approx c_0(1 - 9/8\alpha\sqrt{2})$; however, the amount of dispersion predicted for $\alpha \geqslant 60$ is negligible, and Anliker et al. (1968a) observed no dispersion at such high frequencies.

The nature of the two types of wave can be determined by solving equations (2.36) for the unknown coefficients. From them we deduce that, when $k' = 1$,

$$\eta_1 = \tfrac{1}{2}ap/\rho c_0^2, \quad (\omega/c_0)\xi_1 = i\eta_1/2a,$$

while when $k' = (3/8\alpha)^{1/2}\, e^{-i\pi/8}$,

$$\eta_1 = \frac{3}{16\alpha i^{1/2}} \cdot \frac{ap_1}{\rho c_0^2}, \quad \frac{\omega}{c_0}\xi_1 = \left(\frac{2\alpha}{3}\right)^{1/2}\frac{i^{5/4}}{a}\eta_1.$$

In the first type of wave, the radial wall displacements are in phase with the pressure pulsations, and the longitudinal wall displacements are (when appropriately non-dimensionalised) of the same order of magnitude as the radial displacements, with a 90° phase lead. This is just what one would expect for the pressure wave analysed in § 2.1. For the second type of wave, however, the radial displacement is much smaller for the same pressure amplitude, and the longitudinal displacements are much larger than the radial. Thus this is a shear wave, in which the principal wall oscillations are longitudinal, and the (small) inertia is provided by the fluid dragged back and forth in the Stokes boundary layers (obviously wall inertia would eventually limit the wave speed as $\alpha \to \infty$).

$\alpha \rightarrow 0$

In this limit, the roots of (2.40) are approximately

$$k'^2 = \frac{6}{i\alpha^2}\left(1 + \frac{i\alpha^2}{6}\right) \quad \text{and} \quad k'^2 = \frac{1}{4}\left(1 - \frac{i\alpha^2}{8}\right).$$

The former represents the pressure wave, with a low propagation speed of $c_0\alpha/\sqrt{3}$ and attenuation per (very short) wavelength given by $k = 2\pi$. The latter represents the shear wave, with a finite wave speed $2c_0$ in this limit, and negligible attenuation, which seems rather surprising until we recognise that the fluid in the tube is carried bodily along with the longitudinal wall motions and the shear-rate within it is negligible. The slowly propagating pressure wave is not really a wave at all in this limit, since inertia is negligible. There is, instead, a balance between the elastic restoring forces and viscosity, which is a diffusive process rather than a wave (see Lighthill's appendix to Caro, Foley & Sudlow (1970)), and this is the mechanism governing the transmission of the pressure pulse through the microcirculation.

Computations for intermediate values of α were made for the pressure wave by Womersley (1957), and for the longitudinal wave by other authors, e.g. Atabek (1968). The results show continuous, smooth variation of wave speed and attenuation between $\alpha = 0$ and $\alpha = \infty$, for each type of wave. The wave speed in each case increases monotonically with α, as does the factor e^{-k} for the pressure waves, although for the longitudinal waves e^{-k} has a minimum at $\alpha \approx 3$ (Atabek, 1968), at least when some wall inertia is present.

Womersley (1957) ignored the existence of the fast shear waves, although the second solution for k'^2 would have arisen from his theory. The reason must be that significant longitudinal wall motions are not observed in real arteries, so no such waves could be measured. The absence of shear waves can be directly attributed to longitudinal wall tethering, as we proceed to show.

2.2.2 The effect of longitudinal tethering

As a preliminary to assessing the effects of the different parameters in (2.38), an estimate of their approximate value in normal arteries will be obtained. We know that the Moens–Korteweg wave speed c_0

is given by

$$c_0^2 = E_\theta h / 2\rho a$$

when E_θ is real; using this relation in equations (2.37) for B'_{11} (incremental) etc. leads to

$$B'_{11} = \frac{2}{1 - \sigma_\theta \sigma_x}, \qquad B'_{12} = \frac{2\sigma_x}{1 - \sigma_\theta \sigma_x},$$

$$B'_{21} = \frac{2\sigma_\theta E_x / E_\theta}{1 - \sigma_\theta \sigma_x}, \qquad B'_{22} = \frac{2 E_x / E_\theta}{1 - \sigma_\theta \sigma_x}.$$

These quantities are all of order 1: if we take $E_x / E_\theta = 1.2$ and $\sigma_\theta = \sigma_x = 0.29$ (cf. Patel & Vaishnav, 1972), we obtain

$$B'_{11} = 2.18, \qquad B'_{12} = 0.63, \qquad B'_{21} = 0.76, \qquad B'_{22} = 2.62.$$

Patel & Vaishnav (1972) also give values for the tethering constants $K_{1,2}$, $L_{1,2}$, $M_{1,2}$ in the aorta. They suggest that the radial and longitudinal tethering constants are roughly equal, and give

$$K_{1,2} \approx 33 \text{ kN m}^{-2} \text{(m)}^{-1}, \qquad L_{1,2} \approx 17 \text{ kN m}^{-2} \text{(m s}^{-1})^{-1},$$

$$M_{1,2} + \rho_w h \approx 4 \text{ N m}^{-2} \text{(m s}^{-2})^{-1};$$

thus the inertia terms are negligibly small except at extremely high frequencies ($\omega \sim 10^2 \text{ s}^{-1}$), and

$$K'_{1,2} \approx (33 + 17 i\omega)(a / \rho c_0^2) \times 10^3.$$

Now in the aorta, $a \approx 10^{-2}$ m, $c_0 \approx 5$ m s^{-1} and $\rho = 10^3$ kg m^{-3}, so $|K'_{1,2}| \approx 0.9$ when $\omega = 4\pi$ s^{-1}, which is markedly smaller than B'_{11} or B'_{22}, so that the terms involving K'_2 in (2.38) can be neglected. However, the terms involving K'_1 are divided by $\omega^2 a^2 / c_0^2$, which is very small, about 6.3×10^{-4} for $\omega = 4\pi$. Thus $|K'_1| / (\omega^2 a^2 / c_0^2) \approx 140$, which is much larger than the other terms, and longitudinal tethering is dominant. Equation (2.38) now reduces approximately to

$$\beta_1 k'^4 + \{\beta_2 K - [\beta_3 / (1 - F)]\} k'^2 - [2K / (1 - F)] = 0, \quad (2.41)$$

where

$$\beta_1 = B'_{22} B'_{11} - B'_{12} B'_{21}, \qquad \beta_2 = B'_{11},$$

$$\beta_3 = -F(B'_{12} + B'_{21} - \tfrac{1}{2} B'_{11}) + 2 B'_{22},$$

and

$$K = K'_1 / (\omega^2 a^2 / c_0^2)$$

so that $|K| \gg 1$. The second term in the square brackets is negligible unless α is very small when $1 - F \sim \frac{1}{8} i \alpha^2$.

When α is not small, so that $1 - F = O(1)$, the two solutions of (2.41) are approximately

$$k'^2 = 2/(1-F)\beta_2 \quad \text{and} \quad k'^2 = -\beta_2 K/\beta_1. \qquad (2.42)$$

The first of these represents the pressure wave, as we can see by putting $F = 0$ ($\alpha \to \infty$) and $\beta_2 = B'_{11} = 2/(1-\sigma^2)$ for an isotropic elastic solid. Then $k'^2 = 1 - \sigma^2$ and $c^2 = c_0^2/(1-\sigma^2)$. This is just the result obtained in an elastic tube when longitudinal wall motions are completely prevented (cf. (2.7b) and Lighthill (1975), chapter 12). Indeed, letting $K \to \infty$ directly in equations (2.36) (with $F \neq 0$, necessarily) shows at once that a consistent solution is one in which

$$\xi_1 \approx 0, \qquad B'_{11}\eta_1 \approx a p_1/\rho c_0^2, \qquad A \approx -k', \qquad k'^2 = 2/(1-F)\beta_2. \qquad (2.43)$$

Thus the phase of η differs from that of p only through the complex nature of the distensibility, proportional to B_{11}^{-1}. Also the velocity profile (2.34) reduces to

$$u_1(r) = (p_1/\rho c_0)k'[1 - J_0(i^{3/2}\alpha r/a)/J_0(i^{3/2}\alpha)]; \qquad (2.44)$$

this is precisely the velocity profile driven by an oscillatory pressure gradient in a *rigid* tube (Womersley, 1955), as we would expect for long waves in which there is no longitudinal wall motion. Only the amplitude and phase depend on wall distensibility and viscoelasticity through the factor k'.

The second of the solutions (2.42) represents the longitudinal wave, but it is no longer necessarily a fast wave, as we can see if we put

$$-iK = (1/\rho a \omega^2)(\omega L_1 - i K_1) = M e^{-i\chi}.$$

Therefore, neglecting for now the fact that β_1 and β_2 are complex, we have the speed of propagation equal to

$$c_0/k'_r = c_0(\beta_1/\beta_2 M)^{1/2} \sec(\tfrac{1}{2}\chi + \tfrac{1}{4}\pi). \qquad (2.45)$$

In general this is very small because $M = |K|$ is very large, but if χ is close to $\frac{1}{2}\pi$, i.e. if $K_1 \gg \omega L_1$, the speed would become large. This is to be expected since increasing the elastic restoring force while keeping the viscous damping force constant should increase the wave speed. For the numbers quoted above, $K_1/L_1 \approx 2$, so ωL_1 will exceed K_1 for frequencies greater than $1/\pi$, and the $M^{-1/2}$ factor in (2.45) will dominate. These longitudinal waves are attenuated by a factor e^{-k} per wavelength, where

$$k = 2\pi \tan \left(\tfrac{1}{2}\chi + \tfrac{1}{4}\pi\right),$$

which is significantly greater than 2π, and increases as χ increases, i.e. as the elastic tethering constant, K_1, increases relative to the viscous constant, ωL_1. The purely numerical results of Atabek (1968) are consistent with this result. Since the wave speed and hence wavelength are in general small, the attenuation per unit *distance* will be extremely great; this is no doubt why such waves are not normally observed, at least in the aorta and femoral arteries. Anliker *et al.* (1968b), however, generated high-frequency axial waves in the exposed canine *carotid* artery, and observed wave speeds of about three times c_0 (with attenuation per wavelength about three times that of the pressure wave); the reason c_0 is not small is presumably because the exposed carotid artery is not tethered, with the consequence that M is not large and the visco-elastic properties of the artery wall itself are once more important.

If α is small enough for $1/\alpha^2$ to be large compared with K, the roots of (2.41) reduce to

$$k'^2 = 8\beta_3/i\alpha^2\beta_1 \quad \text{and} \quad k'^2 = -2K/\beta_3. \tag{2.46}$$

The former has a very low wave speed, and this and the attenuation per wavelength are independent of the tethering constants; this represents the pressure 'wave', as in the absence of tethering. The latter represents the longitudinal wave, and its properties are again dominated by the tethering constants.

2.2.3 *The physiological pressure pulse* (*summary*)

Longitudinal tethering prevents axial wall movements, and causes the physiological pressure wave to satisfy equations (2.43), with a rigid-tube velocity profile (2.44). The quantity β_2 is complex, equal to

$$B'_{11} = (E_{\theta\mathrm{dyn}} + i\eta\omega)h/\rho a c_0^2 (1 - \sigma_\theta\sigma_x),$$

and c_0 is best defined as the speed of propagation, in the absence of tethering, as $\alpha \to \infty$; i.e. as

$$c_0 = |E_{\theta \mathrm{dyn}} + \mathrm{i} \eta \omega|^2 h / E_{\theta \mathrm{dyn}} 2 \rho a. \qquad (2.47a)$$

Thus the value of k'^2 is

$$k'^2 = \frac{(1 - \sigma_\theta \sigma_x)}{1 - (2/\alpha \mathrm{i}^{1/2})} (1 - \mathrm{i} \tan \theta), \qquad (2.47b)$$

where $\tan \theta = \eta \omega / E_{\theta \mathrm{dyn}}$, from (2.9), which has a value of about $\frac{1}{6}$. The feature of this result that has not yet been discussed is the fact that σ_θ and σ_x are complex. Patel $et\ al.$ (1973) measured the dynamic anisotropic elastic properties of the canine aorta, and obtained

$$\sigma_\theta \approx 0.28 + 0.07 \mathrm{i}, \qquad \sigma_x \approx 0.16 + 0.02 \mathrm{i},$$

so that the factor

$$(1 - \sigma_\theta \sigma_x) \approx 0.91 - 0.03 \mathrm{i}.$$

This means (a) that the wave speed c_0/k'_r is increased above c_0, but by a smaller factor than when $\sigma_\theta = \sigma_x = 0.5$, and ($b$) that the complex Poisson's ratios have a negligible influence on the damping, all of which must come from the complex Young's moduli of the wall, at least for high-frequency waves when α is very large as in the experiments of Anliker $et\ al.$ (1968a).

2.2.4 Flow-rate and wall shear

As was explained in § 1.2, two aspects of the velocity field are of particular interest. One is the volume flow-rate,

$$Q = \int_0^a 2 \pi r u \, dr,$$

and the other is the shear stress on the wall,

$$\tau = -\mu \frac{\partial u}{\partial r} \bigg|_{r=a}.$$

When the velocity profile is given by the rigid-tube equation, (2.44), for a pressure-wave amplitude of p_1, the complex amplitudes of these quantities turn out to be

$$Q_1 = (\pi a^2 p_1 / \rho c_0) k' (1 - F) \qquad (2.48)$$

and

$$\tau_1 = (\nu p_1/ac_0)k' \cdot \tfrac{1}{2}i\alpha^2 F. \tag{2.49}$$

In these equations $k'/c_0 = 1/c$, and $-i\omega p_1/c$ is the (complex) pressure-*gradient* amplitude, which we may denote by $-G_1$. For large and small values of α, (2.48) and (2.49) become

$$\left.\begin{array}{l} Q_1 \sim (-\pi a^2/\rho\omega)G_1 i(1-2/\alpha i^{1/2})\\ \tau_1 \sim (-\nu/a\omega)G_1 i^{3/2}\alpha(1+1/2\alpha) \end{array}\right\} \text{ as } \alpha \to \infty, \tag{2.50}$$

and

$$\left.\begin{array}{l} Q_1 \sim (\pi a^2/\rho\omega)G_1(\alpha^2/8)(1-i\alpha^2/6)\\ \tau_1 \sim (-\nu/a\omega)G_1(i\alpha^2/2)(1-i\alpha^2/8) \end{array}\right\} \text{ as } \alpha \to 0. \tag{2.51}$$

As Lighthill (1975) has pointed out, good accuracy is achieved by using (2.50) for $\alpha > 4$ and (2.51) for $\alpha < 4$. Note that the amplitude of the centre-line velocity, $u_1(0)$, differs from that of the average velocity, $Q/\pi a^2$, by a factor $[1 - 1/J_0(i^{3/2}\alpha)]/(1-F)$, which is close to 1 for large α, and close to 2 for small α, when the flow is approximately Poiseuille flow. Note too that at large α, the flow-rate is $\tfrac{1}{2}\pi$ out of phase with the pressure gradient, as one would expect for an inviscid fluid (§ 2.1). At very small α on the other hand, the flow-rate is in phase with the pressure gradient, since the flow is quasi-steady.

Womersley and McDonald verified the validity of using the rigid-tube velocity profiles in practice by measuring the pressure-gradient waveform (in a dog's femoral artery), performing a Fourier analysis on it, using equation (2.48) to calculate Q for each term (with $c_0/k' = c$ assumed real), and then recombining the answers to obtain a predicted flow-rate waveform. They compared the results with the actual flow-rate waveform, measured directly with an electromagnetic flowmeter. Agreement was very good, as shown in fig. 2.5 (see McDonald, 1974, p. 130).

The results of this section have confirmed the validity of the one-dimensional wave theory presented in § 2.1 to describe linear pressure waves in arteries, as long as for each frequency the wave speed c_0 is replaced by c_0/k'_r (k' given by (2.47b)) and the flow-rate is related to the pressure by (2.48) instead of

$$Q_1 = \pi a^2 p_1/\rho c_0 = (A/\rho c_0)p_1, \tag{2.52}$$

as predicted by the theory of § 2.1. The quantity Q_1/p_1 is called the characteristic *admittance* of the tube. In the high-frequency limit

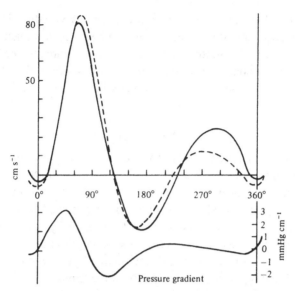

Fig. 2.5. Waveforms of average velocity (upper traces) in a dog's femoral artery; continuous curve, measured with an electromagnetic cuff flowmeter; broken curve, calculated from the measured pressure gradient (lower trace) according to Womersley's rigid-tube theory. (After McDonald, 1974.)

this is equal to $A/\rho c_0$, which is independent of frequency. In this case it is relatively simple to take non-linearities into account by a one-dimensional theory, as shown in § 2.1. At lower frequencies, however, the dispersion of the wave makes non-linear theory more difficult. Womersley (1957) extended his two-dimensional theory to second order in wave amplitude, to take account of small non-linearities. The resulting combinations of Fourier modes became very complicated. Olsen & Shapiro (1967) developed the one-dimensional theory in powers of wave amplitude, and showed, by comparison with model experiments in rubber tubes, that the second-order results were accurate enough to cover normal physiological events. All these theories confirm that the *normal* pressure wave in arteries can be treated as approximately linear.

2.3 Effects of taper and branchings (linear theory)

The theory presented so far is adequate for uniform viscoelastic tubes, but is inadequate for real arteries because they are not uniform. They suffer both from a continuous variation in cross-sectional area and distensibility, and from repeated branchings. In this section we show to what extent the one-dimensional linear theory can be adapted to cope with such non-uniformities. There is little in this section that has not already been outlined by Lighthill (1975).

2.3.1 Taper

If the rate of change of wall properties with distance along the tube is sufficiently gradual, i.e. takes place over a length-scale large compared with the wavelength of the wave, then the modification of a wave as it travels along the tube can be analysed by the WKB-method (Cole, 1968), first applied to arteries by Taylor (1965). Furthermore if, as in arteries, the wavelength is long compared with the tube radius, and if linear theory is applicable, the one-dimensional equations governing the wave, from (2.1)–(2.3), are

$$\left.\begin{array}{l} \partial Q/\partial x + (Y/c)\partial p/\partial t = 0, \\ \partial Q/\partial t + cY(\partial p/\partial x) = 0, \end{array}\right\} \tag{2.53}$$

where $Q = Au$ is the volume flow-rate and $Y = A/\rho c$ is the characteristic admittance of the tube, which, with the wave speed c, may be complex, but is only weakly frequency-dependent. We assume that both c and Y are slowly varying functions of x, and introduce a new variable $\tilde{x} = \varepsilon x$ ($\varepsilon \ll 1$) so that c and Y are functions only of $\tilde{x}$. We shall seek an asymptotic expansion of the solution of (2.53) that is uniformly valid for all x and t as $\varepsilon \to 0$, and present the theory in such a way that subsequent terms in the expansion can be readily constructed, in principle.

The equations (2.53) can be combined into the following single equation for p:

$$\frac{1}{c^2}\frac{\partial^2 p}{\partial t^2} = \frac{\partial^2 p}{\partial x^2} + \frac{\varepsilon(cY)'}{cY}\frac{\partial p}{\partial x},$$

where a prime denotes differentiation with respect to $\tilde{x}$, the 'slow' variable. We introduce a new 'fast' variable

$$x_1 = \int_0^x \frac{dx}{c(\varepsilon x)},$$

in terms of which the equation for p is

$$\frac{\partial^2 p}{\partial t^2} = \frac{\partial^2 p}{\partial x_1^2} - \varepsilon c' \frac{\partial p}{\partial x_1} + \frac{\varepsilon (cY)'}{Y} \frac{\partial p}{\partial x_1}. \tag{2.54}$$

Now suppose p is expanded in a power series:

$$p(x_1, t, \tilde{x}, \varepsilon) = p_0(x_1, t, \tilde{x}) + \varepsilon p_1(x_1, t, \tilde{x}) + \cdots.$$

Substitute into (2.54) and equate like powers of ε. The zeroth-order equation is

$$\partial^2 p_0/\partial t^2 - \partial^2 p_0/\partial x_1^2 = 0,$$

so

$$p_0 = B(\tilde{x})f(t - x_1)$$

for waves travelling in the positive-x direction, where f is arbitrary (depending on initial conditions) and $B(\tilde{x})$ is an as yet undefined function. The first-order equation is

$$\frac{\partial^2 p_1}{\partial t^2} - \frac{\partial^2 p_1}{\partial x_1^2} = 2c \frac{\partial^2 p_0}{\partial \tilde{x} \partial x_1} - c' \frac{\partial p_0}{\partial x_1} + \frac{(cY)'}{Y} \frac{\partial p_0}{\partial x_1}$$

$$= f'(t - x_1)[-2cB' + c'B' - (cY)'B/Y].$$

Now $f'(t - x_1)$ will, in general, be an oscillatory function, so that the solution for p_1 will exhibit resonance and hence no longer be small, unless the term in the square brackets is identically zero. In order that the proposed expansion be uniformly valid, we therefore require that this bracket is identically zero, which determines B. In fact we obtain

$$B(\tilde{x})/B(0) = [Y(0)/Y(\tilde{x})]^{1/2},$$

and the final zeroth-order solution for p (in terms of the old x-variable) is

$$p = \left[\frac{Y(0)}{Y(x)}\right]^{1/2} f\left(t - \int_0^x c^{-1} \, dx\right). \tag{2.55}$$

The corresponding solution for Q, from either of equations (2.53), is

$$Q = Y(x)p. \tag{2.56}$$

These results show that, at every location along the slowly varying tube, the flow-rate is equal to the *local* characteristic admittance multiplied by the pressure, but the inverse of the wave speed is equal to the *average* value of the characteristic inverse wave speed over the length of tube traversed by the wave. In arteries, $|c|$ increases with distance from the heart, while A, and hence $|Y|$, decreases. Thus the amplitude of the pressure pulse will increase ($\propto |Y|^{-1/2}$) while that of the flow-rate pulse will decrease ($\propto |Y|^{1/2}$). Furthermore, the magnitude of the (complex) wave speed will not be as large as would be expected from purely local measurements of distensibility.

The above theory is very attractive, but unfortunately cannot be applied directly to the pressure pulse generated by the heart in mammalian arteries. This is because of the requirement that the length-scale of tube variation should be large compared with the wavelength, whereas we have already seen that the wavelength of the fundamental wave (frequency = heart-rate ≈ 2 Hz in dogs) is several metres, while the length of the longest vessel, the aorta, is of the order of half a metre. The qualitative agreement between the predicted peaking of the pressure pulse (and the corresponding decrease in flow-rate amplitude) and the observations (fig. 1.14) can be no more than suggestive.

The theory should, however, be more applicable to experiments such as those of Anliker *et al.* (1968*a*) and of Histand & Anliker (1973), in which high-frequency waves are generated at one location in an artery and measured at another. The frequency range used was 40–120 Hz, so the wavelength would have been between 4 and 12 cm in the aorta, which is reasonably small compared with the length of that vessel. The ratio between the amplitude of the pressure wave at one position, x, compared with that at another position closer to the origin of the wave, 0, is predicted to be

$$\left[\frac{Y(0)}{Y(x)}\right]^{1/2} \exp\left(\omega \int_0^x \frac{k_1'}{c_0}\,\mathrm{d}x\right), \qquad (2.57)$$

where c_0 and $k'(=k_r' + \mathrm{i}k_i')$ are given by (2.47*a, b*) in which the $1/\alpha$ term is negligible. Histand & Anliker (1973) fitted their measurements in the aorta to the curve $e^{-kx/\lambda}$, where λ is the wavelength, and found $k = 0.7$–1.0 for waves propagated peripherally and

$k = 1.3-1.5$ for waves propagated centrally. Values of x of about 5 cm were generally used. If we assume that the ratio of the viscous to the elastic components of the distensibility (i.e. $\tan \theta$) remains roughly uniform, then the contribution to k from the integral in (2.57) will be approximately the same for both peripheral and central waves, and equal to

$$2\pi \int_0^x \frac{E_\theta^{1/2} \sin \frac{1}{2}\theta}{|E|^{3/2}} \, dx \bigg/ \int_0^x \frac{E_\theta^{1/2} \cos \frac{1}{2}\theta}{|E|^{3/2}} \, dx$$

which is close to $2\pi \tan \frac{1}{2}\theta$. The explanation for the observations presumably lies in the amplitude term $[Y(0)/Y(x)]^{-1/2}$. A constant difference between the values of k for the two directions of wave propagation is consistent with an exponential variation of $Y(x)$, proportional to $e^{-\gamma x/\lambda}$, where γ must be about 0.55 to explain the discrepancy between $k = 0.85$ peripherally and $k = 1.40$ centrally. However, the average, $k \approx 1.13$, is significantly larger than that predicted from the estimates of $\tan \theta = \eta\omega/E_{\theta\text{dyn}}$ reported in § 1.1, which give $2\pi \tan \frac{1}{2}\theta \approx 0.52-0.62$. Therefore those estimates do not apply to the vessels examined by Histand & Anliker (1973); the reason for this discrepancy is not readily apparent.

For $Y = A/\rho c$ to vary with wavelength in the manner predicted above is also inherently unlikely, since we already know, from (1.4), that A is roughly proportional to $e^{-\beta x/a}$, where $\beta \approx 0.02-0.05$. Thus

$$c \propto e^{(x/a)(\gamma a/\lambda - \beta)}, \tag{2.58}$$

which increases with peripheral distance, only if λ is small enough. This is contrary to observation, which suggests that c is independent of frequency (i.e. wavelength) and increases approximately linearly with x:

$$c \approx c_0(1 + nx), \qquad n \approx 0.032 \text{ cm}^{-1},$$

(McDonald, 1974). A linear approximation to (2.58) gives $n \approx \gamma a/\lambda - \beta$, which is consistent with the quoted numbers only for $\lambda \approx 10$ cm (frequency ≈ 50 Hz), and not for the whole range of frequencies used by Histand & Anliker (1973). These inconsistencies have no obvious explanation; it would seem that further experiments designed to test (2.55) critically are required.

The above theory shows how the linear wave is modulated by gradual changes in wall properties. A more general theory of a modulated simple wave has been developed by Seymour (1975) to describe the additional changes brought about by small non-linearities and by weak viscous dissipation. The former tends to steepen the pressure pulse and the latter, like wall viscoelasticity, tends to attenuate it. The development of shocks is predicted if the steepening effects outweigh the dissipative ones. However, like the above elementary theory, Seymour's analysis is restricted in its practical application by the requirement that all amplitude and wave-speed changes should be small over one wavelength of the wave, whereas, in fact, both the taper and the non-linear steepening (§ 2.1) take place over distances short compared with one wavelength of the fundamental cardiac wave. Seymour also makes the unwarranted assumption that the viscous forces in the one-dimensional equation of motion can be represented by a function of the instantaneous average velocity, whereas, in fact, the two are not in phase; this is unimportant in arteries, but only because the fluid viscosity is virtually negligible anyway.

The amplitude gradations predicted by (2.55), in the non-dissipative case where Y and c are real, are the same as those predicted from an assumption that the mean energy flux $\overline{pQ} = Y\overline{p^2}$ down the vessel should be uniform, i.e. no energy is reflected by the gradual taper. Such an assumption is the natural consequence of treating the tapered tube as a sequence of very small isolated discontinuities (Schoenberg, 1968; Lighthill, 1975). To see this and, more importantly, to analyse the effect of arterial junctions, we proceed to analyse the reflection and transmission of waves at junctions.

2.3.2 Isolated wave reflections

We suppose that a linear wave, with $p = P_1 f(t - x/c_1)$ is incident from $x = -\infty$ on a junction at $x = 0$. This may take the form of a branching into two or more daughter vessels (fig. 2.6), or may be only a discontinuity in cross-sectional area or elastic properties, which may be either natural (e.g. an arterial stenosis) or artificial (e.g. a cuff round the artery or a junction with a plastic tube). The dimensions of the junction are taken to be small compared with the wavelength of the wave (or any of its Fourier components) so that the vessels can be considered uniform except at $x = 0$. Suppose that

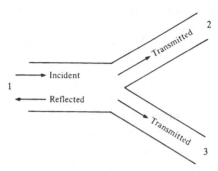

Fig. 2.6. Sketch of an arterial bifurcation. The incident wave is partially reflected in the parent tube 1 and partially transmitted in the daughter tubes 2 and 3.

the parent tube has intrinsic wave speed c_1 and area A_1 so that its characteristic admittance is $Y_1 = A_1/\rho c_1$, and let the daughter tubes have wave speed c_j, area A_j and admittance $Y_j = A_j/\rho c_j$, $j = 2, 3 \ldots N$. Note that, in the case of sinusoidal waves, the wave speeds and admittances can be complex. There will be a reflected wave, with associated pressure fluctuation $p = P_R g(t + x/c_1)$ say, and transmitted waves in each daughter tube, with $p = P_{Tj} h_j(t - x/c_j)$. The net flow-rate in the parent tube is

$$Q = Y_1[P_I f(t - x/c_1) - P_R g(t + x/c_1)];$$

expressions similar to the first term apply in the daughter tubes.

The conditions to be applied at $x = 0$ are continuity of flow-rate, to satisfy mass conservation, and of pressure, to satisfy energy (not momentum) conservation. The latter condition requires that viscous losses at the junction are negligible, and that the kinetic energy term proportional to $\rho Q^2/A^2$ in the expression for the total head is negligible (i.e. that the waves are linear). The conditions can be satisfied only if all waves have the same waveform at $x = 0$, so we can take $g(t) \equiv h_j(t) \equiv f(t)$. Then continuity of pressure and flow-rate require

$$P_I + P_R = P_{Tj} \quad (j = 2, 3 \ldots)$$

$$Y_1(P_I - P_R) = \sum_{j=2}^{N} Y_j P_{Tj},$$

which lead to the following expressions for the amplitudes of the reflected and transmitted waves:

$$\left. \begin{array}{c} P_R/P_I = \left(Y_1 - \sum_{j=2}^{N} Y_j \right) \Big/ \left(Y_1 + \sum_{j=2}^{N} Y_j \right), \\[12pt] P_{Ti}/P_I = 2Y_1 \Big/ \left(Y_1 + \sum_{j=2}^{N} Y_j \right) \quad (i = 2 \ldots N) \end{array} \right\} \qquad (2.59)$$

(as presented by Lighthill (1975), among others).

Restricting attention for now to real values of c_j and Y_j (i.e. to non-dissipative waves), we see that if $\sum_{j=2}^{N} Y_j < Y_1$, then the reflected wave has the same sign as the incident wave, the pressures in the two waves are in phase at $x = 0$ and therefore combine additively to form a large-amplitude fluctuation there, and the effect of the junction is similar to that of a closed end (where $P_R = +P_I$). On the other hand, if $\sum_{j=2}^{N} Y_j > Y_1$, there is a phase change at $x = 0$, the smallest-amplitude pressure fluctuation occurs there, and the junction resembles an open end (where $P_R = -P_I$). If, however, $\sum_{j=2}^{N} Y_j = Y_1$, there is no reflected wave, and the junction is said to be perfectly matched. If the net cross-sectional area increases at a junction, then there should also be an increase in wave speed if the junction is to be perfectly matched. It is interesting to note that the wave speed does indeed increase peripherally, and so does the cross-sectional area. However, this may be misleading since the wave speed also increases with distance down the aorta, while the net cross-sectional area of the circulation does not increase at first (fig. 1.4).

When the reflection is of closed-end type, a sinusoidal incident wave of the form $p = P_I \cos\left[\omega(t - x/c_1)\right]$ will lead to a total pressure fluctuation in the parent tube of

$$p = (P_I - P_R) \cos\left[\omega(t - x/c_1)\right] + 2P_R \cos \omega t \cos (\omega x/c_1).$$

Thus the amplitude of the pressure fluctuation will diminish over the first quarter wavelength from the junction (to $\omega x/c_1 = -\frac{1}{2}\pi$) from $P_I + P_R$ to $P_I - P_R$. Now we have seen that in man there tends to be a reduction of area at the iliac bifurcation, suggestive of a closed-end reflection. Therefore, independently of any taper in the aorta, one would expect an increase in amplitude with distance from the heart, as observed. However, similar observations are made in dogs (fig. 1.14), whose aortic trifurcation tends to be quite well matched, but the amplitude is further increased by occluding the sacral artery in order to create a mis-match (Newman *et al.*, 1973). Thus both taper and the single reflection site at the end of the aorta are implicated in the increase of the pressure pulse along the aorta. Further direct evidence of the isolated reflection site in man can be gathered from fig. 1.17, where a step in the descending part of the velocity waveform can be seen. This step occurs at a time after the start of systole that increases in proportion to the distance *up* the aorta from the iliac bifurcation, reaching a value of about 0.2 s in the innominate artery. The distance from heart to bifurcation and back is about 1.2 m, and the average aortic wave speed is about 6 m s^{-1}, so it is consistent to interpret the step as a mark of the wave reflected from the bifurcation. The obvious notch at the end of systole marks the closure of the aortic valve, and is propagated away with the basic pulse, although it is rapidly attenuated because of its high-frequency content. There are so many fluctuations in the pressure and flow-rate waveforms of fig. 1.17 that it is difficult to give any further interpretation in terms of individual waves reflected at isolated sites.

The rate of energy flux associated with a wave is $pQ = Yp^2$. Thus the proportion of incident energy reflected is P_R^2/P_I^2, while the proportion transmitted is $\sum_{j=2}^{N} Y_j P_{Tj}^2/Y_1 P_I^2$. It can be verified using (2.59) that the sum of these is unity; the former is called the *reflection coefficient*, and the latter the *transmission coefficient*. Note that if a junction is closely, but not perfectly, matched, so that $P_R/P_I = \varepsilon$, say, then the reflected energy is only ε^2, which can be negligible. This shows that there can be an observable effect on the incident wave without a diminution of the rate at which energy is transmitted down the system, and explains why the quantity $Y|p|^2$ remains approximately constant in a very slowly varying vessel (see § 2.3.1).

2.3.3 *Multiple wave reflections*

A full description of the arterial system must include an analysis of the interactions between junctions in a complicated branched network. Consider, for example, a double junction, as illustrated in fig. 2.7, with the single junction already analysed (B) attached to the daughter tube of a previous junction (A) located at $x = -l$. A wave

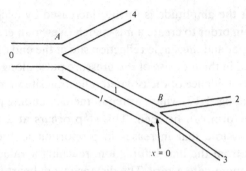

Fig. 2.7. Sketch of a double junction. Wave incident in tube 0 is reflected and transmitted at A. The transmitted wave in tube 1 is reflected and transmitted at B, etc.

incident on A will be partly transmitted there, then partly reflected at B, re-reflected at A, and so on (but with ever decreasing amplitude because the reflection is only partial and there is attenuation). The junction at A can be analysed in exactly the same way as the junction at B if it is possible to write down a general relation between the pressure and flow-rate in tube AB at $x = -l$, i.e. if there is an *effective* admittance $Y_{1\text{eff}}$ of the system downstream of A. Now from the above theory, the ratio of Q to p at $x = -l$ in AB is

$$Y_{1\text{eff}} = \frac{Q}{p} = \frac{Y_1[f(t+l/c_1) - P_{\text{R}}f(t-l/c_1)]}{f(t+l/c_1) + P_{\text{R}}f(t-l/c_1)}; \qquad (2.60)$$

this is very difficult to use because it depends on t. However, the waves under consideration are periodic and linear, and can therefore be subjected to Fourier analysis. Then, as long as only one frequency is analysed at a time, (2.60) does lead to a constant $Y_{1\text{eff}}$: if $f(t) = e^{i\omega t}$, then use of (2.59) leads to

$$Y_{1\text{eff}} = Y_1 \left[\sum_{j=2}^{N} Y_j + i Y_1 \tan(\omega l/c_1) \right] \Big/ \left[Y_1 + i \sum_{j=2}^{N} Y_j \tan(\omega l/c_1) \right]. \qquad (2.61)$$

Now the junction at A can be analysed like that at B, with the *characteristic* admittance of a daughter tube, Y_j, replaced by its *effective* admittance, $Y_{j\text{eff}}$.

Various conclusions can be drawn from (2.61) (Lighthill, 1975). First, if l is much less than a wavelength, $\omega l/c_1 \ll 1$, and $Y_{1\text{eff}} \approx$

$\sum_{j=2}^{N} Y_j$ as if the tube AB were not there. In that case the intermediate tube could be considered as part of the junction. If first
order in $\omega l/c_1$ is retained, we find that

$$Y_{1\text{eff}} \approx \sum_{j=2}^{N} Y_j + \frac{i\omega l}{c_1} Y_1 \left\{ 1 - \left[\left(\sum_{j=2}^{N} Y_j \right) \middle/ Y_1 \right]^2 \right\},$$

so that the intermediate tube does not alter the modulus of the
effective admittance, but introduces a slight phase lead (or lag) of Q
over p in the transmitted wave, according as Y_1 is greater or less
than $\sum_{j=2}^{N} Y_j$. Note that a similar result is obtained if l is a whole
number of half-wavelengths, so that $\omega l/c_1$ is an integer multiple of
π. There is a resonance, or 'window', effect, because standing waves
are set up in AB (in the absence of attenuation), and the incident
wave is 'handed-on' unaltered from A to B. In a similar way we can
deduce that if l is an odd number of quarter-wavelengths, so that
$|\tan (\omega l/c_1)|$ is extremely large, then

$$Y_{1\text{eff}} \middle/ \sum_{j=2}^{N} Y_j \approx \left(Y_1 \middle/ \sum_{j=2}^{N} Y_j \right)^2, \tag{2.62}$$

and the relative importance of Y_1 and $\sum_{j=2}^{N} Y_j$ is accentuated.

The wavelength of the fundamental mode of the normal pressure
pulse in the aorta of man (frequency 1.2 Hz) is about 5 m, roughly
eight times the length of the aorta; in a dog (frequency 2 Hz)
the wavelength is about 3 m, again about eight times the length
of the vessel. Resonance is expected to occur for a mode for
which the vessel length is an integer multiple of half the wavelength; in the aorta, therefore, the mode with a frequency of four
times the fundamental heart-rate is expected to be closest to
resonance. As we shall see, there is normally little evidence of this,
presumably because of wave attenuation. The mode with a
frequency of twice the heart-rate should be one for which (2.62) is
valid, and there is evidence that the modulus of the effective
admittance at the entrance to the aorta has a maximum at about
twice the fundamental frequency, consistent with (2.62) if $Y_1 >$
$\sum_{j=2}^{N} Y_j$ as for a closed-end type of reflection.

The existence of a constant effective admittance for waves of
a given frequency, (2.61), means that a complete network can
be constructed, starting from pure resistances in vessels of the

microcirculation, and building up junction by junction until a single effective admittance for the whole circulation, appropriate to waves of a given frequency, is arrived at. This is equal to the (complex) ratio between the components of flow-rate and pressure corresponding to the same frequency in the ascending aorta. Its reciprocal is called the *effective* or *input impedance* of the arterial tree; such a quantity is relatively simple to measure for a wide range of naturally occurring frequencies, because all that is needed is Fourier analysis of measured waveforms such as those shown in figs. 1.14 and 1.17. Because of the lack of detailed measurements of lengths and diameters of all arteries in a systemic vascular bed, no complete theoretical predictions have been made. However, Taylor (1966) computed the input admittance for a seven-generation branching network of tubes, whose cross-sectional area decreased and whose wave speed increased peripherally in a physiological way, but in which the distances between branches were arranged at random to simulate the great variety of tube lengths present in the actual circulation. Taylor's results show that the modulus of the input admittance, when plotted against frequency, has a maximum at a frequency for which the length of the 'aorta' is rather less than a quarter-wavelength, and thereafter oscillates about a value equal to the characteristic admittance of the 'aorta'; this value is far above the value at zero frequency, which is the inverse of the overall (peripheral) resistance. The *phase* of the admittance also oscillates about zero, decreasing through zero when the modulus is a maximum (as expected for closed-end reflections). The vigorous oscillations are virtually abolished when wave attenuation is properly accounted for.

McDonald (1974, chapter 13) reports a number of studies in which the effective admittance of different arteries is measured for a range of frequencies. Most measurements are in dogs, some in man. The main conclusions seem to be as follows. The modulus of the admittance in the *ascending* aorta of some dogs has two maxima, one at a frequency of about 2 Hz, and one at about 6 Hz. In other dogs, however, there is a rise in the admittance modulus to a frequency of about 2 Hz, after which there is little change; this observation is roughly consistent with the predictions of Taylor (1966). In the descending *thoracic* aorta, there is never more than one maximum (at about 2.5 Hz), followed by a marked minimum which itself is absent in more peripheral arteries. In each aortic case the first maximum of the admittance modulus, combined with a zero and falling phase, is interpreted as indicating the presence of a closed-end type of reflection site a quarter-wavelength away. This puts it some distance beyond the end of the aorta in dogs, which suggests (a) that the aortic trifurcation is quite well matched (the maximum occurs at a smaller frequency when the sacral artery is blocked (Newman & Bowden, 1973)), and (b) that the combined effect of the many peripheral branches is that of a single effective reflection site at about the knee. This is consistent with Taylor's predictions. The subsequent minimum in the admittance modulus presumably reflects the tendency to resonance, so that $Y_{1\text{eff}}$ is closer to the sum of the admittances of peripheral vessels. The second maximum in the proximal aorta of some dogs is thought to represent another effective site of closed-end reflection in the shorter arteries to the head and forelimbs, which would not have an effect in the descending thoracic aorta (O'Rourke & Taylor, 1967). It is not interpreted as a mode for which the first reflection site is three-quarters of a wavelength away, because (a) attenuation would effectively eliminate that, and (b) it would be observed in the descending aorta if that were the cause. The absence of the second maximum in $|Y_{1\text{eff}}|$ in some dogs is thought merely to indicate a less marked difference between the effective impedances of the two arterial systems to the hind- and fore-quarters.

In the ascending aorta of *man* there is just one maximum in $|Y_{1\text{eff}}|$, followed by a marked minimum (Mills *et al.*, 1970). The total absence of the second maximum is interpreted as being the result of much greater reflection from the 'hind-quarters' system in man relative to the 'fore-quarters', largely because of the much smaller blood flow to the latter. Furthermore, the reflection site inferred from the first maximum in $|Y_{1\text{eff}}|$ is much closer to the aortic bifurcation than in dogs, because this is much less well matched in man.

Measurements in the *femoral artery* of a dog also imply a closed-end type of reflection site just below the knee (O'Rourke & Taylor, 1966). Also, the general level of the effective admittance for the first few harmonics of the pulse wave is lower than in the aorta, because the characteristic admittance is lower on account of the smaller area and lower distensibility of more peripheral arteries.

Mention should finally be made of the *pulmonary circulation*, for which both predictions and measurements of the effective input admittance, as well as of wave speed, have been made. Useful reviews are provided by Milnor (1972) and McDonald (1974). The main results are: (i) A typical value of the wave speed in human pulmonary arteries is about $1.75 \, \text{m s}^{-1}$, while the corresponding figure in a dog is about $2.5 \, \text{m s}^{-1}$ (see table 1.1). These are roughly consistent with static distensibility measurements. Pulse-wave velocities of $4-8 \, \text{m s}^{-1}$ have been measured in patients with pulmonary hypertension in which there is pathological thickening and stiffening of vessel walls. (ii) Attenuation studies such as those of Anliker *et al.* (1968*a*) are much more difficult in the short and relatively inaccessible pulmonary arteries, and have not been performed. (iii) The modulus of the input admittance of the pulmonary circulation of a dog has a maximum at a frequency of 3-4 Hz, when the phase becomes zero, which suggests an effective closed-end reflection site in the pulmonary microcirculation. Interpretation in terms of an individual reflection site is even more difficult than in the systemic circulation, because there is no dominant arterial pathway such as that leading through the aorta to the legs. There is a minimum in the admittance modulus at about twice the frequency of the maximum, as one expects if attenuation is not very strong. These results are in qualitative agreement with the theoretical predictions of Wiener *et al.* (1966). The admittance at zero frequency, equal to the reciprocal of the pulmonary vascular resistance, depends strongly on the degree of lung inflation, and reflects a complicated interaction between blood pressure and alveolar air pressure in pulmonary capillaries, which is beyond the scope of this monograph.

2.4 Non-linear models of a complete arterial pathway

We have seen that linear theories can give a reasonable description of the propagation, reflection and attenuation of the pulse wave in a branched network of arteries, as long as the area and distensibility of the arteries vary only very gradually with distance between the junctions. However, this condition is not normally satisfied. Furthermore, there are a number of phenomena, especially abnormal ones, that can be accounted for only by non-linearities (§ 2.1). (This is also true of the development of the mean flow, which according to linear theory is Poiseuille flow everywhere; this aspect is further examined in chapter 3.) In this section, therefore, a method that has been used to analyse non-linear phenomena in an arterial pathway that is not necessarily very slowly varying is outlined; certain drawbacks in the method, at least as practised hitherto, are discussed.

A single arterial pathway, from the heart to a capillary bed, is modelled as a single tube whose cross-sectional area and distensibility vary with distance along it. It is assumed that a one-dimensional theory such as that of § 2.1 can be applied; this means that $u(x, t)$ is the average velocity across a cross-section and that $p(x, t)$ is the average pressure. It is not necessary that a length-scale of longitudinal variation in tube properties be large compared with a wavelength, as in § 2.3, but it is desirable that the length-scale be significantly greater than the tube diameter, both in order that p should approximately represent the pressure measured in the centre of the vessel as well as the average, and so that the numerical methods to be employed do not need absurdly small step lengths. The branching of the arterial tree is modelled by letting the tube have 'porous' walls, with a volume outflow ψ per unit length. In reality ψ will be a function of position with sharp discontinuities, and users of the porous-tube model have invariably smoothed it out considerably. However, wave reflections can still be analysed nearly as accurately as in § 2.3, because there it was assumed that the junction, which really occupies a finite length of tube, was concentrated at a point. In this model it is spread out again, in such a way as to leave the reflection coefficient of the junction unaltered (in principle, at least; I know of no-one who has actually verified this aspect of their outflow model). The idea of modelling an artery as a tapered, porous tube seems to have originated with Streeter, Keitzer & Bohr (1963), and was developed further by Skalak & Stathis (1966). The most thorough investigation of such a model is the simulation of the canine systemic circulation by Anliker et al. (1971), and it is to their work that I shall refer most often.

The equations relating velocity $u(x, t)$, excess pressure $p(x, t)$ and area $A(x, t)$ are similar to, but more general than, (2.1)–(2.3), as follows.

Mass conservation:

$$A_t + (uA)_x + \psi = 0. \tag{2.63}$$

Momentum equation:

$$u_t + uu_x + (1/\rho)p_x = f. \tag{2.64}$$

Wall properties:

$$A = A(\tilde{p}, \tilde{p}_t, x), \tag{2.65}$$

where

$$\tilde{p} = p - P_0(x, t). \tag{2.66}$$

In (2.65) and (2.66) $\tilde{p}$ is the transmural pressure, which may not be equal to the excess pressure p if either there is a variable external pressure or there is a significant variation of hydrostatic pressure within the tube. These effects are represented by the function $P_0(x, t)$; in the latter case $P_0 \equiv -\rho g'x$, where g' is the component of the gravitational acceleration in the x-direction. The distensibility, viscoelasticity and taper of the vessel are represented by the dependence of A on $\tilde{p}$, $\tilde{p}_t$ and x, respectively. Equation (2.64) is derived from the integral momentum equation (because u is really the average cross-sectional velocity, $\bar{u}$) on the assumptions (a) that the difference between $\bar{u}^2$ and $(1/A) \int u^2 \, dA$ is negligible, which is reasonable when the velocity profile in the core of the tube is flat with thin boundary layers outside it, as for oscillatory flow at large α (§ 2.2); and (b) that the x-momentum convected from the tube by the outflow is equal to $\rho u \psi$ per unit length (Schaaf & Abbrecht, 1972). The quantity f in (2.64) is the viscous retarding force, approximately equal to $-\tau S/\rho A$ where τ is the average shear stress at the wall (assumed approximately parallel to the x-direction) and S is the tube perimeter.

The equations can be reduced to characteristic form as in § 2.1. Define the intrinsic wave speed, c, by

$$c^{-2} = (\rho/A) \, \partial A/\partial \tilde{p}, \tag{2.67}$$

and then add $\pm c/A$ times (2.63) to (2.64) to obtain

$$u_t + (u \pm c)u_x \pm (1/\rho c)[\tilde{p}_t + (u \pm c)\tilde{p}_x]$$
$$= f - (1/\rho)P_{0x} \mp (c/A)(\psi + uA_x) \mp (c\tilde{A}_{\tilde{p}t}/A)(\tilde{p}_{tt} + u\tilde{p}_{tx}). \quad (2.68)$$

Thus on the characteristic curves $C_{\pm}$, defined by $dx/dt = u \pm c$, the quantities $(d/dt)(u \pm \int d\tilde{p}/\rho c)$ are given by the right-hand side of (2.68). If the right-hand side is known in terms of u, p, x, t at all values of x and t, the equations can be solved by a straightforward numerical integration forward in time along the characteristics, and this is what almost all workers hitherto have done. Assuming that the functions $P_0(x, t)$ and $A(\tilde{p}, \tilde{p}_t, x)$ are given, the terms to be considered further are ψ, f, and the last term on the right-hand side of (2.68), which represents the viscoelasticity of the wall.

Outflow
The outflow function ψ can clearly not be given as a function of x and t, because the outflow through side-tubes depends on the pressure and/or flow-rate in the parent (§ 2.3). The best choice for ψ would be one that incorporated measurements or predictions of the input admittance of each side-branch. However, the admittance is normally frequency-dependent, so implementation of this choice would require time Fourier analysis of the pressure waveform at each value of x; but in the conventional technique of integrating along characteristics, p is known only for times past. This specification of ψ would therefore require an iterative solution of (2.68): a first guess at the solution would yield an estimate for ψ, which would be used in the conventional integration of the equations; the result of that integration would be used to supply the next estimate of ψ, and so on. This approach would be relatively expensive in computer time (although it would converge rapidly if the effect of ψ were small), and I know of no-one who has followed it. An alternative approach would be to abandon the characteristic equations, and solve the complete problem by a finite-difference method; this has been done by Raines, Jaffrin & Shapiro (1974) for the arteries in the leg. These authors also proposed a useful model for outflow, in which each of the most important side-branches was simulated by means of a linear lumped-parameter system with two

given resistances and a compliance in series. These branches occurred at points and were not smoothed out. The results of this analysis suggested rather surprisingly (but in agreement with experiment) that occlusion of the side-branches has little effect on the propagating pressure pulse. This is not true of the model of Anliker et $al.$ (1971), described further below.

What most authors using the characteristic method have done is to specify ψ in terms either of p or of u in the parent tube. Anliker et $al.$ (1971), following Streeter et $al.$ (1963), assumed a constant peripheral resistance (and no compliance) in each branch, and proposed that

$$\psi = f_1(x)(p - p_c), \tag{2.69}$$

where p_c is a given end-capillary pressure and $f_1(x)$ is a function chosen to simulate approximately the measured distribution of blood flow to different regions of the body. Anliker et $al.$ took

$$f_1(x) = \gamma[1.1 + \cos(5\pi x/2x^*)] \quad \text{for } x \leqslant x^*$$
$$= \gamma \times 1.1 \times e^{-0.08(x - x^*)} \quad \text{for } x \geqslant x^*,$$

where $x^* \approx 70$ cm (for a 30-kg dog) and $1/\gamma$ is a measure of outflow resistance, which can be varied to simulate different experimental conditions.

Skalak & Stathis (1966) analysed a symmetric dichotomously branching network, in which all branches were geometrically similar (quite a good model of the $pulmonary$ arterial tree), by recognising that every branch in a given generation of branching would receive the same flow-rate, equal to half the flow-rate in a tube of the preceding generation. They therefore took

$$\psi = \alpha m A u/(1 - mx), \tag{2.70}$$

which is equal to $d(Au)/dx$ when $Au = q_1(1 - mx)^\alpha$. Here q_1 is the steady flow-rate in the parent tube of a similar $rigid$ network, $\alpha = \frac{1}{2} \log 2/\log \beta$, where β is the ratio of the cross-sectional area of a daughter tube to that of the parent at a junction, and m is the constant in the linear relation between the square root of area and distance down the system: $A^{1/2} \propto 1 - mx$. From the point of view of integrating (2.68), both (2.70) and (2.69) are equally simple.

Viscous friction
The analysis of § 2.2 suggests that the quantity f would be accurately given by Womersley's (1955) analysis of oscillatory flow in a rigid tube. Thus if the pressure gradient at any x were Fourier analysed in time to give

$$\frac{dp}{dx} = \sum_{n=0}^{\infty} G_n(x) \, e^{in\omega t},$$

where $G_n(x)$ is complex, the frictional drag f, equal to $-2\tau/a\rho$ for a circular tube (which we assume) would be

$$f = \frac{1}{\rho} \sum_{n=1}^{\infty} G_n(x) F(\alpha_n) \, e^{in\omega t} + \frac{G_0}{\rho} \qquad (2.71)$$

from (2.49), where G_0 is the mean pressure gradient (<0) and $\alpha_n^2 = n\omega a^2/\nu$; $F(\alpha)$ is given by (2.37). For large α, which is appropriate here for at least the largest arteries in the pathway studied, $F \sim 2/\alpha i^{1/2}$. However, the use of (2.71), like that of a proper outflow function, requires Fourier analysis of the pressure gradient, and therefore an iterative numerical method is needed. Instead, most authors have assumed that f is in phase with u and depends on it alone, and many, including Anliker *et al.* (1971) and Schaaf & Abbrecht (1972) have taken it to be quasi-steady, either for laminar flow ($f = -8\pi\nu u/A$) or for turbulent flow ($f = -0.14\nu^{1/4}|u|^{3/4}u/A^{5/8}$). Olsen & Shapiro (1967) used the quasi-steady relation for turbulent flow when the Reynolds number exceeded 3000, but used the real part of the complex relation between τ and $Q(=uA)$, obtained from (2.48) and (2.49), for laminar flow; however, they were dealing with only small departures from a single sinusoidal mode. Since the effect of viscosity on wave propagation is small when α is large (§ 2.3; see also Olsen & Shapiro, 1967), these approximations are unlikely to cause great inaccuracy in predictions made for large arteries. However, for pathways containing significant portions of small arteries, where α may be relatively small (less than about 4), the effect of friction will be greater and these models inaccurate. This is particularly true of the model of the coronary circulation developed by Rumberger & Nerem (1977).

Viscoelasticity
This is known to be the principal factor causing wave attenuation in arteries, and is likely to be important in all arterial pathways except those so short that attenuation is unimportant (the aorta alone, for example). However, the last term in (2.68) is very complicated, and I know of no integrations in which it has been taken into account. Indeed, it is not possible to take it into account and still to be able to march forward in time, integrating once along the characteristics, because in even the simplest, linear, model of wall viscoelasticity, in which

$$A = A_1(\tilde{p}, x) - \tilde{p}_t B(\tilde{p}, x)$$

(cf. (2.19) and (2.20)), the term $\tilde{p}_{tt} + u\tilde{p}_{tx}$ is still present. Thus to take wall viscoelasticity into account, it is *necessary* to solve the problem either iteratively along the characteristics, which is suitable if this term is relatively small, or by a finite-difference method. This, together with a more adequate treatment of outflow, is the most important next step in the non-linear numerical modelling of wave propagation in arteries.

Elastic properties
Anliker *et al.* (1971) defined the elastic properties of their aortic pathway by specifying c as an empirically determined function of $\tilde{p}$ and x:

$$c(\tilde{p}, x) = (c_0 + c_1\tilde{p})(1 + nx), \qquad (2.72)$$

where $c_0 = 97 \text{ cm s}^{-1}$, $c_1 = 15.3 \text{ cm s}^{-1}$ per kN m^{-2} and $n = 0.02 \text{ cm}^{-1}$. They also specified the area as a function of x at a particular value of $\tilde{p}$ ($p_0 = 13.3 \text{ kN m}^{-2}$):

$$A(p_0, x) = 4.63 \text{ e}^{-0.045x} \qquad \text{for } x \leqslant 54 \text{ cm}$$

$$= 0.41 \text{ e}^{-0.089(x-54)} \qquad \text{for } x \geqslant 54 \text{ cm}.$$

Thus they obtained

$$A(\tilde{p}, x) = A(p_0, x) \exp\left[(\tilde{p} - p_0)/\rho c(\tilde{p}, x) c(p_0, x)\right].$$

They ignored $P_0(x, t)$ and viscoelasticity in their model, and could integrate directly.

Note that a particularly simple integration of (2.68) is possible if the simple elastic properties given by (2.11) are chosen. Then the left-hand side is

$$(d/dt)(u \pm c) \equiv \tfrac{1}{2}(d/dx)[(u \pm c)^2]$$

on the characteristics $C_\pm$. Riemann invariants can be found if the right-hand side can also be expressed as d/dx of something on $C_\pm$, or as $u \pm c$ times d/dx of something. No such case naturally arises, except when attenuation, outflow and taper are negligible, and the effect of an external pressure $P_0(x)$, which is constant in time but relatively rapidly varying in x, is to be analysed. Then $\tfrac{1}{2}(u \pm c)^2 + (1/\rho)P_0(x)$ is constant on $C_\pm$.

Boundary conditions
The conditions at $x = 0$ are those at the exit of the heart, in the aorta or in the pulmonary artery according to which circulatory bed is being modelled. In principle, one may specify (by measurement) either $p(0, t)$ or $u(0, t)$ (or a combination of the two involving the measured input admittance of the system), but Anliker *et al.* (1971) found that much more satisfactory results were obtained by specifying $u(0, t)$, because the numerical procedure was less sensitive to small variations in this quantity. Downstream conditions, at $x = L$ (=150 cm) can most conveniently be expressed in terms of a constant peripheral resistance R_L, so that

$$A[p(L, t), L]u(L, t) = [p(L, t) - p_c]/R_L,$$

where p_c is the end-capillary pressure (as in (2.69)). R_L was given a value that yielded the correct mean pressure at $x = 0$.

Various initial conditions at $t = 0$ may be chosen. The standard case in Anliker *et al.*'s computation was one in which conditions at the start of systole were close to those at the end of diastole (periodicity condition), requiring more than one integration of (2.68). Another possibility is to suppose that, at $t = 0$, the blood is at rest with a given pressure.

Results
The results of computer models such as this always show a broad qualitative agreement with experiment, even without an adequate treatment of wall viscoelasticity, fluid viscosity, or outflow. The standard results of Anliker *et al.* (1971) clearly show peaking and

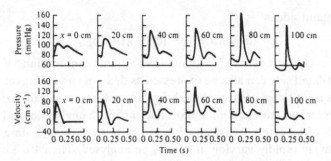

Fig. 2.8. Predicted pressure and velocity waveforms at six different locations along the artery; $x = 0$ is the aortic valve. (After Anliker et al., 1971.)

steepening of the pressure pulse, the development of a second peak in the pressure pulse, and the development of considerable reverse flow in the abdominal aorta (fig. 2.8). However, the usual fall-off in peak velocity with distance and the persistence of backflow into more peripheral arteries are not seen. These features of the results reveal the shortcomings of the model; in my view, its chief defect lies in the excessively smooth and inaccurate form chosen for the outflow function $f_1(x)$ in (2.69). The only direct evidence for this, however, is that varying the outflow constant γ over a reasonable physiological range had a more marked effect on the results than varying the friction factor f. No different shape for $f_1(x)$ was used; neither was any viscoelasticity taken into account. We may note further that the cumulative effect of non-linearities was not negligible (fig. 2.9), as found also by Schaaf & Abbrecht (1972) and by Ling et al. (1973) (the latter authors used a different method; see chapter 3).

Perhaps the greatest value of models such as this is the simulation of different clinical conditions, so that either observed abnormalities may be explained or possible abnormalities can be predicted. Anliker et al. (1971) considered a number of examples. One of particular interest is the modelling of aortic valve incompetence by the imposition of a very-large-amplitude velocity waveform at $x = 0$. The model predicts that very sharp pressure and velocity peaks are formed for x lying between 50 and 100 cm, the thickness of the wavefront being as little as 6 cm. The authors interpret these results (a) as explaining the arterial 'pistol-shot' pulse observed in

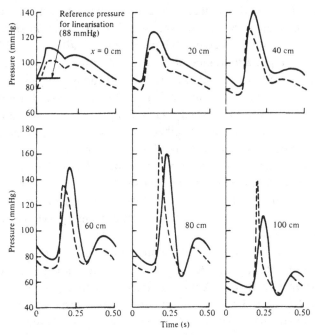

Fig. 2.9. Comparison of pressure waveforms predicted from non-linear (broken curves) and linearised (continuous curves) theory. (After Anliker *et al.*, 1971.)

the limbs of patients with incompetent valves and (*b*) as indicating the presence of shocks. While it is probable that shocks do occur when such sharp pressure pulses are formed, they cannot arise out of a numerical integration along characteristics unless the integration breaks down and two characteristics cross. In that case jump conditions have to be applied across the shock, involving extra dissipation (§ 2.1). Anliker *et al.* did not refer to such numerical breakdown, and it is possible that shocks did not actually appear in their model. Any errors would no doubt be largely due to the assumption that outflow was linearly related to local pressure, by a peripheral resistance, and this would be particularly inaccurate for such sharp pressure peaks since shocks would be formed in side-branch arteries also. The absence of viscoelasticity would also make the model markedly inaccurate when such steep waves are predicted. These authors also noted, but did not model, a 'venous pistol shot', heard over the jugular and femoral veins when the tricuspid valve (between right atrium and ventricle) is incompetent; this phenomenon is described by Hultgren (1962). Anliker *et al.* did not report the input admittance of the arterial system as predicted by their model; Schaaf & Abbrecht (1972) did compute it, and that of the femoral artery, although their model had a less-well-founded description of arterial elasticity. They were able to reproduce some of the measured features described above, such as a more marked maximum of the admittance modulus in the femoral bed than in the aortic bed, but qualitative similarity with observation is to be expected. Future models must either show how to incorporate real features hitherto absent (such as viscoelasticity) or be able to describe quantitatively phenomena that have not previously been accurately modelled.

An arterial bed of particular clinical importance is the coronary circulation. Rumberger & Nerem (1977) have applied a model very similar to that of Anliker *et al.* (1971) to a pathway in the horse, starting at the left coronary ostium and incorporating the left common coronary artery and the left anterior descending artery. There are two important differences between this model and one starting in the aorta. First, the vessels are smaller, so fluid viscosity will have a greater influence (and should therefore be modelled more accurately, but this was not done). Secondly, the pathway plunges into the heart muscle at about 25 cm from the entrance, and this muscle exerts a strong, but variable, external pressure $P_0(x, t)$ (see (2.66)). The authors appreciated this, and used left ventricular pressure (times a known function of x) as the external pressure, which is a reasonable first approximation. However, they still used an equation similar to (2.72) to describe arterial elasticity, which ignores the greater distensibility expected in blood vessels when the transmural pressure is low (cf. fig. 1.10). Much less information on wave propagation is available in coronary arteries than in the aorta, and Rumberger & Nerem derived the elastic constants for their model from a few experiments of their own. They also ignored viscoelasticity, like everyone else. Nevertheless, despite the shortcomings of the model, they were able to reproduce pressure and velocity waveforms that exhibit the principal observed features: (i) a much greater flow-rate in diastole when the squeezing of the peripheral part of the pathway is lifted, and (ii) high-frequency ($\sim$10 Hz) oscillations in pressure and velocity (fig. 1.19). Since the oscillations occur in the model, they must represent a resonance phenomenon involving a Fourier mode for which the length of the pathway to the major reflection site (perhaps where the artery enters the muscle at $x = 25$ cm) is a half-wavelength. A frequency of 10 Hz and a wavelength of 50 cm means a wavespeed of 5 m s^{-1}, which is reasonably consistent with the wave-speed values put into the model by Rumberger & Nerem.

CHAPTER 3

FLOW PATTERNS AND WALL SHEAR
STRESS IN ARTERIES
I STRAIGHT TUBES

The next three chapters, which form the core of this monograph, have two main purposes. One is to give a theoretical explanation for some of the arterial velocity profiles described in § 1.2.4. The other, of greater potential importance in the analysis of arterial disease, is to make predictions of the detailed distribution of wall shear stress in arteries, which is related to the rate of mass transport across artery walls and hence (presumably) to atherogenesis (§ 1.2.6). The second purpose is particularly important because no method has yet been devised to measure wall shear stress accurately as a function of time *in vivo*. This is rather surprising, considering the probable importance of wall shear, and the first section of this chapter is devoted to an explanation of why it is so difficult to measure. The second section begins the analysis of viscous flow in arteries with a discussion of unsteady entry flow (with flow reversal) in a straight tube. In chapters 4 and 5 respectively, curved and branched tubes are considered, and chapter 5 concludes with a discussion of flow instability in arteries.

3.1 The difficulty of measuring wall shear stress

3.1.1 *The need for a good frequency response*

Since the mechanism by which the wall shear stress influences mass transport across the artery wall is unknown, with the consequence that the relevant features of the wall shear distribution cannot be identified, it is important to understand as many features as possible. In particular, the time-dependence of the wall shear at various sites on the artery wall should be accurately recorded, because, as will be argued in § 3.2.3, the unsteady components are likely to be at least as important as the mean shear. We have seen in § 1.2.3 that there are devices that can measure the flow-rate through an artery, and the local blood velocity at points within it, with adequate accuracy. That is, the time variation of these velocities can be faithfully recorded, except at times close to those at which the flow reverses its direction (see appendix). However, the requirements of a device to measure wall shear are more stringent for two reasons. One is that the shear on a tube wall may reverse its direction when the average and centre-line velocities do not; furthermore, it always has a phase lead over those velocities (except when the Womersley parameter α (see (2.35)) is so small that the flow is quasi-steady). This is because the slow-moving fluid near the wall responds more readily to variations in the driving pressure gradient than the faster-moving fluid in the core. Thus the problems associated with reversal are more pronounced in the case of wall shear measurement. The less obvious, but probably more significant, difficulty is that a wall shear probe must respond accurately to higher frequencies than a centre-line-velocity probe whereas, as we shall see, the probes hitherto designed are less well able to respond to high frequencies.

The need for a very good frequency response in wall shear measurement can be seen from Womersley's solution for oscillatory flow in a long, rigid, circular tube (§ 2.2). If the applied pressure gradient is

$$\frac{dp}{dx} = - \sum_{n=0}^{\infty} G_n \, e^{in\omega t}, \tag{3.1a}$$

then the average velocity is

$$\bar{u} = \frac{G_0 a^2}{8\mu} + \frac{a^2}{i\mu} \sum_{n=1}^{\infty} \frac{G_n}{\alpha_n^2} [1 - F(\alpha_n)] \, e^{in\omega t}, \tag{3.1b}$$

the centre-line velocity is

$$u(0) = \frac{G_0 a^2}{4\mu} + \frac{a^2}{i\mu} \sum_{n=1}^{\infty} \frac{G_n}{\alpha_n^2} \left[1 - \frac{1}{J_0(i^{3/2}\alpha_n)} \right] e^{in\omega t}, \tag{3.1c}$$

and the wall shear stress is

$$\tau = \frac{G_0 a}{2} + \frac{a}{2} \sum_{n=1}^{\infty} G_n F(\alpha_n) \, e^{in\omega t}, \tag{3.1d}$$

where $\alpha_n^2 = nwa^2/\nu$ and $F(\alpha)$ is given by (2.37). Now we saw in § 1.2 that up to 10 harmonics are required to reproduce accurately the average velocity waveform in the canine aorta; this neglects Fourier modes whose amplitudes are less than about 5 cm s^{-1}, 10% of the amplitude of the largest mode (Patel *et al.*, 1963*a*). Equations (3.1*b* and *c*) indicate that the same accuracy is required for centre-line velocity, and thus that a hot-film velocity probe for use in a dog has to have a good response to frequencies up to about 20 Hz. However, the terms in equation (3.1*d*) for τ have (for large α_n) an amplitude equal to $\alpha_n (\propto n^{1/2})$ times those in (3.1*b* and *c*). Thus (*a*) G_n must diminish faster than $n^{-1/2}$ for this to be convergent, and assuming that this is satisfied (*b*) the sum must be taken to significantly higher values of n to achieve the same accuracy. The precise value of n required cannot be predicted without a knowledge of G_n for $n > 10$, but it can be seen clearly that a shear-stress probe must have a much better frequency response than one measuring centre-line velocity.

An example of what is required in practice can be derived by calculating the wall shear from a known velocity waveform, either measured or predicted. Fig. 3.1(*a*) shows the waveform of average velocity in the left anterior descending coronary artery of a *horse*, as calculated from the computer programme of Rumberger & Nerem (1977) (a similar measured waveform was shown in fig. 1.18). The heart-rate was taken to be 1 Hz. Fig. 3.1(*b*) indicates how many Fourier components are needed for an accurate representation of the root-mean-square (r.m.s.) velocity and the peak velocity; it can be seen that 8 harmonics (8 Hz) are adequate for the r.m.s. velocity, while 30 harmonics (30 Hz) are needed for the peak. Fig. 3.2(*a*) shows the corresponding waveform of wall shear-rate, calculated from (3.1*b* and *d*) using (*a*) over 100 and (*b*) 50 Fourier components; fig. 3.2(*b*) shows the peak and r.m.s. values of wall shear-rate as functions of the number of Fourier components used. From these we can see that even 50 components are not enough to represent the peak shear, although 20 are adequate for the r.m.s. Note that while the twentieth component corresponds to a frequency of 20 Hz in the horse, it corresponds to 40 Hz in the dog. The peak shear stress in this example is about 9 N m^{-2} and the r.m.s. is 1.8 N m^{-2}, compared with the mean value of about 0.6 N m^{-2}.

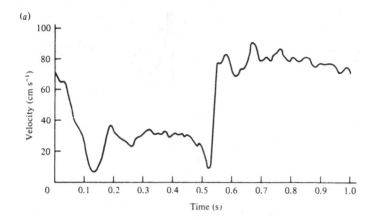

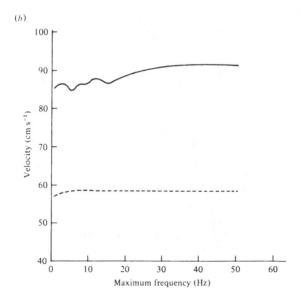

Fig. 3.1. (a) Calculated waveform of average velocity in the left anterior descending coronary artery of a horse. $x = 25$ cm. (After Rumberger, 1976.) (b) Plots of peak (continuous curve) and r.m.s. (broken curve) velocities in a waveform reconstituted from a Fourier series of the waveform in (a), terminated after terms of a given frequency (abscissa).

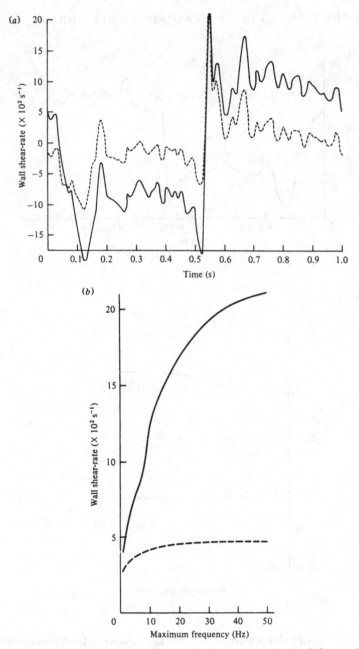

Fig. 3.2. (a) Calculated waveform of wall shear-rate in the left anterior
descending coronary artery of a horse (continuous curve), together with the
waveform reconstituted from the Fourier series, terminated at a frequency
of 50 Hz (broken curve). (b) As for fig. 3.1(b), but for wall shear-rate not
velocity. (I am most grateful to Mr T. Benson of the Ohio State University
for performing the computations from which figs. 3.1 and 3.2 have been
plotted.)

3.1.2 *The limitations of the hot-film anemometer*

One method of estimating wall shear in a tube is to measure the velocity profile at internal points and to extrapolate it to zero at the wall. However, this cannot be accurate unless there are several points within the inner part of the boundary layer on the wall, so that a smooth monotonic curve can be drawn through them. In arteries the unsteady boundary layers have thickness approximately equal to $4(\nu/\omega)^{1/2}$, where ω is the angular frequency of the oscillations, and this takes a value of about 2.3 mm when the frequency is 2 Hz. The inner part of the boundary layer, in which the velocity profile is monotonic, is at most about 1 mm thick. Furthermore, both these numbers would be diminished for higher harmonics of the pulse. Now the hot-film anemometers used hitherto for velocity measurement are about 1 mm across, so it is difficult to make accurate measurements within the boundary layer. The only chance is to insert the probe in one wall of an artery and traverse the artery with the hot-film towards the far wall so that the film can enter the far boundary layer. Even so, accurate measurements cannot be made because of the radial excursion of the artery wall: the probe must remain far enough from the wall in systole not to be struck by it in diastole. Thus in the aortae of dogs (Seed & Wood, 1971; Clark & Schultz, 1973) measurements are not made in the boundary layer. The situation is somewhat better in horses, because of their lower heart-rate and, hence, thicker boundary layers, and Nerem *et al.* (1974*a*) have reported some measurements in the far boundary layer (without commenting on the problem of wall movement).

Other methods of measuring velocity profiles, e.g. with a catheter-tip electromagnetic flowmeter (Mills *et al.*, 1970), or by ultrasonic Doppler anemometry (Peronneau *et al.*, 1970), suffer from the same limitation of poor resolution, to within 1 mm. Similar limitations apply to flow in rigid transparent models, but in these the scale of the flow can be increased so that the probe width is a small fraction of the boundary layer thickness (also laser Doppler anemometry, which has somewhat finer resolution, is nowadays available). Even when the velocity profile is known quite accurately, extrapolation to estimate wall shear is notoriously inaccurate. Brech & Bellhouse (1973) reported a 50% discrepancy between such estimates and values measured directly by a hot-film probe (but see below); Minton & Selvalingam (1970) have also shown a 25% discrepancy between this method and theoretical deduction for oscillatory flow in a long straight tube. Thus one should attempt to measure wall shear stress another way. The only method at present available for use *in vivo* employs a constant-temperature hot-film embedded in the vessel wall, heat transfer from it being related to the local shear-rate. A variation of this method involving *mass* transfer from an electrode in a flowing electrolyte (the 'electrochemical technique') can also be used in model experiments (Reiss & Hanratty, 1962). Such probes have been used to measure the wall shear in steady flow in rigid models, ranging from two-dimensional branched channels (Smith, Colton & Freedman, 1974) to three-dimensional casts of the canine aorta (Lutz *et al.*, 1977), and yield satisfactory results. These authors also used their probes in sinusoidally oscillating flow, assuming that the quasi-steady calibration was applicable.

The behaviour of such shear probes is investigated theoretically in the appendix, on the assumption that two-dimensional boundary layer theory is applicable. This requires *both* that the Péclet number $Pe = S_0 l^2/\kappa$ is large (greater than about 400), where S_0 is the mean wall shear-rate, l is the length of the hot film (or electrode) in the direction of motion and κ is the thermal (or concentration) diffusivity of the fluid ($\kappa \approx 1.4 \times 10^{-7}$ m^2 s^{-1} for heat in blood), *and* that l is much less than the width of the film in the perpendicular direction. Neither condition is well satisfied in practice: for example, when $S_0 = 320$ s^{-1} (as for a 0.5 cm artery with a mean velocity of 20 cm s^{-1} such as the horse coronary artery) and when $l = 0.1$ mm (as for the smallest hot-films hitherto made), the Péclet number is only about 23; both larger and smaller values of S_0 are encountered in arteries. Furthermore, hot-films are often made only 2.5 times as wide as they are long, so the three-dimensionality of the temperature field over the film may be important. Nevertheless, the conclusions of the theory are expected to give a qualitative indication of the frequency response of a hot-film shear probe.

Use of the probe requires that there is a unique relation between the rate of heat transfer from the film, Q, and the wall shear-rate, S. Steady boundary layer theory indicates that

$$Q = 0.81(\kappa^2 l^2 S)^{1/3}(T_1 - T_0)\rho c_p, \qquad (3.2)$$

where T_1 is the temperature of the film and T_0, ρ, c_p are the temperature, density and specific heat of the ambient fluid (Lévêque, 1928), although in any experiment the steady relation between Q and S would of course be measured directly. In order that the probe can be used in unsteady flow, therefore, its response must be known to be quasi-steady, i.e. (3.2) or its empirical equivalent must be known to be true at every instant. One of the main conclusions of the theory outlined in the appendix is that a hot-film shear probe will respond quasi-steadily to an unsteady shear $S(t)$ as long as the quantity λ is small enough, where λ is defined by

$$\lambda = \left| \frac{3^{1/3} l^{2/3} \, dS/dt}{\kappa^{1/3} S^{5/3}(t)} \right|. \tag{3.3}$$

Numerical results suggest that 0.1 is the critical value above which λ should not rise. This makes it clear that as the shear approaches zero, prior to a flow reversal, the probe becomes completely inaccurate. Such inaccuracy is inevitable since the heat transfer from the film will be positive whatever the sign of the shear over it (and even in the complete absence of flow), so that it cannot fall to zero with the shear (see fig. A.10 in the appendix).

The requirement that $\lambda < 0.1$ also demonstrates how the response will cease to be quasi-steady, in a flow of sufficiently large frequency and not too small amplitude, well before reversal of the shear, and even in the absence of shear reversal. Consider a sinusoidal shear variation,

$$S(t) = S_0(1 + \alpha_1 \cos \omega t), \tag{3.4}$$

for which

$$\lambda = \left| \frac{3^{1/3} l^{2/3} \omega \alpha_1 \sin \omega t}{\kappa^{1/3} S_0^{2/3} (1 + \alpha_1 \cos \omega t)^{5/3}} \right|. \tag{3.5}$$

The predicted heat transfer as a function of time is shown for various values of the amplitude parameter α_1 (both greater and less than 1) and the frequency parameter $\omega_1 = \omega l^{2/3}/\kappa^{1/3} S_0^{2/3}$ in figs. A.11(a)–(c). The increasing departure from the quasi-steady response as ω_1 (and hence λ for given values of α_1 and ωt) increases is most marked. As a practical example, we once more take the horse coronary artery, with $S = 320 \text{ s}^{-1}$ and $\alpha_1 \approx 5$ (Nerem et al.,

1974a): the value of α_1 was estimated from the quoted values of the ratio of maximum to mean velocity and of the Womersley parameter, using equations (3.1); l will again be taken to be 0.1 mm. A reasonable criterion for the usefulness of the hot-film output is that it should be quasi-steady for at least a quarter of a cycle around the time of peak shear, i.e. $-\frac{1}{4}\pi \leqslant \omega t \leqslant \frac{1}{4}\pi$. Equation (3.5) shows that, when $\omega t = \frac{1}{4}\pi, \lambda \approx 0.0036\omega$; thus $\lambda < 0.1$, and the probe can be used, only as long as the frequency $f(=\omega/2\pi)$ is below 4.5 Hz. We have seen that this is not good enough for accurate wall shear measurements in the coronary arteries, even discounting the greater inaccuracy at times closer to flow reversal. If we were to require comparable accuracy for reversed shear, at $\omega t = \frac{3}{4}\pi$, the critical frequency would be reduced to 2 Hz. If we took $\alpha_1 = 0.5$, the critical frequency for accuracy at $\omega t = \frac{1}{4}\pi$ would be 6 Hz, although in that non-reversing case the maximum value of λ would be less than 0.5 only if $f < 2.5$ Hz. For a different example we consider a sinusoidal oscillation typical of the aorta of a dog, at a distance of 5 cm from the entrance (as analysed in § 3.2 below). Here S_0 is smaller ($=80\ \mathrm{s}^{-1}$) but α_1 is greater (≈ 50): in this case the critical frequency for accuracy at $\omega t = \frac{1}{4}\pi$ is 5.6 Hz. Once more, this does not represent an adequate frequency response. The only way open to an experimentalist to improve the frequency response, i.e. to reduce λ for a given shear distribution at a given time, is to make l smaller. However, the value of 0.1 mm used here for l is the smallest length of hot-film employed to date; new methods of probe construction are therefore required. The claim by Ling et al. (1968) to have measured wall shear in vivo, in the aorta of a dog, and to be accurate throughout the cycle despite the fact that no reversed shear was recorded when it is invariably predicted (cf. § 3.2), is not substantiated by a description of unsteady calibration experiments and must be regarded with some scepticism. We may also note here that the electrochemical technique is even less appropriate than the hot-film for unsteady studies, in models or casts of arteries, since κ is much smaller for electrolytes in water than for heat ($\kappa \approx 10^{-9}\ \mathrm{m}^2\ \mathrm{s}^{-1}$), making λ larger.

Further problems in using a hot-film shear probe in vivo are the difficulty in setting the film flush with the pulsating vessel wall and the three-dimensional character of flow in most arteries, leading to *two* unsteady components of wall shear stress. The former is important because even a slight protrusion into the vessel can seriously disturb the local shear distribution (Lutz *et al.*, 1977). The latter may actually be an advantage because it means that the shear stress *vector* does not fall to zero at most points on the artery wall, so, if the probe is insensitive to direction, it can record the magnitude of the vector at all times in the cycle without the problems associated with flow reversal. This could perhaps be achieved by using a film either of circular or of spiral cross-section, and not placing it on a plane of symmetry of the artery.

We have seen that the frequency response of a wall shear probe is expected to be worse than that of a centre-line- (or average-) velocity probe, while a better frequency response is needed. Thus the most accurate way of obtaining a 'measured' waveform of wall shear, at least in a fairly straight segment of artery far from its entrance, is probably to measure the centre-line velocity and calculate the wall shear from Womersley's equations (3.1*a–d*), since the velocity waveform is known to be related to the pressure gradient as in a rigid tube (fig. 2.5). The improvement in frequency response required of velocity probes is not as dramatic as, and is much more feasible than, that which would be needed from direct wall shear measurement. This method will be inaccurate in arteries so small that the profile in the core is not flat, and near the entrance of an artery where Womersley's theory is invalid (especially for the mean shear: see § 3.2). It will also, of course, be inaccurate near branches and bends.

Ling & Atabek (1972) have proposed that the way to derive a 'measured' waveform of wall shear is to measure the pressures at two nearby stations, calculate the pressure gradient from them, and then use a non-linear theory (based on approximate integration of the axisymmetric equations of motion, including radial wall displacement and fluid viscosity, but not wall viscoelasticity) to compute the velocity and wall shear waveforms. They have used this method in the aorta and in the coronary arteries (Ling *et al.*, 1973; Atabek, Ling & Patel, 1975), and there is no reason to doubt the accuracy of the method. However, there is no evidence that it is more accurate than that proposed above, it requires two simultaneous measurements (of pressure) instead of only one

(of velocity), and the computations would be laborious for a new-
comer to the field to reproduce, while use of (3.1b and d) needs only
a standard Fourier-series routine.

The next section and the following two chapters are concerned
with the theoretical prediction of wall shear at arterial sites where
the above semi-empirical method of determining it cannot be used.

3.2 Entry flow in a straight tube

The aorta has been the site of a high proportion of velocity-profile
measurements; different parts of the aortic wall also have different
mass-transport properties and different susceptibilities to atheroma
(§ 1.2); it is therefore a vessel in which prediction of the wall shear,
as a function of position and time, is desirable. The geometry of the
aorta is complicated, as shown in fig. 1.4, and in order to isolate
different effects we make several drastic simplifications. The
thoracic aorta is modelled as a uniform, rigid, curved tube of
circular cross-section, with a flat, but unsteady, velocity profile
entering from the aortic valve. Either the pressure gradient or the
centre-line (or average) velocity can be regarded as given, and the
neglect of elasticity is justified by the correspondence between
rigid-tube theory and experiment (fig. 2.5). Kuchar & Ostrach
(1966) calculated entry flow in an elastic tube. They found that
elasticity was important only within about one tube radius of the
entrance, and depended strongly on the boundary condition that
the tube radius was rigidly fixed there. We take the origin $x = 0$ at
the downstream end of the aortic valve, where expansion is possible; thus the tube remains effectively parallel-sided at all times,
since the wavelength of the pressure wave greatly exceeds the
longitudinal excursion of any fluid element. The gradual taper of the
real aorta can be neglected for similar reasons.

The branches from the arch of the aorta will clearly have a
significant influence on the flow near their entrances. However,
their effect on the flow nearer to the heart is likely to be less
pronounced; as we shall see, the skewed velocity profiles in this
region of the canine aorta can be accounted for by the curvature of
the vessel. The flows in certain types of branch are discussed in
chapter 5. The two effects remaining to be analysed are those of

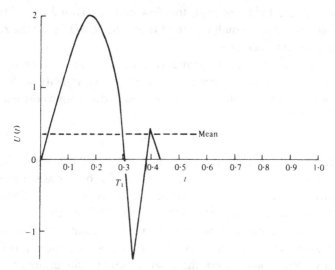

Fig. 3.3. Dimensionless velocity, $U(t)$, in the core of the aorta, simplified from a measured waveform presented by Nerem *et al.* (1972). Peak velocity $(U = 2)$ is 1 m s^{-1}, mean velocity $(U = 0.34)$ is 0.17 m s^{-1}; duration of beat $(t = 1)$ is 0.4 s. (After Pedley, 1976*a*.)

unsteadiness and of curvature. Curvature is discussed in the next chapter; in this section we simplify the problem still further and consider unsteadiness alone, analysing unsteady entry flow in a straight tube. The incoming waveform of centre-line velocity is taken to have a shape typical of that measured in the canine ascending aorta, as shown in fig. 1.14(*b*) and modelled in fig. 3.3, where the peak velocity is 1 m s^{-1}, where there is flow reversal with a maximum negative velocity of -0.7 m s^{-1}, and where the mean velocity is 0.17 m s^{-1} (Nerem, *et al.*, 1972). Note that the neglect of curvature is partly justified by the fact that the initial part of the ascending aorta is more or less straight, curvature developing only after 2 or 3 cm.

3.2.1 *Steady flow*

Steady axisymmetric entry flow in a tube, with a more or less flat velocity profile at $x = 0$, has been the subject of theoretical study for many years (see Goldstein, 1938; Schlichting, 1968). In the limit of large Reynolds number Re $(=2Ua/\nu$, where a is tube radius and U

is the average fluid velocity), the flow can be divided into three regions, excluding a small region of length-scale ν/U near the rim of the tube entrance at $x = 0$.

(a) For $x/a = O(1)$, a boundary layer on the wall $r = a$, of thickness $\delta \propto (\nu x/U)^{1/2}$, in which the longitudinal velocity profile is that of a Blasius boundary layer to leading order, with a correction of $O(Re^{-1/2})$:

$$u = U[f'_0(\eta) + O(Re^{-1/2})], \tag{3.6}$$

where $\eta = y(U/2\nu x)^{1/2}$, $y = a - r$, and $f_0(\eta)$ is the Blasius function.

(b) Also for $x/a = O(1)$, there is an inviscid core, consisting of a uniform velocity plus a correction due to the displacement effect of the Blasius boundary layer (Van Dyke, 1970; Wilson, 1971); for the case when the entry condition is that of irrotational, but not necessarily exactly parallel, flow ($u = U$, $\partial v/\partial x = 0$, where v is the radial velocity component), the axial velocity profile in the core is

$$\frac{u}{U} = 1 + \frac{\beta_0}{(\pi Re)^{1/2}} \int_0^\infty \frac{\sigma^{-1/2} I_0(\sigma r/a)}{I_1(\sigma)} \sin(\sigma x/a) \, d\sigma + O(Re^{-1}),$$

where $\beta_0 = \lim_{\eta \to \infty} (\eta - f_0) \approx 1.217$ (Singh, 1974). The $O(Re^{-1/2})$ correction to the core flow satisfies the boundary condition

$$v/U \sim -\beta_0/(2Rex/a)^{1/2} \quad \text{at } r = a,$$

and as $x/a \to \infty$ the axial velocity component has the asymptotic expansion

$$\frac{u}{U} \sim 1 + \frac{2\beta_0}{Re^{1/2}} \left[\left(\frac{2x}{a}\right)^{1/2} + O\left(\frac{x}{a}\right)^{-1/2} \right] + o(Re^{-1}),$$

representing a flat profile. Thus the expansion breaks down when $x/a = O(Re)$ at which point the boundary layers fill the tube ($\delta = O(a)$).

(c) There is therefore a third region, in which $\xi = x/aRe = O(1)$, where the boundary layer equations apply (to leading order) for all r, and the flow develops from a flat core surrounded by a Blasius boundary layer at $\xi \ll 1$ to Poiseuille flow as $\xi \to \infty$. This development has been calculated numerically (at least for two-dimensional flow) by Bodoia & Osterle (1961).

Despite the displacement effect, and the non-uniform core profile at finite values of Re, the velocity profile near the wall, and in particular the wall shear, in the region $x/a \ll Re$ is given by the Blasius profile (3.6). This means that the boundary layer is so thin that it behaves, to leading order in $Re^{-1/2}$, as if it were on a flat plate and not in a tube: the curvature of the walls is irrelevant. We assume that this is also the case in unsteady entry flow, and calculate the structure of the unsteady boundary layer on a semi-infinite flat plate; the procedure is expected to be valid for x/a much smaller than the mean Reynolds number.

3.2.2 Non-reversing unsteady flow

We consider a flat plate occupying the half-plane $\hat{y} = 0$, $\hat{x} > 0$ (where a hat over a variable means that it is dimensional; this convention will be used from now on), with a viscous incompressible fluid flowing over it in the $\hat{x}$-direction. The velocity far from the plate is $\hat{U}(\hat{t})$. Suppose that a velocity-scale is U_0 and a time-scale is Ω^{-1}, and introduce dimensionless variables,

$$\left. \begin{array}{l} x = \hat{x}\Omega/U_0, \quad y = \hat{y}(\Omega/\nu)^{1/2}, \quad t = \Omega\hat{t}, \quad U(t) = \hat{U}(\hat{t})/U_0, \\ u = \hat{u}/U_0, \quad v = \hat{v}/(\Omega\nu)^{1/2}, \quad \psi = (\hat{\psi}/U_0)(\Omega/\nu)^{1/2}, \end{array} \right\} \quad (3.7)$$

where $\hat{u}$, $\hat{v}$, $\hat{\psi}$ are the $\hat{x}$- and $\hat{y}$-components of velocity, and the stream function, respectively. The velocity field is related to ψ by

$$(u, v) = (\psi_y, -\psi_x).$$

Boundary layer theory is applicable as long as the Reynolds number, $U_0^2/\Omega\nu$, is much greater than 1. The boundary layer equations reduce to

$$\psi_{yt} + \psi_y\psi_{xy} - \psi_x\psi_{yy} = U_t + \psi_{yyy}, \quad (3.8)$$

which is to be solved subject to boundary conditions:

$$\left. \begin{array}{ll} \psi = \psi_y = 0 & \text{on } y = 0 \text{ (no slip)} \\ \psi_y \sim U(t) & \text{as } y \to \infty. \end{array} \right\} \quad (3.9)$$

The function $U(t)$ should eventually be taken to be of the form shown in fig. 3.3; however, this subsection is devoted to non-reversing flow, and when it is necessary to specify $U(t)$, we take as a prototype

$$U(t) = 1 + \alpha \sin t, \quad (3.10)$$

where the amplitude parameter, α, is less than 1.

This problem has been investigated by several authors. The best known early work is that of Lighthill (1954), who restricted himself to $\alpha \ll 1$, and used a Karman–Pohlhausen method to compute the skin friction for both large and small values of x. Moore (1951, 1957) showed that the small-α restriction is unnecessary for the small-x expansion, and Lin's (1956) expansion for large x did not require the restriction either. This seems not to have been generally realised, however, until Pedley (1972b) combined the two approaches, and extended them to cover a free stream whose velocity varies as a power of x. The following account is a summary of the large-amplitude theory for the flat plate; its object is to calculate the dimensionless wall shear stress

$$S(x, t) = \psi_{yy}|_{y=0}$$

for all t and for as wide a range of values of x as possible.

Small x

At a fixed value of $\hat{x}$ we would expect the flow to be a quasi-steady Blasius boundary layer if the frequency Ω is low enough. This suggests that an expansion in powers of the dimensionless coordinate x should have the quasi-steady solution as its first term. We therefore introduce appropriate similarity variables, η_1 and ϕ, such that

$$y = [2x/U(t)]^{1/2}\eta_1, \qquad \psi = [2xU(t)]^{1/2}\phi,$$

and the governing equation, (3.8), becomes

$$\phi_{\eta_1\eta_1\eta_1} + \phi\phi_{\eta_1\eta_1} + 2x[\phi_x\phi_{\eta_1\eta_1} - \phi_{\eta_1}\phi_{x\eta_1}$$
$$+ (\dot{U}/U^2)(1 - \phi_{\eta_1} - \tfrac{1}{2}\eta_1\phi_{\eta_1\eta_1}) - (1/U)\phi_{\eta_1 t}] = 0. \quad (3.11)$$

The boundary conditions are

$$\phi = \phi_{\eta_1} = 0 \quad \text{on} \quad \eta_1 = 0, \qquad \phi_{\eta_1} \to 1 \text{ (exponentially)} \quad \text{as } \eta_1 \to \infty.$$

Formally putting $x = 0$, we see that the problem reduces to that of the Blasius boundary layer, with solution $\phi = f_0(\eta_1)$. We therefore seek a solution in powers of x:

$$\phi = f_0(\eta_1) + \sum_{m=1}^{\infty} x^m \phi_m(\eta_1, t), \qquad (3.12)$$

where the functions ϕ_m satisfy homogeneous boundary conditions in η_1, but inhomogeneous differential equations. Substitution into (3.11) and equating first and second powers of x show that

$$\phi_1 = \frac{\dot{U}}{U^2} f_{11}(\eta_1), \qquad \phi_2 = \frac{\dot{U}^2}{U^4} f_{21}(\eta_1) + \frac{\ddot{U}}{U^3} f_{22}(\eta_1),$$

where $f_{mk}(\eta_1)$ satisfies the ordinary differential equation

$$f'''_{mk} + f_0 f''_{mk} - 2m f'_0 f'_{mk} + (2m+1) f''_0 f_{mk} = \mathscr{F}_{mk}(\eta_1) \qquad (3.13)$$

and

$$\mathscr{F}_{11} \equiv \eta_1 f''_0 + 2(f'_0 - 1), \qquad \mathscr{F}_{22} \equiv 2 f'_{11},$$

$$\mathscr{F}_{21} \equiv 2 f'^2_{11} - 3 f_{11} f''_{11} + \eta_1 f'''_{11} - 2 f'_{11}.$$

These equations have been solved numerically, and the values of $f''_{mk}(0)$ and of $\beta_{mk} \equiv \lim_{\eta_1 \to \infty} (-f_{mk})$ are given in table 3.1. The dimensionless skin friction is

$$S = \frac{U^{3/2}(t)}{(2x)^{1/2}} \left\{ f''_0(0) + \frac{x\dot{U}}{U^2} f''_{11}(0) \right.$$

$$\left. + x^2 \left[\frac{\dot{U}^2}{U^4} f''_{21}(0) + \frac{\ddot{U}}{U^3} f''_{22}(0) \right] + O(x^3) \right\}. \qquad (3.14)$$

Another quantity that proves to be of importance is the displacement thickness $\hat{\delta}_1$ defined by

$$\hat{\delta}_1 = \int_0^\infty \left(1 - \frac{\hat{u}}{\hat{U}} \right) d\hat{y},$$

Table 3.1. *Values of* $f''_{mk}(0)$
and of $\beta_{mk} = \lim_{\eta_1 \to \infty} (-f_{mk})$,
$(f_{00} \equiv f_0 - \eta)$

m	k	$f''_{mk}(0)$	β_{mk}
0	0	0.470	1.217
1	1	1.200	−0.727
2	1	0.383	−0.541
2	2	−0.664	0.845

so that the dimensionless displacement thickness $\delta_1 = (\Omega/\nu)^{1/2}\hat{\delta}_1$ is equal to

$$\delta_1 = \left(\frac{2x}{U}\right)^{1/2}\left[\beta_0 + \frac{x\dot{U}}{U^2}\beta_{11} + x^2\left(\frac{\dot{U}^2}{U^4}\beta_{21} + \frac{\ddot{U}}{U^3}\beta_{22}\right) + O(x^3)\right]. \quad (3.15)$$

The above expansion shows that the quasi-steady solution is accurate at a given value of x as long as

$$\varepsilon = |x\dot{U}/U^2| \quad (3.16)$$

is much less than 1 for all t. The series itself will be a useful asymptotic expansion if ε (and $|x^2\ddot{U}/U^3|$) is small enough for the $O(x^2)$ term to be small compared with the $O(x)$ term for all t. In that case the first two terms will be a reasonable approximation to the almost quasi-steady solution. Pedley (1972b) made a number of computations, and a useful rule of thumb to emerge is that the first two terms are quite accurate as long as $\varepsilon < 0.5$. An example is given in fig. 3.4, where the values of S calculated from one, two and three terms of (3.14) are plotted against t for the case when U is sinusoidal, given by (3.10) with $\alpha = 0.5$, and $x = 0.6$. The three-

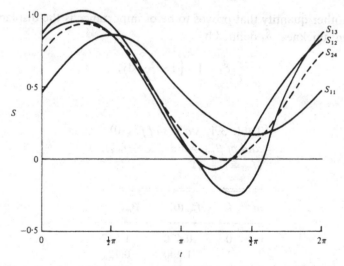

Fig. 3.4. Graphs of S against t as calculated from one, two and three terms of (3.14) (continuous curves, S_{11}, S_{12}, S_{13}) and as calculated from (3.30) with $\gamma = 0$ (broken curve, S_{24}). $\alpha = 0.5$, $x = 0.6$. (After Pedley, 1972b.)

term expansion differs significantly from the two-term expansion (and both are therefore inaccurate) for t between about $\frac{5}{4}\pi$ and $\frac{7}{4}\pi$, at which times $\varepsilon = 0.51$. Note, incidentally, that although the free stream does not reverse its direction, the shear at the wall is predicted to do so in the example shown. This is because the slowly moving fluid near the wall, with little inertia, tends to respond instantaneously to the pressure gradient U_t, which is negative for half the cycle. This also accounts for the phase lead of the actual wall shear over the free-stream velocity (i.e. over the quasi-steady wall shear).

Large x

If the flow were purely oscillatory, so that for large x the influence of the leading edge could not be felt, we would expect the flow to be that of the classical Stokes oscillatory boundary layer. This is the solution of (3.8) and (3.9) that is independent of x; the departure from the free stream would be confined to a boundary layer of (non-dimensional) thickness 1. On the other hand, if the amplitude of the oscillations in the free stream were very small compared with the mean velocity ($\alpha \ll 1$ in (3.10)), the flow should be approximately given by the Blasius boundary layer, with thickness $(2x)^{1/2}$. This suggests that for large x (when $(2x)^{1/2} \gg 1$) there are two length-scales for variation in the y-direction, one primarily associated with the mean flow and one with the oscillations. Together with the observation that the coefficient of the highest-order derivative ($\phi_{\eta_1\eta_1\eta_1}$) in (3.11) is small compared with that of other terms, this makes it clear that a solution should be sought by the method of matched asymptotic expansions (Van Dyke, 1975). We introduce outer and inner variables as follows:

Outer:

$$\eta = y/(2x)^{1/2}, \quad \tilde{\psi} = (2x)^{-1/2}\psi,$$

Inner:

$$\zeta = y/\sqrt{2} = x^{1/2}\eta, \quad \Psi = \psi/\sqrt{2} = x^{1/2}\tilde{\psi},$$

where the $\sqrt{2}$ is retained in the inner variables so that they are the same as those used by Pedley (1972b). From now on $U(t)$ is restricted for convenience to take the form (3.10).

Outer expansion. In terms of the outer variables, (3.8) becomes

$$2\tilde{\psi}_{\eta t} - (1/x)[\tilde{\psi}_{\eta\eta\eta} + \tilde{\psi}\tilde{\psi}_{\eta\eta} + 2x(\tilde{\psi}_x\tilde{\psi}_{\eta\eta} - \tilde{\psi}_{x\eta}\tilde{\psi}_\eta)] = 2\alpha \cos t.$$
(3.17)

The outer boundary condition is that

$$\psi_\eta \to 1 + \alpha \sin t \quad \text{as } \eta \to \infty,$$
(3.18)

while the inner condition is that of matching to the inner solution. We would also like to impose the condition that, as $\alpha \to 0$, $\tilde{\psi}$ takes its steady form $f_0(\eta)$. We seek an expansion for $\tilde{\psi}$ of the form

$$\tilde{\psi} = \sum_{m=0}^{\infty} x^{-m/2}\tilde{\psi}_m(\eta, t),$$

and substitute it into (3.17), equating like powers of x, to obtain a series of equations for $\tilde{\psi}_m$. The zeroth-order equation is

$$\tilde{\psi}_{0\eta t} = \alpha \cos t,$$

and the general solution satisfying (3.18) is

$$\tilde{\psi}_0 = \alpha\eta \sin t + f_0(\eta) + F_0(\eta) + \dot{T}_0(t),$$
(3.19)

where the last two terms tend to zero as $\alpha \to 0$ but are otherwise arbitrary functions. $\dot{T}_0$ is expected to be oscillatory with zero mean; any constant part can be incorporated into F_0. The solution of the first order equation is

$$\tilde{\psi}_1 = F_1(\eta) + \dot{T}_1(t),$$
(3.20)

where F_1 and $\dot{T}_1$ are arbitrary (a dot and a prime denote differentiation with respect to t and η respectively). The equation for $\tilde{\psi}_2$ is

$$2\tilde{\psi}_{2\eta t} = \bar{f}_0''' + (\eta\alpha \sin t + \bar{f}_0' + \dot{T}_0)\bar{f}_0'',$$
(3.21)

where (3.19) has been used, with $\bar{f}_0 \equiv f_0 + F_0$. In order that there should be no secular terms, $\bar{f}_0(\eta)$ must satisfy the same equation as the Blasius function f_0, which also satisfies both the wall boundary conditions and the steady part of the outer boundary condition, but we cannot yet rule out the existence of functions $F_0(\eta)$ such that $F_0(0)$ or $F_0'(0)$ is non-zero. After the time-independent part of (3.21) is removed, that equation can be integrated to give

$$\tilde{\psi}_2 = -\tfrac{1}{2}\alpha \cos t(\eta\bar{f}_0' - \bar{f}_0) + \tfrac{1}{2}\bar{f}_0'T_0(t) + F_2(\eta) + \dot{T}_2(t),$$
(3.22)

where again the last two terms are arbitrary. In all cases we expect the arbitrary functions occurring in $\tilde{\psi}_m$ to be determined by the equation obtained from prohibiting secular terms in $\tilde{\psi}_{m+2}$ and by boundary conditions derived from the inner expansion. Eigenfunctions may occur (solutions that satisfy both the outer and the wall boundary conditions identically), and they cannot be determined without taking into account conditions at the leading edge (see p. 147).

The problems for $\tilde{\psi}_3$ and $\tilde{\psi}_4$ can be solved similarly. For secular terms to be absent, $F_1(\eta)$ and $F_2(\eta)$ must satisfy the equations

$$F_m''' + \bar{f}_0 F_m'' + m\bar{f}_0' F_m' + (1-m)\bar{f}_0'' F_m = -\delta_{m2}F_1'^2 \quad (m = 1, 2) \tag{3.23}$$

and F_m' should tend to zero as $\eta_1 \to \infty$. The solution for $\tilde{\psi}_3$ is

$$\tilde{\psi}_3 = -\tfrac{1}{2}[\alpha\eta \cos t - T_0(t)]F_1' + F_3(\eta_1) + \dot{T}_3(t), \tag{3.24}$$

while that for $\tilde{\psi}_4$ is

$$\tilde{\psi}_4 = -\tfrac{1}{4}\alpha\bar{\alpha}_2 \sin t$$
$$+ \tfrac{1}{16}\alpha^2 \cos 2t \, (\eta^2\bar{f}_0'' + \eta\bar{f}_0' - \bar{f}_0)$$
$$- \tfrac{1}{2}\alpha \cos t \, (\eta F_2' + F_2) - \tfrac{1}{2}T_2(t)\bar{f}_0' + F_4(\eta) + \dot{T}_4(t), \tag{3.25}$$

where $\bar{\alpha}_2 = \bar{f}_0''(0)$ and a number of terms containing $T_0(t)$ have been omitted because that function is subsequently shown to be zero. In each case the last two terms are arbitrary functions. The outer boundary conditions on the terms in the inner expansion are obtained by rewriting the outer expansion in terms of the inner variable $\zeta = x^{1/2}\eta$ and expanding in powers of $x^{-1/2}$. Each term of that expansion can be used as the large-ζ condition on a corresponding term of the inner solution.

Inner expansion. In terms of inner variables, (3.8) is unchanged, as follows:

$$2\Psi_{\zeta t} - \Psi_{\zeta\zeta\zeta} + 2(\Psi_{x\zeta}\Psi_\zeta - \Psi_x\Psi_{\zeta\zeta}) = 2\alpha \cos t,$$

with inner boundary condition

$$\Psi = \Psi_\zeta = 0 \quad \text{at } \zeta = 0;$$

we seek a solution of the form

$$\Psi = \sum_{m=-1}^{\infty} x^{-m/2}\Psi_m(\zeta, t).$$

The problem for Ψ_{-1} is then

$$2\Psi_{-1\zeta t} - \Psi_{-1\zeta\zeta\zeta} = 0;$$

$$\Psi_{-1}(0, t) = \Psi_{-1\zeta}(0, t) = 0, \qquad \Psi_{-1}(\infty, t) = F_0(0) + \dot{T}_0(t),$$

which has no steady or periodic (non-diffusing) solution unless $\Psi_{-1}(\infty, t) = 0$, when $\Psi_{-1} \equiv 0$. Thus $T_0(t) \equiv 0$, $F_0(0) = 0$. The problem for Ψ_0 is

$$2\Psi_{0\zeta t} - \Psi_{0\zeta\zeta\zeta} = 2\alpha \cos t,$$

with outer boundary condition

$$\Psi_0(\infty, t) \sim F_1(0) + \dot{T}_1(t) + \alpha\zeta \sin t + \zeta F_0'(0),$$

from (3.19) and (3.20). This has no periodic solution unless

$$F_0'(0) = F_1(0) = 0, \qquad T_1(t) = \tfrac{1}{2}\alpha(\sin t + \cos t), \qquad (3.26)$$

and then

$$\Psi_0 = \alpha\zeta \sin t - \text{Im}\left\{\frac{\alpha\, e^{it}}{1+i}[1 - e^{-(1+i)\zeta}]\right\}. \qquad (3.27)$$

This represents the Stokes layer, and has zero mean. Furthermore, we now have three, homogeneous, boundary conditions on $F_0(\eta)$, which is therefore identically zero since $f_0(\eta)$ is the unique solution of the equation and boundary conditions defining the Blasius function (Coppel, 1960). Hence the leading term in the inner expansion is the *oscillatory* Stokes layer, with no contribution from the mean flow, while the leading term in the outer expansion is the *steady* Blasius boundary layer corresponding to the mean velocity, plus the inviscid oscillations of the free stream (the term $\alpha\eta \sin t$ in (3.19)). To leading order in $x^{-1/2}$, therefore, the mean and oscillatory components of the flow are completely uncoupled and do not interact.

Subsequent terms in the inner expansion, and boundary conditions on the unknown terms of the outer expansion, are obtained similarly. We require the expansion

$$f_0(\eta) = \tfrac{1}{2}\alpha_2\eta^2 - (\alpha_2^2/5!)\eta^5 + O(\eta^8),$$

where $\alpha_2 = f_0''(0)$ is given in table 3.1. The results for the next three terms are given below. First

$$\Psi_1 = \tfrac{1}{2}\alpha_2\zeta^2, \qquad (3.28)$$

the innermost part of the steady Blasius layer, requiring

$$F_1'(0) = F_2(0) = T_2(t) = 0.$$

Then
$$\Psi_2 = \tfrac{1}{2}F_1''(0)\zeta^2,$$

requiring
$$F_3(0) = F_2'(0) = T_3(t) = 0.$$

Finally
$$\Psi_3 = \tfrac{1}{2}F_2''(0)\zeta^2 + \mathrm{Im}\,\{\alpha\alpha_2\,e^{it}[\tfrac{13}{32} - \tfrac{1}{4}i\zeta^2$$
$$- e^{-k\zeta}(\tfrac{13}{32} + \tfrac{13}{32}k\zeta + \tfrac{5}{16}i\zeta^2 - \tfrac{1}{24}\bar{k}\zeta^3)]\}, \tag{3.29}$$

where $k = 1 + i$, $\bar{k} = 1 - i$, and we require
$$F_3'(0) = F_4(0) = 0, \qquad T_4(t) = -\tfrac{21}{32}\alpha\alpha_2 \cos t.$$

This last term is the first in which interaction between the oscillatory flow and the mean becomes apparent, through both Ψ_3 and $T_4(t)$.

Note that all the boundary conditions for the functions F_1, F_2 and F_3 have now been obtained. Unless there is an eigensolution, they will all be identically zero. It is a simple matter to show that F_1 and F_3 are identically zero, but the problem for F_2 is satisfied by an arbitrary multiple of $\eta f_0' - f_0$ (Stewartson, 1957; Libby & Fox, 1963). The appearance of eigenfunctions is to be expected on physical grounds because the large-x expansion must in some way be influenced by different conditions near the leading edge. The eigensolutions that emerge in the above way are also eigensolutions of the steady boundary layer equations; unsteadiness could come in only at subsequent terms in the expansion where $F_2(\eta)$, say, may be multiplied by functions of time. The above procedure also yields only those eigensolutions that correspond to integer powers, m, of $x^{-1/2}$, whereas all powers will in general be required to take account of upstream conditions. However, Libby & Fox (1963) showed numerically that $m = 2$ is the first eigenvalue, and that the second is $m = 3.77$; the above expansion was therefore terminated at $m = 3$.

There is also the possibility of eigensolutions that are intrinsically unsteady. Brown & Stewartson (1973) have investigated these, and found a class of very complicated eigensolutions, whose structure suggests that disturbances are propagated downstream near the edge of the Blasius boundary layer with approximately the free-stream velocity, and their effect then diffuses both inwards to the wall and out into the free stream. At any given time, the variation with x of the disturbance to the boundary layer has the form of a decaying oscillation, which was also found in a numerical solution to the small-amplitude equations by Ackerberg & Phillips (1972). Both groups emphasised the marked influence of these eigensolutions on the unsteady displacement thickness of the boundary layer, while the influence on the wall shear is small since the eigensolutions are greatest at the edge of the boundary layer.

If we ignore this class of eigensolutions, we may write down the large-x expansion for the dimensionless wall shear, S:

$$S = \frac{1}{\sqrt{2}} \Psi_{\zeta\zeta}\bigg|_{\zeta=0} = (\alpha/\sqrt{2})(\cos t + \sin t) + (2x)^{-1/2}\alpha_2$$
$$+ (x^{-3/2}/\sqrt{2})\gamma\alpha_2 - \tfrac{5}{16}\alpha\alpha_2 \cos t + O(x^{-2}). \qquad (3.30)$$

The large-x expansion for the dimensionless displacement thickness, again ignoring the eigensolutions other than $F_2(\eta)$, is

$$\delta_1 = (2x)^{1/2}/(1 + \alpha \sin t)[\beta_0 - (\alpha/2x^{1/2})(\cos t - \sin t)$$
$$+ (\beta_0/x)(\tfrac{1}{2}\alpha \cos t - \gamma) + O(x^{-2})]; \qquad (3.31)$$

in both these equations γ is defined by

$$F_2(\eta) = \gamma(\eta f_0' - f_0).$$

Matching. Pedley (1972b), reviewing the numerical results of a number of papers, showed that the amplitudes and phases of the wall shear oscillations calculated from the small- and large-x expansions, (3.14) and (3.30), are close together for x in the range 0.5–0.7 for several values of α up to 0.8. Agreement was closer when only two terms of (3.14) were used than when three terms were used, presumably because at this value of x there is a range of values of t for which $\varepsilon = |x\dot{U}/U^2|$ exceeds 0.5. Fig. 3.4 shows the close correspondence between the wall shear calculated from (3.30) and that from two terms of (3.14) when $x = 0.6$ and $\alpha = 0.5$. In the large-x expansion, Pedley arbitrarily took $\gamma = 0$, and fig. 3.4 indicates that no significant improvement can be expected by choosing any other value for γ, which would merely alter the mean shear slightly (see (3.30)): the means with $\gamma = 0$ differ only because the mean value of $(1 + \alpha \sin t)^{3/2}$ is not precisely 1. Similar conclusions are reached if we match the means of the two expansions of the displacement thickness ((3.31) with $\gamma = 0$ and two terms of (3.15)).

The expansion (3.30) shows that the phase lead of the shear over the free stream tends to the limit $\frac{1}{4}\pi$ as $x \to \infty$. In fact, if we write

$$S = \bar{S} + \alpha S_1 \sin (t + \tfrac{1}{4}\pi - \chi),$$

where $\bar{S}$ is the mean shear and $\frac{1}{4}\pi - \chi$ is the phase lead, we obtain

$$\chi \approx 1 - S_1 \approx 0.073 x^{-3/2}.$$

This indicates that the asymptotic limit, in which the mean flow is uncoupled from the oscillation, is accurate to within 10% for x as low as 0.81.

Finally, we may notice that the large-x expansion does not formally require the restriction that $\alpha < 1$, i.e. that the free stream is non-reversing. Thus, where the influence of the leading edge is manifest only in the development of the mean flow, the amplitude of the oscillation can be as large as we please, although the small-x expansion is clearly not valid throughout the cycle in a reversing flow, because ε becomes infinite when U passes through zero.

3.2.3 Reversing flow

The aim of this subsection is to investigate the flow near the leading edge of a semi-infinite flat plate (i.e. the entry flow in a tube) when the free stream reverses its direction. The work described is largely contained in the paper by Pedley (1976a). We shall consider a free stream whose velocity varies with time like that shown in fig. 3.3. Since this represents a periodic motion, we expect that far from the leading edge the whole flow can be described as a developing mean flow superimposed on a number of oscillatory terms, each representing the Stokes layer corresponding to one Fourier component of the waveform. From the work described above, this is expected to be valid for dimensional distances $\hat{x}$ greater than approximately 1.0 U_0/Ω, where U_0 is the mean velocity, and Ω is the lowest angular frequency of the motion. In the example shown in fig. 3.3, this distance is only about 1.1 cm, but, as we shall see, the large amplitude of the oscillations means that the influence of the leading edge (other than in the development of the mean flow) is felt up to $\hat{x} = 4$–5 cm at some stage during the cycle.

The main additional assumption to be made is that the flow in the aorta (or over the flat plate) comes completely to rest at the end of each cycle. This is based on the fact that the free-stream velocity (fig. 3.3) is zero for more than half the cycle; but it cannot be strictly true since there will inevitably be residual motions in the boundary layer when the free stream has come to rest, and these will take a time proportional to a^2/ν to decay (here a is the radius of the aorta, so the decay time is about 20 s). However, these residual motions turn out to be very small, and it is consistent to neglect them in describing the large oscillatory wall shear during systole. This assumption is very useful, because it means that the thickness of the boundary layer that develops on the tube wall during every beat never exceeds a value proportional to $(\nu T)^{1/2}$, where T is the period of the beat. For blood in the dog aorta, $(\nu T)^{1/2} \approx 0.13$ cm, which is less than a tenth of the vessel diameter. Therefore we are still justified in neglecting the curvature of the tube wall.

The development with time of the flow at a given value of $\hat{x}$ quite near the leading edge can be described qualitatively as follows. Initially vorticity diffuses out from the wall in a time-dependent but $\hat{x}$-independent manner, and is restricted to a Rayleigh layer of thickness proportional to $(\nu \hat{t})^{1/2}$. This persists until fluid particles that have passed the leading edge arrive at $\hat{x}$; from then on the diffusive flow is continuously modified until eventually an approximately quasi-steady boundary layer is set up, described by the first one or two terms of the expansion (3.12). This modification is quite rapid in the case of the impulsively started flat plate (Hall, 1969), and we may suppose that it is quite rapid in the present case too. The approximately quasi-steady boundary layer is expected to persist until after the peak velocity has passed, but must break down again well before the free-stream velocity becomes zero. Near the time of zero velocity the convective inertia terms will be small, and the flow is expected once more to represent a diffusive balance between unsteady inertia and viscosity. Subsequent oscillations could be treated similarly, but it turns out that for values of $\hat{x}$ that are of interest the influence of the leading edge is not felt after the first reversal, and the flow remains diffusive until the next cycle. The main approximation of the theory to be outlined here is to treat the transitions between diffusive and approximately quasi-steady flows as instantaneous, not gradual.

The variables are non-dimensionalised as in (3.7), with Ω replaced by $1/T$, where T is the period of the cycle. The boundary layer equation for the longitudinal velocity u, (3.8), can be written as

$$(u_t - U_t) + \left(uu_x - u_y \int_0^y u_x \, dy \right) = u_{yy}, \tag{3.32}$$

with boundary and initial conditions $u = 0$ on $y = 0$, $u \to U(t)$ as $y \to \infty$, and $u = 0$ at $t = 0$. We assert that the initial diffusive solution, a balance between the first term in brackets on the left-hand side of (3.32) and the viscous term on the right-hand side, holds exactly until a certain time, $t_1'(x)$, which is to be determined (it will be the time at which the influence of the leading edge is first felt at x). That is,

$$u = U(t) - 2\pi^{-1/2} \int_{\eta_0}^{\infty} U\left[t - \frac{\eta_0^2(t - t_0')}{\mu^2} \right] e^{-\mu^2} \, d\mu \quad \text{for } 0 < t \leqslant t_1'(x),$$

$$\tag{3.33}$$

where

$$\eta_0 = \tfrac{1}{2} y (t - t_0')^{-1/2}$$

and the time origin of the diffusion is $t_0' = 0$. After this time, we assert that there is a sudden transition to the approximately quasi-steady solution represented by (3.12) which holds until another transition time $t_2'(x)$; that is,

$$u = U(t)[f_0'(\eta_1) + (x\dot{U}/U^2)f_{11}'(\eta_1)] \quad \text{for } t_1'(x) < t < t_2'(x)$$

$$\tag{3.34}$$

where $\eta_1 = y[U(t)/2x]^{1/2}$, f_0 is the Blasius function and f_{11} is the solution of (3.13) with $m = k = 1$. (The primes on the times t_0', t_1', t_2' are put there because these are not the same as the times t_0, t_1, t_2 defined by Pedley (1976a).) The expansion (3.34) in no way takes account of the flow before $t = t_1'(x)$, so there will inevitably be a discontinuity at that time; the removal of that discontinuity involves an extremely complicated exercise in multi-layered boundary layer theory, even in the apparently simple case of the impulsively started flat plate (Stewartson, 1973), and will not be attempted here. As in the case of the periodic boundary layer considered above, the influence of the leading edge is convected along the outer edge of the boundary layer, at approximately the free-stream velocity, and its effect then diffuses into the inner parts of the layer, affecting the wall shear somewhat later than, say, the displacement thickness.

A new diffusive flow takes over from the quasi-steady flow at $t = t_2'(x)$, and persists until the end of the cycle, $t = 1$. This will again represent a solution of (3.32) without the convective inertia terms, which should be solved subject to an initial condition derived from (3.34). However, such a solution would be rather cumbersome, and we make the further approximation that the new diffusive solution also takes the form (3.33), with the disposable function $t_0'(x)$ chosen to make the displacement thickness of the boundary layer continuous at $t = t_2'$. The displacement thickness is chosen because it is thought to be the most fundamental single parameter representing the motion in a boundary layer (in any case, choosing another parameter, such as momentum thickness, makes little difference to the results (Pedley, 1976a)). Continuity of displacement thickness (see (3.15)) at $t = t_2'$ requires that

$$[2xU(t_2')]^{1/2}\left[\beta_0 + \frac{x\dot{U}(t_2')}{U^2(t_2')}\beta_{11}\right]$$

$$= 2\left(\frac{t_2' - t_0'}{\pi}\right)^{1/2}\int_0^1 U[t_2' - \lambda^2(t_2' - t_0')]\,d\lambda, \qquad (3.35)$$

where $\beta_0 \ (=\beta_{00})$ and β_{11} are given in table 3.1.

It still remains to determine the take-over times $t_1'(x)$ and $t_2'(x)$. In Pedley (1976a) the former was chosen to be the time at which information about the leading edge is first convected to x at the edge of the boundary layer, i.e. t_1' is chosen so that

$$x = \int_0^{t_1'} U(t)\,dt, \qquad (3.36a)$$

although it is recognised that the effect on wall shear is not felt until a little later (Hall, 1969; Dennis, 1972). The second take-over time is chosen in a similar way:

$$x = \int_{t_2'}^{T_1} U(t)\, dt, \qquad (3.36b)$$

where T_1 is the first zero of $U(t)$ (≈ 0.3 in fig. 3.3). This means that a fluid particle passing the leading edge at t_2' only just reaches x before being swept back by the reversed flow; this seems a reasonable choice, by symmetry with (3.36a), but there is no strong physical reason for it. A better choice in this case (and perhaps also in the case of t_1') may be to take t_2' (and t_1') as the times at which ε, from (3.16), takes the value 0.5, since for smaller values of ε, (3.34) is expected to be very accurate (see p. 142). This would also be the only possible choice for a free stream that did not reverse its direction at $t = T_1$, but merely came close to zero, so that the quasi-steady expansion broke down. The results to be quoted were based on the choices (3.36a, b), but it was verified by Pedley (1976a) that ε lay between 0.4 and 0.6 for all take-overs, so the two methods of evaluating t_1' and t_2' are almost equivalent. This result follows from the fact that for a uniformly accelerating (or decelerating) flow, when $U \equiv \pm t$, then t_1' (or t_2') $\equiv (2x)^{1/2}$, and ε is exactly equal to 0.5 at these times. The equivalence of the two methods would break down only for flows whose acceleration varied rapidly as the velocity passed through zero.

Somewhat different approximate methods have been used in the past to analyse unsteady boundary layers. Atabek & Chang (1961), in a discussion of unsteady entry flow in a tube, used the Oseen approximation, replacing the convective inertia terms in (3.32) by $U(t)u_x$. This makes the equation linear and therefore relatively easy to solve. However, while this might be expected to give an accurate description of the flow at the outer edge of the boundary layer, there are no grounds for supposing it to be a good approximation near the wall, where $u \ll U$ and where the shear stress is calculated. The present method is certainly much more accurate in the approximately quasi-steady regime; it is unlikely to be significantly more inaccurate during the diffusive periods when U is small, and it is just as easy to use. A modified Oseen approximation was proposed by Lewis & Carrier (1949), who replaced the convective inertia terms by cUu_x for some constant c such that $0 < c < 1$. The quantity cU

was to represent an average across the boundary layer of the convection velocity. These authors found that if c is taken equal to about 0.35, the Blasius value of wall shear was obtained for steady flow over a flat plate (although the predicted displacement thickness is in error by 11%). Carrier & Di Prima (1956) subsequently applied the method to small-amplitude oscillatory flow over a semi-infinite flat plate; it has the merit of giving the correct answer in steady flow, and of predicting a continuous velocity field. However, the method has not been used for large-amplitude oscillations, and there is no *a priori* way of knowing whether a constant value of c is appropriate throughout a flow reversal, or for different geometries (e.g. with an x-dependent outer flow).

Under the modified Oseen approximation, a change of variables to

$$w(\xi, y, t) = u(x, y, t) - U(t), \qquad \xi = x - \int_0^t cU(t')\, dt'$$

reduces the problem of an unsteady flow starting from rest at $t = 0$ to the solution of the one-dimensional diffusion equation for $t > 0$:

$$w_t = w_{yy}$$

with

$$w(\xi, \infty, t) = 0, \quad w_y(\xi, 0, t) = 0, \quad w(\xi, y, 0) = 0,$$

$$w(\xi, 0, t) = -U(t)H\left[\xi + \int_0^t cU(t')\, dt'\right],$$

where H is the Heaviside step function. This has the solution (in terms of u and x again)

$$u = U(t) - \frac{2}{\sqrt{\pi}} \int_{\eta_0}^{\infty} U\left[t\left(1 - \frac{\eta_0^2}{\mu^2}\right)\right]$$

$$\times H\left[x - \int_{t(1-\eta_0^2/\mu^2)}^t cU(t')\, dt'\right] e^{-\mu^2}\, d\mu, \qquad (3.37)$$

where

$$\eta_0 = y/2\sqrt{t}.$$

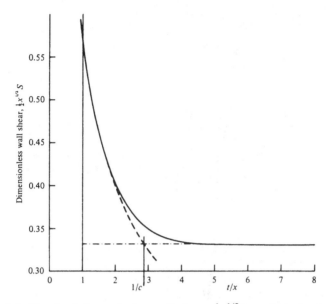

Fig. 3.5. Graph of dimensionless wall shear, $\tfrac{1}{2}x^{1/2}S$, against t/x, for the impulsively started flat plate ($U = 1$): continuous curve, exact numerical solution, calculated by Hall (1969); dashed curve, Rayleigh solution; dot–dash curve, Blasius solution. The method of Pedley (1976a) is equivalent to jumping from the Rayleigh to the Blasius solution at $t/x = 1$; the modified Oseen solution makes the jump at $t/x = 1/c$.

In the case of the impulsively started flat plate, for which $U \equiv H(t)$, this reduces to

$$u = \operatorname{erf}\left[\tfrac{1}{2}y(1/t)^{1/2}\right] \quad \text{for } 0 < t < x/c$$

$$= \operatorname{erf}\left[\tfrac{1}{2}y(c/x)^{1/2}\right] \quad \text{for } t > x/c.$$

This solution consists of the pure Rayleigh layer formed initially, but taken to extend until $t = x/c$ not $t = x$ as in the method proposed here; at $t = x/c$, the Rayleigh layer is taken over by an approximation to the Blasius layer. The values of wall shear predicted in this case are

$$S = (\pi t)^{-1/2} \qquad \text{for } 0 < t < x/c$$

$$= (\pi x/c)^{-1/2} \qquad \text{for } t > x/c,$$

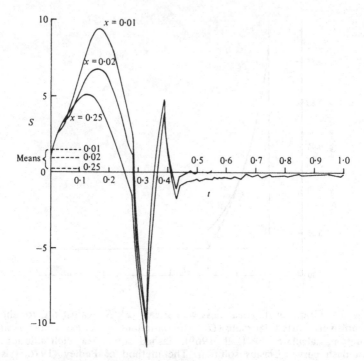

Fig. 3.6. Dimensionless shear-rate on the wall of the aorta, plotted against time, at three distances from the inlet. Motion in the boundary layer has almost completely died out by the end of the cycle. (After Pedley, 1976a.)

and agree more closely with the exact numerical values than the present method (fig. 3.5; I am grateful to Dr S. Farthing for pointing this out to me). The improvement essentially lies in choosing a more appropriate time for the take-over of the quasi-steady solution from the diffusive one, equivalent to multiplying the right-hand sides of equations (3.36) by c. The shear stress agrees with the exact solution in the quasi-steady region because c was chosen to make it do so. Whenever the modification to the quasi-steady shear (represented by the second term in (3.34)) is significant, the agreement will not be exact. On the other hand, in all cases, use of the modified Oseen solution means that the influence of convection will not be completely lost in the diffusive region, as in the present method. An improvement in the present method could therefore be made by keeping (3.34) in the quasi-steady regions, but using (3.37) instead of (3.33) in the diffusive regions, the time origin of the diffusion still being determined by continuity of the displacement thickness. I would propose that the take-over times still be determined either by equations (3.36) or by the condition $\varepsilon = 0.5$ (see (3.16)), except in cases of impulsively started steady motion when a different condition may lead to an improvement (cf. fig. 3.5).

According to the approximate theory of Pedley (1976a), the dimensionless wall shear is predicted to be

$$S = 2\pi^{-1/2}t^{1/2}\int_0^1 \dot{U}[t(1-\lambda^2)]\,d\lambda \quad \text{for } 0 < t < t_1', \tag{3.38a}$$

$$S = [U^{3/2}(t)/(2x)^{1/2}][f_0''(0) + (x\dot{U}/U^2)f_{11}''(0)] \quad \text{for } t_1' < t < t_2', \tag{3.38b}$$

$$S = [\pi(t-t_0)]^{-1/2}\Big\{ U(t_0') + 2(t-t_0')$$

$$\times \int_0^1 \dot{U}[t - \lambda^2(t-t_0')]\,d\lambda \Big\} \quad \text{for } t_2' < t < 1, \tag{3.38c}$$

where (3.38b) consists of the first two terms of (3.14). This quantity has been calculated for the case in which the free-stream velocity takes the form shown in fig. 3.3. S is plotted as a function of time for three different values of x in fig. 3.6. Very near the entrance ($x = 0.01$) there is a large, quasi-steady peak in S, almost in phase with the peak velocity. This diminishes rapidly as x increases, and has an increasing phase lead, as one would expect from the theory of § 3.2.2. However, at all values of t or x, the shear-rates with the greatest magnitudes are negative. These high reversed shear-rates are a consequence of the large adverse pressure gradient associated with the rapid deceleration of the aortic core flow as the aortic valve closes. It can also be seen that in all cases the wall shear has almost completely died away at the end of the beat, confirming that the residual motions are very small and can be neglected in calculations of the oscillatory components of wall shear.

If x is too large, it is predicted that no quasi-steady regime exists, even at peak systole, because no fluid element that has passed the leading edge at any time during the cycle can arrive at x before the end of the cycle (i.e. (3.36a) has no solution). In the case considered here, the critical value of x above which no quasi-steady solution exists is about 0.21 (corresponding to $\hat{x} \approx 4.2$ cm). Thus the curve for $x = 0.25$ in fig. 3.6 is calculated entirely from the first diffusive solution (3.38a), and the same curve would be predicted for all x greater than 0.21. This cannot be strictly true, because if the flow were independent of x the equations would be linear, and the mean flow (uncoupled from the oscillations) would be Poiseuille flow. But the steady entrance length for Poiseuille flow is approximately 0.06 a Re (where a is the tube radius and Re is the Reynolds number), equal to about 33 cm in the present case. So the mean flow cannot be Poiseuille flow for all $\hat{x}$ greater than 4.2 cm. The error in the present method lies in neglecting the residual motions at the end of each cycle; after many cycles these would build up (differently for each x) and would lead to an x-dependent mean flow. The remedy consists simply in assuming that, for values of x greater than 0.21, the oscillatory and mean flows do not interact, as predicted by the theory of § 3.2.2. The oscillatory flow would be independent of x, consisting of a number of superimposed Stokes layers, and the associated wall shear could be calculated using either (3.1d) or (3.38a) (the two are equivalent if the boundary layers are thin, i.e. if the Womersley parameter, α, is large). The mean flow would be steady entry flow, as described in § 3.2.1. Thus the present method is expected to predict the oscillatory components of wall shear accurately for all x, while giving increasingly inaccurate predictions of the mean shear as x increases up to the value 0.21; for x greater than 0.21, steady entry flow should predict the mean shear accurately.

We can see from fig. 3.6, as we saw in the case of fully developed flow in coronary arteries, that the amplitude of wall shear stress oscillations is predicted to be much larger than the mean shear. At $x = 0.01$, the mean shear is about 15% of the (negative) peak, compared with a mean velocity that is 17% of the (positive) peak velocity. However, this is predicted to fall rapidly, being 9% at $x = 0.02$, and reaching as little as 2% in fully developed flow. The r.m.s. value of S, on the other hand, falls from about 40% of the peak to about 30%, and remains considerably larger than the mean. In view of the large amplitudes of the oscillations in S, it seems much more likely *a priori* that the permeability of the artery wall, and thus the generation of atherosclerosis, is correlated with some measure of wall shear that is independent of direction, like the peak or r.m.s. value, rather than the arithmetic mean. Finally, we note that S is non-dimensionalised with respect to $U_0(\nu T)^{-1/2}$, so in this case the peak wall shear-rate is about 4000 s^{-1}. This corresponds to a wall shear *stress* of 16 N m^{-2}, which is below that which causes endothelial damage (40 N m^{-2}; see § 1.2) but not by a huge factor.

CHAPTER 4

FLOW PATTERNS AND WALL SHEAR
STRESS IN ARTERIES
II CURVED TUBES

We now turn to the second main feature of the thoracic aorta: its curvature. The aim is to describe flow near the entrance of a curved tube in the same way that the previous section described flow near the entrance of a straight tube. However, we immediately come up against the major difficulty that the fully developed flow to which the entry flow tends, and which in a straight tube is Poiseuille flow (the mean flow) plus an easily calculated oscillatory component, is very complicated, and even the steady component is not yet completely understood. In the next three sections, therefore, we concentrate on fully developed flow in curved tubes, leaving a discussion of entry flow to §§ 4.4 and 4.5.

The reason why the flow in a curved tube is difficult to calculate lies in the fact that the motion cannot be everywhere parallel to the curved axis of the tube, but transverse (or secondary) components of velocity must be present. This follows because in order for a fluid particle to travel in a curved path of radius R with speed w it must be acted on by a lateral force (provided by the pressure gradients in the fluid) to give it a lateral acceleration w^2/R. Now the pressure gradient acting on all particles will be approximately uniform, but the velocity of those particles near the wall will be much lower than that of particles in the core, as a result of the no-slip condition. Therefore the radius of curvature of the path of particles in the core must be greater than that of particles near the wall. In other words, fluid in the core is swept to the outside of the bend, and that near the wall returns towards the inside; a secondary circulation, like that depicted in fig. 4.1, is set up. These secondary motions themselves influence the distribution of axial velocity and result in a complicated interaction of the two.

We begin by writing down the full equations of motion for unsteady viscous flow in a curved tube, without yet imposing the

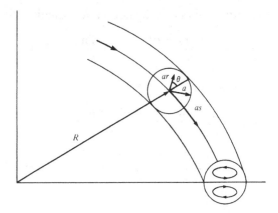

Fig. 4.1. Coordinate system to be used in analysing flow in curved tubes (a is the radius of the tube, R is the radius of curvature of the axis). A qualitative sketch of the secondary motions to be expected in steady flow is included.

restriction that the flow should be fully developed. We also retain the possibility that the curvature of the centre line is non-uniform, although we do restrict it to lie in a single plane. It is not difficult, but is even more cumbersome, to include the effect of torsion (Farthing, 1977). Coordinates (ar, θ, as) are chosen, where a is the radius of the tube cross-section, assumed uniform (so the tube boundary is $r = 1$); as is the distance measured along the centre line (equivalent to $\hat{x}$ in the straight tube considered in chapter 3), and (ar, θ) are polar coordinates in the cross-section. We take the velocity field to be $\hat{W}_0(u, v, w)$, where $\hat{W}_0$ is a suitable velocity-scale, so that the dimensionless longitudinal velocity is w, not u as heretofore; this is done so that the present notation should be consistent with that of the other authors referred to subsequently. Let the radius of curvature of the centre line at a particular cross-section be R, and define the dimensionless ratio $\delta(s) = a/R$, which depends on s if R is not uniform. Finally, we define the parameter

$$h = 1 + \delta(s)r \cos \theta,$$

which is $1/R$ times the distance from the centre of curvature of the centre line (at a particular s) to the projection on the plane containing the centre line of a general point (r, θ, s). In this orthogonal coordinate system, the continuity equation div $\mathbf{u} = 0$ is

$$u_r + \frac{u(1+2\delta r \cos \theta)}{rh} + \frac{v_\theta}{r} - \frac{v\delta \sin \theta}{h} + \frac{w_s}{h} = 0, \qquad (4.1)$$

and the three momentum equations are:

$$u_t + uu_r + \frac{v}{r}u_\theta + \frac{w}{h}u_s - \frac{v^2}{r} - \frac{\delta \cos \theta}{h}w^2$$

$$= -p_r + \frac{Re^{-1}}{rh}\left\{\frac{\partial}{\partial s}\left[\frac{r}{h}(u_s - hw_r - \delta \cos \theta\, w)\right]\right.$$

$$\left. - \frac{\partial}{\partial \theta}\left[h\left(v_r + \frac{v}{r} - \frac{u_\theta}{r}\right)\right]\right\}, \qquad (4.2)$$

$$v_t + uv_r + \frac{v}{r}v_\theta + \frac{w}{h}v_s + \frac{uv}{r} + \frac{\delta \sin \theta}{h}w^2$$

$$= -\frac{1}{r}p_\theta + \frac{Re^{-1}}{h}\left\{\frac{\partial}{\partial r}\left[h\left(v_r + \frac{v}{r} - \frac{u_\theta}{r}\right)\right]\right.$$

$$\left. - \frac{\partial}{\partial s}\left[\frac{1}{rh}(hw_\theta - \delta r \sin \theta\, w - rv_s)\right]\right\}, \qquad (4.3)$$

$$w_t + uw_r + \frac{v}{r}w_\theta + \frac{w}{h}w_s + \frac{\delta \cos \theta}{h}uw - \frac{\delta \sin \theta}{h}vw$$

$$= -\frac{1}{h}p_s + \frac{Re^{-1}}{r}\left\{\frac{\partial}{\partial \theta}\left[\frac{1}{rh}(hw_\theta - \delta r \sin \theta\, w - rv_s)\right]\right.$$

$$\left. - \frac{\partial}{\partial r}\left[\frac{r}{h}(u_s - hw_r - \delta \cos \theta\, w)\right]\right\}. \qquad (4.4)$$

The time has been non-dimensionalised with respect to $a/\hat{W}_0$ and the pressure with respect to $\rho \hat{W}_0^2$; the Reynolds number $Re = \hat{W}_0 a/\nu$. These equations will be used in § 4.2 as well as in this section.

In the next three subsections we shall consider: (i) Steady fully developed flow in a tube with small curvature ($\delta \ll 1$); most of the section will also assume that δ is uniform so that the velocity field $\mathbf{u}$

and pressure gradient $(-\partial p/\partial s)$ are both independent of s. When δ is allowed to vary, it is supposed to do so on a length-scale large compared with a, so $\partial/\partial s = O(\varepsilon) \ll 1$. (ii) Fully developed flow with an oscillatory mean pressure gradient, $\delta \ll 1$ and δ uniform (so $\partial \mathbf{u}/\partial s = 0$). (iii) Unsteady flow starting from rest with a given waveform of average axial velocity; here either δ is uniform and not necessarily much less than 1, or is much less than 1 but need not be uniform $(\partial/\partial s = O(\varepsilon))$. When δ is uniform, and hence the velocity field independent of s, the only way that the transverse velocities are affected by the axial velocity is through the 'centrifugal force' terms that drive it; these are the last terms on the left-hand sides of (4.2) and (4.3), proportional to δw^2. The axial velocity is itself influenced by the transverse motions through the convective inertia terms in (4.4).

In §§ 4.1 and 4.2 it is convenient to take the velocity-scale $\hat{W}_0$ equal to ν/a, so $Re = 1$. In §§ 4.3–4.5, $\hat{W}_0$ is an actual order of magnitude for the axial velocity; there we take $Re \gg 1$.

4.1 Fully developed steady flow, $\delta \ll 1$

In the case where δ is uniform and hence the velocity field independent of s, (4.4) shows that the pressure gradient $-p_s$ is independent of s. Thus the pressure can be written in the form $-Gs + p'(r, \theta)$; from (4.2) and (4.3) G is seen to be independent of r and θ and is therefore a constant. When δ is not uniform, but $\partial/\partial s = O(\varepsilon) \ll 1$, we write $-p_s = G - \varepsilon p_s'''$. We also set $\delta \ll 1$ and neglect all terms that are $O(\delta)$ or $O(\varepsilon^2)$ compared with the leading term in any equation. First, however, we rescale w so that the centrifugal force terms in (4.2) and (4.3) are of the same order of magnitude as the viscous and inertial terms (this is necessary because the centrifugal force terms *drive* the secondary motion). To this end we replace w by $(2\delta)^{-1/2} w'$ (the factor of $2^{-1/2}$ is introduced for consistency with other authors); we also replace s by s/ε. Equations (4.1)–(4.4) with $Re = 1$ then reduce to

$$u_r + \frac{u}{r} + \frac{v_\theta}{r} + \varepsilon(2\delta)^{-1/2} w_s' = 0, \qquad (4.5)$$

$$uu_r + \frac{v}{r}u_\theta - \frac{v^2}{r} - \tfrac{1}{2}w'^2 \cos\theta + \varepsilon(2\delta)^{-1/2}w'u_s$$

$$= -p_r' - \frac{1}{r}\frac{\partial}{\partial\theta}\left(v_r + \frac{v}{r} - \frac{u_\theta}{r}\right) - \varepsilon(2\delta)^{-1/2}w_{rs}', \qquad (4.6)$$

$$uv_r + \frac{v}{r}v_\theta + \frac{uv}{r} + \tfrac{1}{2}w'^2 \sin\theta + \varepsilon(2\delta)^{-1/2}w'v_s$$

$$= -\frac{1}{r}p_\theta' + \frac{\partial}{\partial r}\left(v_r + \frac{v}{r} - \frac{u_\theta}{r}\right) - \varepsilon(2\delta)^{-1/2}\frac{w_{\theta s}'}{r}, \qquad (4.7)$$

$$uw_r' + \frac{v}{r}w_\theta' + \varepsilon(2\delta)^{-1/2}w'w_s'$$

$$= D - \varepsilon(2\delta)^{-1/2}p_s'' + \left(w_{rr}' + \frac{1}{r}w_r' + \frac{1}{r^2}w_{\theta\theta}'\right). \qquad (4.8)$$

The boundary conditions are that $u = v = w' = 0$ on $r = 1$, and that there should be no singularity at $r = 0$.

In the last equation $p_s'' = 2\delta p_s'''$, and the parameter D is given by

$$D = (2\delta)^{1/2}G = (2\delta)^{1/2}\frac{\hat{G}a^2}{\mu}\cdot\frac{a}{\nu}, \qquad (4.9)$$

where $-\hat{G}$ is the dimensional pressure gradient. $\hat{G}a^2/\mu$ is four times the peak velocity in Poiseuille flow in a straight tube, driven by the same pressure gradient $-\hat{G}$, so $D = (2\delta)^{1/2} \cdot 4Re'$, where Re' is the Reynolds number of the straight-tube flow. D is called the Dean number, because Dean (1928) was the first to realise its significance in curved-tube flow.

When the curvature is uniform ($\varepsilon = 0$), the flow is entirely governed by the single dimensionless parameter D; a slow variation in curvature will have a small effect on the flow as long as $\varepsilon' = \varepsilon(2\delta)^{-1/2} \ll 1$. We shall examine the flow with $\varepsilon' = 0$, (a) for small D (an analytical solution), (b) for a wide range of D, becoming quite large (numerical solutions), and (c) for very large D (an incomplete asymptotic theory). In (a) we shall also briefly examine the consequences of a non-zero, but small, ε'. When $\varepsilon' = 0$, the continuity equation, (4.5), admits of the existence of a secondary stream function, ψ, such that

$$u = (1/r)\psi_\theta, \quad v = -\psi_r, \qquad (4.10)$$

and then the other equations reduce to the pair of coupled equations:

$$\nabla_1^2 w' + D = (1/r)(\psi_\theta w_r' - \psi_r w_\theta'), \tag{4.11}$$

$$\nabla_1^4 \psi + \frac{1}{r}\left(\psi_r \frac{\partial}{\partial\theta} - \psi_\theta \frac{\partial}{\partial r}\right)\nabla_1^2 \psi = -w'\left(\sin\theta\, w_r' + \frac{\cos\theta}{r} w_\theta'\right), \tag{4.12}$$

where

$$\nabla_1^2 \equiv \partial^2/\partial r^2 + (1/r)\partial/\partial r + (1/r^2)\partial^2/\partial\theta^2.$$

The boundary conditions at $r = 1$ are $\psi = \psi_r = w' = 0$.

4.1.1 Small D

Dean (1928) gave the first few terms of a series solution of (4.11) and (4.12) in powers of D; the leading term for w' $(=O(D))$ is just Poiseuille flow in a straight tube, and the leading term for ψ is $O(D^2)$ from (4.12). The procedure is equivalent to the successive approximation of inertia terms in lubrication theory, only the lubrication theory is particularly simple because the cross-sectional area of the tube is uniform. If we set

$$w' = D \sum_{n=0}^{\infty} D^{2n} w_n(r, \theta), \qquad \psi = \sum_{n=1}^{\infty} D^{2n} \psi_n(r, \theta),$$

then

$$\left.\begin{array}{l} w_0 = \frac{1}{4}(1 - r^2), \\ \psi_1 = [2\sin\theta/(9 \times 8^3)]r(1 - \frac{1}{4}r^2)(1 - r^2)^2, \\ w_1 = [\cos\theta/(45 \times 8^5)]r(1 - r^2)(19 - 21r^2 + 9r^4 - r^6), \end{array}\right\} \tag{4.13}$$

etc. Dean's main application was to calculate the dimensionless flow rate, Q, corresponding to a particular value of D. The term w_1 does not contribute to it, and, in fact, Dean calculated the contributions from all the terms up to $O(D^9)$ (i.e. w_4) to obtain

$$Q = \int_0^{2\pi} \int_0^1 w'r\, dr\, d\theta$$

$$= (\pi D/8)[1 - 0.0306(D/96)^4 + 0.0120(D/96)^8 + O(D^{12})],$$

where $\frac{1}{8}\pi D$ is the Poiseuille flow value in the present scaled variables. The size of the coefficients in this series suggests that the small-D expansion is valid for values of D up to about 100. In the canine aorta, however, where $\delta \approx 0.2$ and the mean Reynolds number is approximately 800, the mean value of D is greater than 2000 (and the peak value is greater still), so this expansion can be useful only for much smaller blood vessels. Note that the leading term in the secondary flow, represented by ψ_1 in (4.13), takes the form of a pair of vortices, as shown schematically in fig. 4.1 (see also fig. 4.2). They are symmetrical about $\theta = \pm\frac{1}{2}\pi$, and the flow is towards the outside of the curve ($\theta = 0$) in the core of the tube and towards the inside near the walls.

The effect of non-uniform curvature can be assessed in a similar way when $\delta \ll \varepsilon' \ll 1$. We calculate merely the first-order correction to u_1, v_1 and w_1. If we suppose that the δ used in the change of variables leading to (4.5)–(4.8) is a constant, δ_0, then the fact that δ is a function of s appears only in the two centrifugal force terms on the left-hand sides of (4.6) and (4.7), involving w'^2; these terms should be multiplied by $\Delta(s) = \delta(s)/\delta_0 = O(1)$. Then the solution with $\varepsilon' = 0$ is altered only in that w_1, ψ_1 are multiplied by $\Delta(s)$ etc.; w_0 is unchanged. To find the first-order correction to u_1, v_1, w_1, write

$$
\left.
\begin{aligned}
u &= D^2[(\Delta(s)/r)\psi_{1\theta} + \varepsilon' D\Delta'(s)u_{11}], \\
v &= D^2[-\Delta(s)\psi_{1r} + \varepsilon' D\Delta'(s)v_{11}], \\
w' &= Dw_0 + D^3[\Delta(s)w_1 + \varepsilon' D\Delta'(s)w_{11}],
\end{aligned}
\right\} \tag{4.14}
$$

substitute into (4.5)–(4.8) and retain only the leading terms. The continuity equation, (4.5), gives

$$
u_{11r} + u_{11}/r + v_{11\theta}/r = -w_1(r, \theta), \tag{4.15}
$$

while the $O(\varepsilon' D^3)$ terms in (4.6) and (4.7) can be combined to yield

$$
\nabla_1^2 \Omega_{11} = (1/r)(\partial/\partial r)(rw_0 v_1) - w_0 u_{1\theta}/r, \tag{4.16}
$$

where

$$
\Omega_{11} = v_{11r} + v_{11}/r - u_{11\theta}/r.
$$

The leading term in (4.8) yields an equation for w_{11}, as follows:

$$
\nabla_1^2 w_{11} = -G_{11} + u_{11}w_{0r} + w_0 w_1, \tag{4.17}
$$

where $-\varepsilon' D^4\Delta'(s)G_{11}$ is independent of r and θ and is the correction to the pressure gradient required to keep the volume flux independent of s; G_{11} is zero, in fact, because w_{11} is proportional to $\cos\theta$, but the corresponding term at the next order would in general not be zero. The solutions of (4.15)–(4.17) that satisfy the boundary conditions are

$$
\left.
\begin{aligned}
u_{11} &= C\cos\theta\,(1-r^2)^2 f(r), \\
v_{11} &= C\sin\theta\,(1-r^2)g(r), \\
w_{11} &= C\cos\theta\,r(1-r^2)h(r),
\end{aligned}
\right\} \tag{4.18}
$$

where

$$
\left.
\begin{aligned}
C &= 1/(270\times 8^5), \\
f(r) &= -13+15r^2-7r^4+r^6, \\
g(r) &= 13-224r^2+266r^4-124r^6+17r^8, \\
h(r) &= (1/3360)(-7338+7362r^2-6498r^4+3903r^6 \\
&\quad -548r^8+40r^{10}).
\end{aligned}
\right\} \tag{4.19}
$$

It is to be expected that, in a tube of increasing curvature ($\Delta'>0$), there will be a delay in setting up the secondary motions, which will therefore be weaker at a given value of s than they would be in a uniformly curved tube, with Δ equal to its value at s. The above solutions confirm this expectation, as we can see by considering the secondary motion at $r=0$, where it reduces to a velocity of magnitude

$$[D^2/(9\times 8^3)][2\Delta(s)-\varepsilon'D\Delta'(s)(13/1920)]$$

in the $\theta=0$ direction.

The quantity of greatest interest in the context of blood vessels is the shear stress on the vessel walls. This has two components which in non-dimensional variables are $-w_r'|_{r=1}$ in the axial (s-) direction and $-v_r|_{r=1}$ in the tangential (θ-) direction. To the order of this calculation, these quantities are:

$$
\begin{aligned}
-w_r'|_{r=1} &= \tfrac{1}{2}D+[D^3\cos\theta/(30\times 8^4)] \\
&\quad \times[\Delta(s)-\varepsilon'D\Delta'(s)(73/3024)], \\
-v_r|_{r=1} &= [D^2\sin\theta/(6\times 8^2)] \\
&\quad \times[\Delta(s)-\varepsilon'D\Delta'(s)(13/5760)].
\end{aligned}
\tag{4.20}
$$

These results show (i) that curvature increases axial wall shear on the outside wall ($|\theta| < \frac{1}{2}\pi$) and decreases it on the inside, (ii) that increasing curvature ($\Delta' > 0$) diminishes this effect, (iii) that curvature generates a positive secondary shear in the θ-direction (consistent with the qualitative diagram in fig. 4.1), and (iv) that increasing curvature reduces this secondary shear slightly (but only slightly: the numerical coefficient of $\varepsilon'D\Delta$ in the second of equations (4.20) is about 0.0023 compared with 0.024 for that in the first equation).

Two final points to note are (v) that the 'centres' of the secondary vortices, i.e. the stagnation points on $\theta = \pm\frac{1}{2}\pi$, which are at $r \approx 0.43$ for uniform curvature, are moved out slightly to $r \approx 0.43 + (\varepsilon'D\Delta'/\Delta) \times 5.1 \times 10^{-4}$ when the curvature is increasing; and (vi) that the position of maximum axial velocity, which is at $r \approx 2.6 \times 10^{-5}\Delta D^2$, $\theta = 0$ for uniform curvature, is moved to $r \approx 2.6 \times 10^{-5}\Delta D^2[1 - (\varepsilon'D\Delta'/\Delta) \times 0.019]$, $\theta = 0$ when the curvature is variable.

Low-Dean-number flow in a tube of variable curvature has also been studied by Murata, Miyake & Inaba (1976). They obtained the same results as above, but with some extra terms arising from a less conventional non-dimensionalisation than that used here. They kept $Re'(\propto \delta^{-1/2}D) = O(1)$ as $\delta \to 0$, and then gave solutions for small and for moderate Re', whereas we have kept $D = O(1)$ as $\delta \to 0$, subsequently considering the small-D expansion. The extra terms, which appear on the right-hand sides of (4.6)–(4.8), constitute kinematic corrections to zero-Reynolds-number flow, and they demonstrate, for example, that when $Re' \ll 1$ the axial flow is shifted to the inside of a bend and not to the outside, in order to minimise the rate of energy dissipation. Murata et al. presented solutions for Re' up to 2000, but because of their scaling this is still equivalent to small Dean number. The great merit of their paper is that they show how to develop numerical solutions of the small-δ and small-D equations for arbitrarily large values of ε', so that tubes of small, but rapidly varying, curvature can be examined.

4.1.2 Intermediate D

Reverting now to the case of uniform curvature, we recall that the small-D expansion is inadequate to describe arteries of interest such as the aorta. Equations (4.11) and (4.12) have been solved

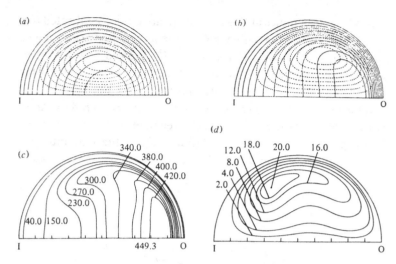

Fig. 4.2. Contour plots of axial velocity (continuous curves) and secondary streamlines (broken curves), (a) $D = 96$, (b) $D = 606$. (c) Contour plots of axial velocity, $D = 5000$. (d) Secondary streamlines, $D = 5000$. I is the inside of the bend, O the outside. (After McConalogue & Srivastava (1968) and Collins & Dennis (1975).)

numerically for values of D up to about 600 by McConalogue & Srivastava (1968), and subsequently up to $D = 5000$ by Collins & Dennis (1975); although the methods used by these two groups were quite different, the results agree very well for $D < 600$. The flow pattern is depicted by plotting curves of constant ψ (secondary streamlines) and of constant w' (axial velocity contours). Results for $D = 96$, 606 and 5000 are shown in fig. 4.2. Fig. 4.2(a) shows that for $D = 96$ the secondary motions are approximately symmetrical about the line $\theta = \frac{1}{2}\pi$ (as predicted by Dean's expansion) and that the point of peak axial velocity ($w'_{max} = 23.4$) is shifted from the tube axis to about $r = 0.20$ (compared with 0.24, which is the value predicted above by the first term in Dean's expansion). Fig. 4.2(b) shows that at the intermediate value of D, a boundary layer appears to be developing (on the outside wall of the bend) in which the axial shear is high, whereas in the core the secondary flow appears to be approximately uniform while the contours of axial velocity are approximately straight lines across the tube. This view of the axial velocity field is confirmed at $D = 5000$ by fig. 4.2(c), but the secondary streamlines at this value of D are quite complicated (fig. 4.2(d)), suggesting that an asymptotic theory for large D may not be straightforward. Collins & Dennis (1975) showed how well their computations agreed with experiment: figs. 4.3(a) and (b) show the agreement between the computed position and value of the maximum axial velocity and the same quantities as measured by Adler (1934), while fig. 4.3(c) shows the agreement between theoretical and experimental values of the ratio $f(D)$ of the flow-rate in a straight tube to that in a curved tube subject to the same pressure

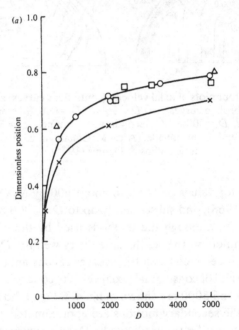

Fig. 4.3. (a) Variation of the position of maximum axial velocity along the plane of symmetry as a function of D. Numerical calculations: open circles, present theory; crosses, filled circles, open triangles, other theories. Open squares, experimental measurements of Adler (1934). (b) Magnitude of maximum dimensionless axial velocity as a function of D. Numerical calculations: continuous curve, present theory; open circles, filled triangles, broken curve, other theories. Open squares, experimental measurements of Adler (1934). (c) Variation of the friction ratio f as a function of κ ($\propto D^{2/3}$). Numerical calculations: open circles, present theory; solid circles and curve joining them, (3.53); open triangles, continuous curve, other theories. Broken curve, experimental measurements of White (1929). (After Collins & Dennis, 1975.)

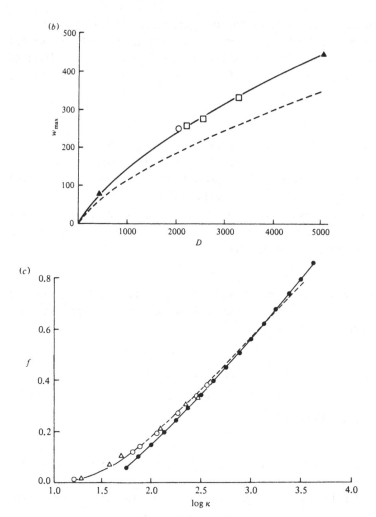

Fig. 4.3. (*continued*)

gradient. Collins & Dennis were also concerned to confirm the asymptotic structure of the solution for large D, as proposed by Ito (1969). This structure is discussed below; suffice it here to say that the numerical solutions confirmed the predicted structure in that (i)

$$f(D) \sim D^{-1/3}[a + bD^{-1/3} + O(D^{-2/3})], \qquad (4.21)$$

where a and b are constants (equal to 8.12 and -16.7 respectively); and that (ii) $D^{-2/3}w'$ is approximately independent of D across the plane of symmetry of the tube (the maximum value of $D^{-2/3}w'$ appears to tend to a constant value around 1.6 as $D \to \infty$), except near the outside wall, where (iii) there is a boundary layer of thickness proportional to D^{-13}, so that the wall shear, $-w_r$ at $r = 1$, $\theta = 0$, is proportional to D ($\approx 0.85D$).

4.1.3 Large D

The numerical results described above suggest that the structure of the solution at large D is that of an inviscid rotational core surrounded by a thin boundary layer. Suppose that a scale for the axial velocity w is D^α, while a scale for ψ is D^β and a scale for boundary layer thickness is $D^{-\gamma}$. In the core, where viscous forces are negligible, the inertia terms in (4.11) must balance the driving pressure gradient D, so we must have

$$\alpha + \beta = 1.$$

In the boundary layer, however, the inertia terms in both (4.11) and (4.12) must balance the viscous terms and (in (4.12)) the centrifugal force terms. Hence

$$\beta + 4\gamma = 2\beta + 3\gamma = 2\alpha + \gamma,$$

so that $\beta = \gamma = \tfrac{1}{2}\alpha = \tfrac{1}{3}$. Thus we can write

$$w' = D^{2/3}\tilde{w}, \qquad \psi = D^{1/3}\tilde{\psi} \qquad (4.22)$$

and, in the boundary layer,

$$\zeta = D^{1/3}(1-r).$$

This scaling corresponds to that introduced at the end of the subsection for intermediate D above.

The inviscid core equations, to leading order in D, now become

$$\tilde{w}[\sin\theta\,\tilde{w}_r + (\cos\theta/r)\tilde{w}_\theta] = 0$$

and

$$(1/r)(\tilde{\psi}_\theta\tilde{w}_r - \tilde{\psi}_r\tilde{w}_\theta) = 1, \qquad (4.23)$$

and their solution is

$$\tilde{w} = w_c(r\cos\theta), \qquad \tilde{\psi} = r\sin\theta/w_c'(r\cos\theta), \qquad (4.24)$$

where w_c is an arbitrary function of $x = r\cos\theta$, the Cartesian coordinate in the direction $\theta = 0$. The axial velocity is constant along lines of constant x, on which the secondary stream function is proportional to $y = r\sin\theta$.

In the boundary layer, the equations are

$$\tilde{w}_{\zeta\zeta} + \tilde{\psi}_\theta \tilde{w}_\zeta - \tilde{\psi}_\zeta \tilde{w}_\theta = 0 \qquad (4.25)$$

and

$$\tilde{\psi}_{\zeta\zeta\zeta\zeta} + (\tilde{\psi}_\theta\, \partial/\partial\zeta - \tilde{\psi}_\zeta\, \partial/\partial\theta)\tilde{\psi}_{\zeta\zeta} = \sin\theta\, \tilde{w}\tilde{w}_\zeta, \qquad (4.26a)$$

with boundary conditions

$$\left.\begin{array}{l} \tilde{\psi} = \tilde{\psi}_\zeta = \tilde{w} = 0 \quad \text{on } \zeta = 0, \\[4pt] \tilde{\psi}_\zeta \to 0, \qquad \tilde{w} \to w_c(\cos\theta) \quad \text{as } \zeta \to \infty, \\[4pt] \tilde{\psi} = \tilde{\psi}_\zeta = \tilde{w}_\theta = 0 \quad \text{on } \theta = 0, \pi. \end{array}\right\} \qquad (4.27)$$

Equation (4.26a) can be integrated once to give

$$\tilde{\psi}_{\zeta\zeta\zeta} + \tilde{\psi}_\theta\tilde{\psi}_{\zeta\zeta} - \tilde{\psi}_\zeta\tilde{\psi}_{\zeta\theta} = -\tfrac{1}{2}\sin\theta[w_c^2(\cos\theta) - \tilde{w}^2]. \qquad (4.26b)$$

The boundary layer problem appears to be completely specified for any reasonable function $w_c(\cos\theta)$. This function is then, in principle, determined by imposing the condition that the (secondary) mass flux in the boundary layer is just enough to balance the normal inflow from the core, which is non-zero and given by $\tilde{\psi}$ in (4.24). That is, we impose

$$\tilde{\psi}(\zeta \to \infty, \theta) = \sin\theta / w_c'(\cos\theta). \qquad (4.28)$$

An apparently self-consistent approximate solution to this problem was given by Ito (1969) using the Pohlhausen momentum integral method. He was able to derive a function $w_c(\cos\theta)$ that decreased monotonically from a given constant (≈ 1.9) at $\theta = 0$ to zero at $\theta = \pi$. Note that $w_c(1)$ should be equal to the limit of the maximum value of $D^{-2/3}w'$ as $D \to \infty$, about 1.6 according to the numerical results of Collins & Dennis (1975). The boundary layer thickness in Ito's calculations remained more or less constant until θ was almost equal to π, when it increased sharply. This increase occurs because the external velocity, proportional to $\sin\theta$, becomes small. In fact, if we put $\phi = \pi - \theta \ll 1$, and suppose that near $\phi = 0$ we have $\psi \propto \phi^\alpha$, $w' \propto \phi^\beta$ and boundary layer thickness $\propto \phi^\gamma$, then (4.28) implies that $\alpha + \beta = 3$. Balance of the viscous, inertial and centrifugal force terms in the boundary layer equation (4.26b) then gives

$$\alpha - 3\gamma = 2\alpha - 2\gamma - 1 = 2\beta + 1,$$

from which we obtain $\alpha = \frac{5}{3}$, $\beta = \frac{4}{3}$, $\gamma = -\frac{2}{3}$. Thus the solution in the core near $\phi = 0$ should be of the form

$$\tilde{w}_c = \sqrt{2}A\phi^{4/3}[1 + O(\phi^2)], \qquad \tilde{\psi}_c = (3/4A\sqrt{2})\phi^{5/3}[1 + O(\phi^2)],$$

$$\tag{4.29}$$

while that in the boundary layer should be of the form

$$\tilde{w} = \sqrt{2}A\phi^{4/3}H(\eta)[1 + O(\phi^2)],$$
$$\tilde{\psi} = A^{1/2}\phi^{5/3}F(\eta), \qquad \eta = A^{1/2}\zeta\phi^{2/3},$$

$$\tag{4.30}$$

where

$$F''' = \tfrac{5}{3}FF'' - \tfrac{7}{3}F'^2 + H^2 - 1, \qquad H'' = \tfrac{5}{3}FH' - \tfrac{4}{3}F'H,$$
$$F(0) = F'(0) = H(0) = F'(\infty) = H(\infty) - 1 = 0.$$

This problem has been solved numerically by Smith (1975), who found

$$F''(0) = 1.595, \qquad H'(0) = 1.265, \qquad F(\infty) = 0.503.$$

This last result, together with (4.29) and (4.30), shows that $A = 1.05$ (a different value from that quoted by Smith, because his D differs from that defined in (4.9) by a factor of $\sqrt{2}$).

This analysis is important because it indicates that the asymptotic structure for large D, consisting of an inviscid core and a $D^{-1/3}$ boundary layer, may have a self-consistent solution for virtually all θ. In fact, the solution breaks down only in a region for which $\phi = O(D^{-1/8})$, $1 - r = O(D^{-1/4})$, because although the boundary layer approximation is still valid there (so that $\nabla_1^2 \approx \partial^2/\partial r^2$) the previously neglected terms D (in (4.11)) and $w'w'_\theta \cos\theta/r$ (in (4.12)) become as important as the others.

Riley & Dennis (1976) and others have attempted accurate numerical solutions of the boundary layer problem defined by (4.25)–(4.28). The procedure is to guess a value of w_c at $\theta = 0$ (e.g. the value of 1.6, derived from the full numerical solution of Collins & Dennis (1975)), and then to integrate forward in θ. For example, if we write

$$w_c(\cos\theta) = \sqrt{2} \sum_{n=0}^{\infty} \theta^{2n} W_n,$$

$$\tilde{w}(\zeta, \theta) = \sqrt{2} \sum_{n=0}^{\infty} \theta^{2n} W_n h_n(\zeta W_0^{1/2}),$$

$$\tilde{\psi}(\zeta, \theta) = W_0^{1/2} \sum_{n=0}^{\infty} \theta^{2n+1} f_n(\zeta W_0^{1/2}),$$

the following problem is obtained to leading order:

$$\left. \begin{array}{l} f_0''' + f_0 f_0'' - f_0'^2 + 1 - h_0^2 = 0, \qquad h_0'' + f_0 h_0' = 0, \\ f_0(0) = f_0'(0) = h_0(0) = f_0'(\infty) = h_0(\infty) - 1 = 0. \end{array} \right\} \qquad (4.31)$$

This was solved by Stewartson (1958), and has the properties $f_0''(0) = 0.953$, $h_0'(0) = 0.463$, $f_0(\infty) = 1.33$ (see Smith, 1975). Then the condition (4.28) implies $W_1 = -0.27/W_0^{1/2}$, so the second problem is completely determined:

$$h_1'' + f_0 h_1' - 2f_0' h_1 + 3(W_0/W_1) f_1 h_0' = 0,$$

$$f_1''' + f_0 f_1'' - 4f_0' f_1' + 3f_0'' f_1 = -2(W_1/W_0)(1 - h_0 h_1) + \tfrac{1}{6}(1 - f_0^2),$$

$$h_1(\infty) = 1, \qquad f_1'(\infty) = 0;$$

this is now linear and can also be solved numerically. Riley & Dennis (1976) preferred a finite-difference integration of the boundary layer equations to the above series solution. They proposed that the correct value of W_0 is that which causes the solution to approach its asymptotic form, (4.29)–(4.30), smoothly as $\theta \to \pi$.

However, Riley & Dennis's numerical results were incomplete, although their iterations appeared to lead to a well-defined value of W_0, because the numerical solution of the boundary layer equations broke down for values of θ considerably smaller than π. This was not because of the development of separation (despite the conclusions of Barua (1963), who wrongly imposed the condition that $w_c'(x) = $ constant; Ito (1969) did not predict separation), but was associated with the appearance of reversed secondary flows near the edge of the boundary layer. This is not inconsistent with the full numerical solutions (fig. 4.2(d)), but makes forward integration of the boundary layer equations impossible, and any accurate integration very difficult. It also means that Ito's (1969) momentum integral solution, which expressly forbids reversed flow in the secondary boundary layer, is unlikely to be accurate. Thus it still has not been demonstrated conclusively that the boundary layer structure outlined above is actually correct for $\theta < \pi - O(D^{-1/8})$. It may, for example, be correct only for θ less than some value $\theta_s = \pi - O(1)$ at which separation develops, but not thereafter; if separation does develop, the core flow will be significantly affected everywhere, and a theoretical prediction of W_0 would then be impossible. It is also possible that the boundary layer structure is quite incorrect: Smith (1976a) has considered curved tubes whose cross-sections have straight sides, and has proposed a plausible structure in which the flow near a flat outer wall could consist of an inviscid layer (thickness $\sim D^{-1/6}$) in which the core flow turns a corner, with a thin viscous region (thickness $\sim D^{-1/4}$) embedded within it to satisfy the no-slip condition. He also showed that no simple solution of the boundary layer problem is possible for a rectangular cross-section, although it is for a triangular cross-section. It is not clear whether this work is relevant to tubes with curved (in particular, circular) cross-sections, but it serves to underline the difficulty of analysing curved-tube flows for large Dean numbers.†

Although the suggested boundary layer structure is not yet certain, the fact that it has a well-behaved limit as $\theta \to \pi$, and the consistency between it and the quoted numerical results of Collins & Dennis (1975), suggest that it can be used with some confidence for predictions of particular quantities, such as the flow-rate (represented by $f(D)$ in (4.21)) and the axial wall shear at the outer wall, $-w'_r$ at $r = 1$, $\theta = 0$, which is $\approx 0.85D$. The values of a and b in (4.21), as obtained by Collins & Dennis, agree closely (to within $\frac{1}{2}$% and 6% respectively) with the corresponding values derived by Ito (1969) from his approximate method, which lends circumstantial support to the latter.

---

† See also Van Dyke (1978), where further circumstantial evidence against the boundary layer structure is adduced.

4.2 Fully developed oscillatory flow, $\delta \ll 1$

In this section we consider only tubes of uniform curvature, but extend the work described above to cover an axial pressure gradient ($-p_s$ in the dimensionless equation (4.4)) that varies sinusoidally with time according to

$$-p_s = G + \alpha^2 W_0 \cos \alpha^2 t. \qquad (4.32)$$

Here W_0 is the (dimensionless) velocity amplitude that would be driven by such a pressure gradient in a straight pipe containing an inviscid fluid, and α is Womersley's parameter (see (2.35)): $\alpha^2 = \Omega a^2 / \nu$, where Ω is the dimensional angular frequency of the oscillation. This class of flows has been analysed by Lyne (1971) for the case $G = 0$, by Smith (1975), and by Blennerhassett (1976).

A secondary stream function can be defined as in (4.10) and the governing equations corresponding to (4.11) and (4.12) are

$$\nabla_1^2 w' + D + \alpha^3 (2R_s)^{1/2} \cos \tau = (1/r)(\psi_\theta w'_r - \psi_r w'_\theta) + \alpha^2 w'_\tau$$
$$(4.33)$$

and

$$\nabla_1^4 \psi + (1/r)(\psi_r \, \partial/\partial\theta - \psi_\theta \, \partial/\partial r)\nabla_1^2 \psi = -w' w'_y + \alpha^2 \nabla_1^2 \psi_\tau,$$
$$(4.34)$$

where $y = r \sin \theta$, $\tau = \alpha^2 t$ and there is a new dimensionless parameter, R_s:

$$R_s = \delta W_0^2 / \alpha^2 = \delta \hat{W}_a^2 / \Omega \nu, \qquad (4.35)$$

where $\hat{W}_a$ is the dimensional axial velocity amplitude. R_s has a large value, of the order of 10^3–10^4, in the aorta. Note that the ratio of the unsteady pressure gradient amplitude to the mean is equal to $\alpha^3 (2R_s)^{1/2}/D$. The present notation differs from that of Lyne, Smith and Blennerhassett in that they used $\beta = \sqrt{2}/\alpha$ instead of α, and they omitted the $\sqrt{2}$ from the definitions of D in (4.9) and of w', with the result that the $w' w'_y$ term in (4.34) was multiplied by 2.

Since the problem has three independent parameters, each of which may be taken to be large or small, there is an enormous variety of different limiting cases for which asymptotic expansions can be sought analytically. Smith (1975) has investigated many of them. Here we restrict attention to those that are potentially of greatest interest in the study of flow in large arteries; that is, we restrict attention from the start to large values of α, and to flows in which the amplitude of the oscillatory motion is at least as large as the mean velocity; small perturbations about the steady flow are relatively easy to analyse (Smith, 1975, § 2).

4.2.1 *Zero mean pressure gradient*

In order to discover the principal effects of oscillations in a curved tube, we follow Lyne (1971) and examine first the case in which the flow is purely oscillatory, i.e. $D = 0$. If the fluid were inviscid, there would be potential flow in which

$$w' = \alpha(2R_s)^{1/2} \sin \tau, \quad \psi = 0; \qquad (4.36a)$$

this is the dimensionless, small-δ version of a potential vortex, centred on the centre of curvature of the tube axis, which has dimensional velocity

$$\hat{w} = R\hat{W}_a \sin (\Omega \hat{t})/(R + \hat{r} \cos \theta). \qquad (4.36b)$$

In a viscous fluid with large α, this oscillatory core flow will be surrounded by a Stokes layer on the boundary $r = 1$. In that Stokes layer the centrifugal forces will drive an oscillatory secondary motion, primarily consisting of a θ-component of velocity, v. Since the centrifugal force term is proportional to w'^2 (see (4.6) and (4.7)), the solution for v will contain both an oscillation with double the fundamental frequency, and a mean flow. This mean secondary flow is similar to the steady streaming generated by the high-frequency oscillation of a cylinder or sphere (Riley, 1965); as in those examples, the mean flow does not fall to zero at the edge of the Stokes layer, but drives steady secondary motions in the core. The details of the core flow will depend on the value of the 'secondary-streaming Reynolds number', which in this case is R_s. This parameter must be allowed to remain $O(1)$ as $\alpha \to \infty$.

Recognising that a secondary motion will be generated in the core, we seek a solution there of the form

$$w' = \alpha(2R_s)^{1/2} \sin \tau,$$

$$\psi = \psi_0 + \alpha^{-1}\psi_1 + \alpha^{-2}\psi_2 + O(\alpha^{-3});$$

substitution into (4.34) shows that $\nabla_1^2\psi_0$ and $\nabla_1^2\psi_1$ are independent of τ, and that

$$\frac{\partial}{\partial \tau} \nabla_1^2 \psi_2 = \nabla_1^4 \psi_0 + \frac{1}{r}\left(\psi_{0r}\frac{\partial}{\partial \theta} - \psi_{0\theta}\frac{\partial}{\partial r}\right)\nabla_1^2 \psi_0. \tag{4.37}$$

The Stokes layer has a dimensional thickness proportional to $(\nu/\Omega)^{1/2}$, so we introduce new variables

$$\eta = (\alpha/\sqrt{2})(1-r), \qquad \Psi = \alpha\psi, \qquad W = \alpha^{-1}w',$$

and seek solutions of the form

$$W = W_0 + O(\alpha^{-1}), \qquad \Psi = \Psi_0 + O(\alpha^{-1}).$$

These must satisfy the no-slip condition on $\eta = 0$, and must match to the core flow as $\eta \to \infty$, which among other things requires

$$W_0 \to (2R_s)^{1/2} \sin \tau \quad \text{as } \eta \to \infty. \tag{4.38}$$

The equations satisfied by W_0 and Ψ_0, from (4.33) and (4.34), are

$$\tfrac{1}{2}W_{0\eta\eta} - W_{0\tau} = -(2R_s)^{1/2} \cos \tau \tag{4.39}$$

and

$$\tfrac{1}{2}\Psi_{0\eta\eta\eta\eta} - \Psi_{0\eta\eta\tau} = \sqrt{2} \sin \theta \, W_0 W_{0\eta}. \tag{4.40}$$

The solution of (4.39) satisfying (4.38) and the wall condition is

$$W_0 = (2R_s)^{1/2}[\sin \tau - e^{-\eta} \sin (\tau - \eta)], \tag{4.41}$$

which is the basic Stokes layer solution. This is then substituted into the right-hand side of (4.40); the most general periodic solution that has frequency 2 and satisfies the wall boundary conditions is

$$\begin{aligned}
\Psi_0 = \sqrt{2}\, R_s \sin \theta &\left[\tfrac{5}{8} - \tfrac{1}{4}\eta - \tfrac{1}{8}e^{-2\eta} - \frac{1}{\sqrt{2}}e^{-\eta} \cos (-\eta + \tfrac{1}{4}\pi)\right.\\
&+ \frac{1}{8\sqrt{2}}(9 - 5\sqrt{2}) \cos (2\tau + \tfrac{1}{4}\pi) - \frac{1}{8\sqrt{2}}e^{-2\eta} \cos (2\tau - 2\eta + \tfrac{1}{4}\pi)\\
&\left.- \frac{1}{\sqrt{2}}e^{-\eta} \cos (2\tau - \eta + \tfrac{1}{4}\pi) + \tfrac{5}{8}e^{-\sqrt{2}\eta} \cos (2\tau - \sqrt{2}\eta + \tfrac{1}{4}\pi)\right]\\
&+ B(\theta)\eta^3 + C(\theta)\eta^2. \tag{4.42}
\end{aligned}$$

As usual in such steady-streaming problems the arbitrary functions $B(\theta)$ and $C(\theta)$ turn out on matching to be zero, but not only does Ψ_0 not tend to zero as $\eta \to \infty$ (a displacement effect, present in all boundary layers) but also $\Psi_{0\eta}$ does not tend to zero as $\eta \to \infty$. In fact, with $B = C = 0$,

$$\Psi_0 \sim \sqrt{2} \, R_s \sin \theta [\tfrac{5}{8} - \tfrac{1}{4}\eta + (1/8\sqrt{2})(9 - 5\sqrt{2}) \cos (2\tau + \tfrac{1}{4}\pi)],$$

$$\Psi_{0\eta} \to -(1/2\sqrt{2}) R_s \sin \theta \qquad (4.43)$$

as $\eta \to \infty$.

The core solution ψ_0 must match with (4.43) as $r \to 1$. Since $\nabla_1^2 \psi_0$ is independent of τ, we can write

$$\psi_0 = \psi_0^{(u)}(r, \theta, \tau) + \psi_0^{(s)}(r, \theta),$$

where $\psi_0^{(s)}$ is steady and $\psi_0^{(u)}$ is periodic with zero mean and satisfies $\nabla_1^2 \psi_0^{(u)} = 0$. In fact, matching with (4.43) leads to the conclusion that $\psi_0^{(u)} \equiv 0$, the unsteady term in (4.43) giving rise only to a displacement effect at higher order in α^{-1}. Since ψ_2 is expected to contain only periodic and mean terms, (4.37) yields the equation satisfied by $\psi_0^{(s)}$:

$$\nabla_1^4 \psi_0^{(s)} + (1/r)(\psi_{0r}^{(s)} \, \partial/\partial\theta - \psi_{0\theta}^{(s)} \, \partial/\partial r)\nabla_1^2 \psi_0^{(s)} = 0;$$

and this must be solved subject to the boundary conditions derived from (4.43):

$$\psi_0^{(s)} = 0, \qquad \partial\psi_0^{(s)}/\partial r = \tfrac{1}{4}R_s \sin \theta \quad \text{on } r = 1.$$

If we write $\psi_0^{(s)} = R_s \chi_0$, χ_0 satisfies

$$\left.\begin{array}{l} \dfrac{1}{R_s}\nabla_1^4 \chi_0 + \dfrac{1}{r}\left(\chi_{0r}\dfrac{\partial}{\partial\theta} - \chi_{0\theta}\dfrac{\partial}{\partial r}\right)\nabla_1^2 \chi_0 = 0, \\[2mm] \chi_0 = 0, \qquad \chi_{0r} = \tfrac{1}{4}\sin \theta \quad \text{on } r = 1, \end{array}\right\} \qquad (4.44)$$

which is the problem of the steady, two-dimensional viscous flow in a circle, driven by a given tangential velocity on the boundary. This steady secondary flow in the core is governed by the single dimensionless parameter R_s, the secondary-streaming Reynolds number.

As usual for viscous flows, analytical progress towards solution of the problem (4.44) can be achieved only in the limits of small and large R_s (Lyne, 1971). The small-R_s limit is a simple Stokes flow problem, with solution

$$\chi_0 = -\tfrac{1}{8}r(1 - r^2) \sin \theta - (R_s/3072)r^2(1 - r^2)^2 \sin 2\theta + O(R_s^2).$$

Note that $(1/r)\chi_{0\theta}$, proportional to the radial velocity, is to leading order *negative* on $\theta = 0$ (and positive on $\theta = \pi$). This means that the steady secondary streaming in the core proceeds across the tube from the outside of the bend to the inside. This is in the opposite direction to the secondary flow in a pipe when the axial pressure gradient is steady, and corresponds to the negative value of the azimuthal velocity (proportional to $-\chi_{0r}$) on $r = 1$.

At large values of R_s (more relevant physiologically) one expects the effects of viscosity to be confined to boundary layers of thickness $R_s^{-1/2}$ near $r = 1$; however, these come together at $\theta = 0$ and erupt across the centre of the tube. Lyne (1971) postulates (and demonstrates the self-consistency of) a scheme in which the wall boundary layers are supplemented by another layer of thickness $R_s^{-1/2}$ across the plane of symmetry ($\theta = 0$, $\theta = \pi$), the bulk of the core flow being inviscid (fig. 4.4). In the inviscid regions the streamlines are closed, and therefore the vorticity is uniform (Batchelor, 1956) and of opposite sign in the two regions; one reason why the shear layer across the middle is required is to smooth out the discontinuity in vorticity between the two inviscid regions. In the upper core region, the stream function (denoted by an overbar) is given by

$$\nabla^2 \bar{\chi}_0 = -\zeta,$$

where ζ is the vorticity, expected to be negative (fig. 4.4). The boundary conditions are

$$\bar{\chi}_0 = 0 \quad \text{on } r = 1 \quad \text{and} \quad \text{on } \theta = 0, \pi.$$

The solution is

$$\bar{\chi}_0 = \frac{\zeta}{2\pi} \left\{ \left[1 - \frac{1}{2}\left(r^2 + \frac{1}{r^2} \right) \cos 2\theta \right] \tan^{-1} \left(\frac{2r \sin \theta}{1 - r^2} \right) \right.$$
$$- \frac{1}{4}\left(r^2 - \frac{1}{r^2} \right) \sin 2\theta \log \left(\frac{1 + 2r \cos \theta + r^2}{1 - 2r \cos \theta + r^2} \right)$$
$$\left. + \left(r - \frac{1}{r} \right) \sin \theta - \tfrac{1}{2}\pi r^2 (1 - \cos 2\theta) \right\}, \tag{4.45}$$

and this gives the scaled tangential velocity at the edge of the boundary layer on $r = 1$ to be

$$\bar{v}_1 = -\bar{\chi}_{0r}|_{r=1} = (\zeta/\pi)(\pi \sin^2 \theta - 2 \sin \theta - \sin 2\theta \log \tan \tfrac{1}{2}\theta),$$

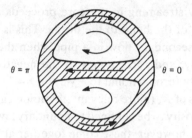

Fig. 4.4. Sketch of cross-section of curved tube showing steady secondary streaming at large values of R_s. Shaded regions are viscous boundary layers.

while that at the edge of the layer on $\theta = 0$, π is

$$\bar{u}_1 = \frac{1}{r}\bar{\chi}_{0\theta}|_{\theta=0,\pi} = \frac{\zeta}{2\pi}\left(1 - \frac{1}{x^2}\right)\left[2 - \left(x + \frac{1}{x}\right)\log\left(\frac{1+x}{1-x}\right)\right],$$

where $x = r\cos\theta$. The boundary condition to be satisfied by the tangential velocity v on $r = 1$ is $v_0 = -\frac{1}{4}\sin\theta$. The boundary condition to be satisfied by the tangential velocity u on $y = r\sin\theta = 0$ is $\partial u/\partial y = 0$; the value of u on $y = 0$ is unknown. The usual two-dimensional boundary layer equations are to be satisfied in each layer. Their exact solution would be a difficult problem, and Lyne (1971) found an approximate solution by assuming (a) that the longitudinal velocity in a shear layer departed by only a little from that in the core just outside, so the equations could be linearised, and (b) that the velocity profile in the shear layers is convected unchanged around the corners at $\theta = 0$, π. He showed that (b) is self-consistent with the linearisation and with earlier work by Harper (1963) and Moore (1963), but that (a) is inaccurate near $\theta = 0$, π. This is not expected to be important because the velocities are small there anyway. Lyne's numerical results showed that $\zeta \approx -0.56$, and that the secondary flow has stagnation points (the vortex centres) at $r \approx 0.48$, $\theta = \pm\frac{1}{2}\pi$. These values should be compared with

$$r \approx 0.58 - 1.04 \times 10^{-6}R_s^2, \qquad \theta = \pm(\frac{1}{2}\pi - 2.00 \times 10^{-3}R_s)$$

when R_s is small: the vortex centre moves off the line $\theta = \pm\frac{1}{2}\pi$ in the direction of the fluid motion at $r = 1$ as R_s increases from zero, but comes back on to $\theta = \pm\frac{1}{2}\pi$ as $R_s \to \infty$. Lyne made qualitative experimental observations of the direction of steady secondary streaming in the core, which was from the outside to the inside of the bend as predicted.

From the physiological point of view, the most important quantities to predict are the axial velocity profile in the core and the wall shear stress. Lyne's theory predicts no departure from a flat profile in the core, because it is limited to infinitesimally small values of δ, and the secondary motions affect the profile only at the next order (see § 4.3). The wall shear has two components, which can be calculated from the boundary layer solutions (4.41) and (4.42). In dimensional terms the axial wall shear-rate is

$$\frac{\nu}{a^2}\frac{1}{(2\delta)^{1/2}}\frac{\alpha^2}{\sqrt{2}}W_{0\eta}\big|_{\eta=0} = \hat{W}_0\left(\frac{\Omega}{\nu}\right)^{1/2}\sin\left(\Omega\hat{t}+\tfrac{1}{4}\pi\right),$$

as in a Stokes layer, and the secondary wall shear-rate is

$$\frac{\nu}{a^2}\frac{1}{2}\alpha\Psi_{0\eta\eta}\big|_{\eta=0} = \frac{\hat{W}_0^2\delta}{(\Omega\nu)^{1/2}a}\frac{\sin\theta}{2\sqrt{2}}[1+(3\sqrt{2}-5)\sin\left(2\Omega\hat{t}+\tfrac{1}{4}\pi\right)].$$

It is interesting to note that the steady component of the secondary shear is in the direction of θ increasing (for $\theta>0$), which is in the opposite sense to the mean velocity at the edge of the Stokes layer, driving the steady streaming in the core.

4.2.2 Non-zero mean pressure gradient

One of the reasons why Lyne's theory is not directly applicable to the aorta is that it requires the mean pressure gradient to be zero. Only if D is very small (i.e. $D \ll \alpha^{-1} \ll 1$) can the flow be expanded in a power series in D with Lyne's solution as the leading term (Smith, 1975, § 3). However, in the canine aorta we have $\alpha \approx 13$, $D \approx 2000$ and $R_s \approx 4200$, so although the ratio between the amplitude and the mean of the pressure gradient, $\alpha^3(2R_s)^{1/2}/D$, is large (about 100), D should be taken to be large not small. If the flows associated with the mean and the oscillations could be uncoupled, one would expect the mean to be described by large-D steady flow (§ 4.1.3) and the oscillations by large-α, large-R_s unsteady flow (§ 4.2.1), except that there we let $\alpha \to \infty$ before allowing $R_s \to \infty$, while the order of magnitude of the numbers here suggests that the limits should be taken in the reverse order. However, the problem is non-linear and the mean and oscillatory parts cannot be separated; we should therefore examine carefully those cases in which α, D, R_s are all large. The values of the above numbers suggest that we should concentrate particularly on values of the parameters such that $1 \ll \alpha \ll D < R_s$ (and in which $D \ll \alpha^3 R_s^{1/2}$, so that the amplitude of the pressure gradient oscillations is much greater than the mean). The regimes of greatest theoretical interest are those in which the secondary streaming in the core changes from being outwards, as it would be if the pressure gradient were steady and equal to the mean, to being inwards, as if driven by the Stokes layer on the wall in the absence of a mean pressure gradient. *A priori* one would expect this transition to occur when $R_s = O(D^{1/3})$, on the assumption that each of the two contributions to the secondary streaming has the same order of magnitude as it would in the absence of the other. However, little work has been done on this case, and the greatest progress in studying such a transition has been made by Blennerhassett (1976) who took the limit $\alpha \to \infty$ and then examined cases in which $D \lesssim R_s$. This is physiologically reasonable, apart from the initial limit $\alpha \to \infty$.

Blennerhassett's limiting procedure ensures that the thinnest boundary layer on the tube wall is the Stokes layer, in which the flow is that already calculated by Lyne (1971). Interest can therefore be centred on the steady component of the flow, driven both by the mean axial pressure gradient and by the slip velocity at the edge of the Stokes layer. Blennerhassett showed that if the axial velocity outside the Stokes layer, non-dimensionalised as in the early part of this section, is written

$$w' = \alpha (2R_s)^{1/2} \sin \tau + w_1 + O(\alpha^{-1}),$$

and if the stream function is again written as

$$\psi = \psi_0 + \alpha^{-1}\psi_1 + O(\alpha^{-2}),$$

then the equations for w_1 and ψ_0 are just the equations for exactly steady flow in a curved tube, (4.11) and (4.12), with (w', ψ) replaced by (w_1, ψ_0). The boundary conditions, however, are different from those in § 4.1, being

$$\psi_0 = w_1 = 0, \qquad \psi_{0r} = \tfrac{1}{4} R_s \sin \theta \quad \text{on } r = 1. \qquad (4.46)$$

The solution of this problem in the limit $R_s \to \infty$ with $D = o(R_s)$ reveals that the secondary stream function ψ_0 $(= O(R_s))$ becomes equal to that calculated by Lyne (1971), so that the flow in the core (away from the steady-streaming boundary layer) is given by $\psi_0 = R_s \bar{\chi}_0$ and (4.45) above. The axial velocity w_1 is then to be calculated from (4.11). This equation suggests that $w_1 = O(D/R_s)$, so that in the core the axial flow results from a balance between the convective and the pressure gradient terms. However, by converting to (secondary) streamline coordinates, Blennerhassett was able to show that such a balance leads to a solution for w_1 that is not a single-valued function of position, and hence w_1 must be $O(D)$. Thus w_1 in the core must satisfy

$$\bar{\chi}_{0\theta} w_{1r} - \bar{\chi}_{0r} w_{1\theta} = 0,$$

which means that w_1 is constant on the secondary streamlines. In fact, examination of the equations for the $O(DR_s^{-1/2})$ and $O(DR_s^{-1})$ corrections to $\bar{\chi}_0$ shows that

$$w_1(r, \theta) = (D/\zeta)\bar{\chi}_0(r, \theta)$$

in the core, where ζ (≈ 0.56) is the magnitude of the axial vorticity in each of the two inviscid core regions, separated by a shear layer across the middle of the tube (fig. 4.4). In this limit, therefore, the secondary flow is inwards across the core, as in purely oscillating flow, and there are peaks of axial velocity at the vortex centres in each half of the tube. Blennerhassett also gave an approximate numerical solution of the equations for w_1 in the wall and centre-plane boundary layers (thickness $\propto R_s^{-1/2}$) using Lyne's solution for ψ_0.

More important than these asymptotic results, however, are Blennerhassett's numerical solutions for the steady components of the flow, obtained directly from (4.11) and (4.12) for arbitrary values of D and R_s, without any boundary layer approximation. He first solved the equations for $D = o(R_s)$, in which case the secondary stream function is not affected by the axial flow and is determined from (4.44) as in the purely oscillatory case. Then w_1 is given by solving the linear equations (4.11). Results for various values of R_s are shown in fig. 4.5. In each part of the figure the secondary streamlines are shown on the left, and contours of axial velocity on

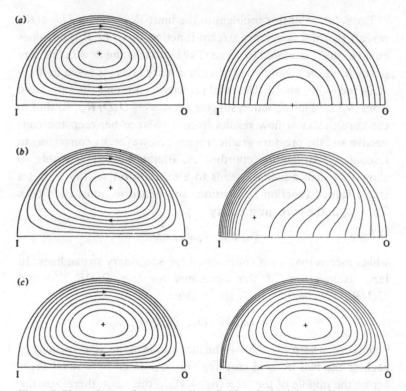

Fig. 4.5. Plots of secondary streamlines (left) and of axial velocity contours (right) representing the steady flow induced by pulsatile motion in a curved tube. I is the inside of the bend, O the outside. Here $D = o(R_s)$, (a) $R_s = 10$, (b) $R_s = 150$, (c) $R_s = 1800$. (After Blennerhassett, 1976.)

the right. The secondary streamlines show no qualitative changes with R_s, the velocity being inwards across the centre of the tube, but the axial velocity distribution changes considerably. At fairly small R_s (e.g. $R_s = 10$, fig. 4.5(a)) the axial velocity represents only a small deviation from axial symmetry (Poiseuille flow). As R_s is increased, however, axial momentum is advected towards the inside wall of the bend, and is swept round sideways, so that the peak axial velocity occurs much closer to the inside wall. Furthermore, at a value of R_s between 100 and 150, the position of peak axial velocity comes off the centre plane (fig. 4.5(b), for $R_s = 150$), and, as $R_s \to \infty$, this position tends to coincide with the secondary vortex centre (fig. 4.5(c), for $R_s = 1800$). These calculations confirm the asymptotic

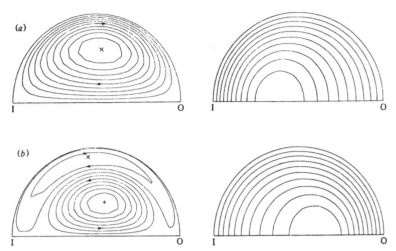

Fig. 4.6. Plots of secondary streamlines (left) and of axial velocity contours (right) representing the steady flow induced by pulsatile motion in a curved tube. I is the inside of the bend, O the outside. Here $R_s = 30$, (a) $D = 40$, (b) $D = 160$. (After Blennerhassett, 1976.)

result that the contours of constant w_1 coincide with the secondary streamlines when $R_s \to \infty$.

When D is not much smaller than R_s, (4.11) and (4.12) are not uncoupled, and must be solved simultaneously. Their solution yields some very interesting results. At fairly small values of R_s they are not unexpected, as shown by fig. 4.6 for $R_s = 30$, and (a) $D = 40$, (b) $D = 160$ (the Dean number defined by (4.9) and used here is eight times that used by Blennerhassett). When D is small, the secondary streaming is directed towards the inside wall (fig. 4.6(a)), and the peak axial velocity occurs near the inside wall. As D is increased, however, the influence of the steady axial pressure gradient becomes stronger, and the secondary flow driven by the slip velocity at the edge of the Stokes layer is squeezed to the sides, to be replaced in the core by a secondary flow directed outwards, as in steady flow (fig. 4.6(b)). The peak axial velocity then moves towards the outside of the tube. This sequence of solutions is consistent with that predicted by Blennerhassett for creeping flow ($R_s \ll 1$).

For somewhat larger values of R_s, the transition from inwards to outwards secondary flow continues to arise in much the same way.

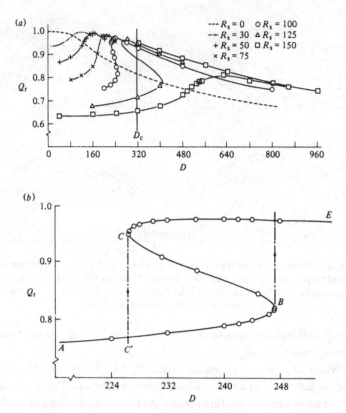

Fig. 4.7. (a) Plots of the flow-rate ratio, Q_r, against D, for various values of R_s. Q_r is the actual flow-rate through the tube at a given mean pressure gradient divided by the flow-rate in a straight tube driven by the same mean pressure gradient. (b) Enlargement of the S-shaped section of the curve for $R_s = 100$. (After Blennerhassett, 1976.)

However, for values of R_s above a certain critical value, between 75 and 100, a most unexpected result is obtained. This is best exhibited in graphs of mean volume flow-rate through the tube, expressed as a fraction of the mean flow-rate in a straight tube with the same mean pressure gradient, against D for various values of R_s. Fig. 4.7(a) shows that for R_s below its critical value the flow-rate ratio, Q_r, is a single-valued function of D with a maximum at a value of D that is predicted to be approximately $22R_s^{1/2}$ when R_s is small. This value is slightly above that at which the transition in the nature of the secondary flow takes place.

For R_s equal to 100 or more, however, an S-shaped portion appears in the graph of Q_r against D, with the implication that the value of Q_r actually achieved at a given D depends on how the flow is started. If D is gradually increased from zero, with R_s fixed, the flow-rate will follow the lower branch of the curve, jumping discontinuously to the upper branch when D reaches the value at which the slope of the curve becomes infinite (point B on fig. 4.7(b)). Similarly, if D is reduced from a very large value, Q_r will follow the upper branch, jumping discontinuously to the lower branch at the point C. Blennerhassett was able to compute flow patterns corresponding to points on each section of the curve. He found that the flows do represent a one-parameter family of solutions of the equations, with the axial velocity on the centre line increasing monotonically as the curve is followed continuously from point A to point E (fig. 4.7(b)). Some of his computations are also shown in fig. 4.8, and confirm that the steady secondary flow corresponding to points on the lower branch of the curve has the same direction as for a purely oscillatory pressure gradient (fig. 4.8(a)), while for points on the upper branch the secondary flow in the core resembles that for a steady pressure gradient (fig. 4.8(c) and (d)). The middle branch represents an intermediate type of flow (fig. 4.8(b)), but this is presumably unstable and unrealisable in practice.

Even more unexpected results were obtained for $R_s = 150$ (the largest value considered by Blennerhassett). Here it seems as if the point B of fig. 4.7(b) has gone off to infinity, and at least two solutions of the problem exist for all D above D_c (fig. 4.7(a)). In this case the flow on the upper branch would be achieved only if a secondary motion with outwards velocity on the centre plane were imposed at the start (i.e. if R_s were increased from zero with D fixed), and that on the lower branch only if an inwards secondary motion were imposed at the start (or if D were gradually increased from zero). The flow patterns corresponding to the lower branch have features in common with the large-R_s, $D = O(R_s)$ results, in that the positions of maximum axial velocity do not lie on the centre plane (compare fig. 4.9 with figs. 4.5(b), (c)). The upper-branch flow patterns are qualitatively the same as other upper-branch patterns, such as that shown in fig. 4.8(d).

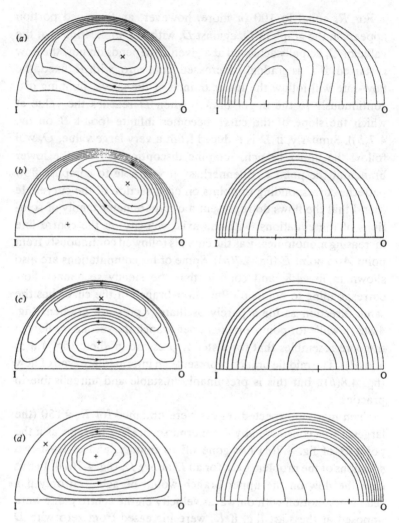

Fig. 4.8. Plots of secondary streamlines (left) and of axial velocity contours (right) representing the steady flow induced by pulsatile motion in a curved tube. I is the inside of the bend, O the outside. Here $R_s = 100$, (a) $D = 244$ (lower branch of curve in fig. 4.7(b)), (b) $D = 245$ (middle branch), (c) $D = 244$ (upper branch), (d) $D = 400$. (After Blennerhassett, 1976.)

These results are too novel and too incomplete for a genuine physical or mathematical understanding to be available as yet. It is clear both that (4.11) and (4.12), with boundary conditions (4.46),

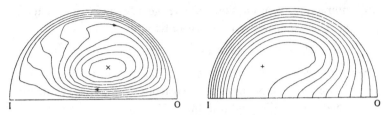

Fig. 4.9. Plots of secondary streamlines (left) and of axial velocity contours (right) representing the steady flow induced by pulsatile motion in a curved tube. I is the inside of the bend, O the outside. Here $R_s = 150$, $D = 512$ (lower branch of curve in fig. 4.7(b)). (After Blennerhassett, 1976.)

contain a wealth of fascinating information and that the problem of pulsatile flow in a curved tube is far from completely solved, even in the limit $\alpha \to \infty$.

Returning now to cases in which the Stokes layer is not necessarily the thinnest layer present, i.e. the limit $\alpha \to \infty$ is not taken before any other, we recall that the thickness of the boundary layer in steady flow at large D is $O(D^{-1/3})$, while the Stokes layer has thickness $O(\alpha^{-1})$. Thus important interactions in the boundary layer can be expected when $\alpha = O(D^{1/3})$, and again the numbers quoted on p. 183 suggest that this is a physiologically important regime $(2000^{1/3} \approx 12.6)$. As in § 4.1.3, let $w' = D^{2/3}\tilde{w}$, $\psi = D^{1/3}\tilde{\psi}$ and also set $\alpha = \tilde{\alpha}D^{1/3}$. Then in the inviscid core, the equations of motion are

$$\left. \begin{aligned} D^{1/3}\tilde{\alpha}^2\tilde{w}_\tau + (1/r)(\tilde{\psi}_\theta\tilde{w}_r - \tilde{\psi}_r\tilde{w}_\theta) &= \tilde{\alpha}^3(2R_s)^{1/2}\cos\tau + 1, \\ D^{-1/3}\tilde{\alpha}^2\nabla_1^2\tilde{\psi}_\tau - \tilde{w}\tilde{w}_y &= (D^{-2/3}/r)(\tilde{\psi}_\theta\,\partial/\partial\theta - \tilde{\psi}_\theta\,\partial/\partial r)\nabla_1^2\tilde{\psi}, \end{aligned} \right\} \quad (4.47)$$

where $y = r\sin\theta$. In the boundary layer, we make the substitution $\zeta = D^{1/3}(1-r)$ and obtain to leading order (cf. (4.25) and (4.26a))

$$\left. \begin{aligned} -\tilde{\alpha}^2\tilde{w}_\tau + \tilde{w}_{\zeta\zeta} + \tilde{\psi}_\theta\tilde{w}_\zeta - \tilde{\psi}_\zeta\tilde{w}_\theta &= -\tilde{\alpha}^3 D^{-1/3}(2R_s)^{1/2}\cos\tau, \\ -\tilde{\alpha}^2\tilde{\psi}_{\zeta\zeta\tau} + \tilde{\psi}_{\zeta\zeta\zeta\zeta} + \tilde{\psi}_\theta\tilde{\psi}_{\zeta\zeta\zeta} - \tilde{\psi}_\zeta\tilde{\psi}_{\zeta\zeta\theta} &= \sin\theta\,\tilde{w}\tilde{w}_\zeta. \end{aligned} \right\} \quad (4.48)$$

In order that the unsteady terms in (4.48) should formally be of the same order of magnitude as the others, we set $(2R_s)^{1/2} = SD^{1/3}$, although we recognise that S must be allowed subsequently to be large.

The second of equations (4.47) shows that the leading term in the expansion for the axial velocity in the core represents a non-interacting combination of the steady and the unsteady motion. The expansion can therefore be written

$$\tilde{w} = \tilde{\alpha}S \sin \tau + w_{c0}(x) + D^{-1/3} w_{c1}(x, y, \tau) + O(D^{-2/3}),$$

where $x = r \cos \theta$. The leading term in the stream function expansion, however, cannot be uncoupled from the unknown function w_{c1}; equations (4.47) yield (Smith, 1975, § 8):

$$\tilde{\psi} = \tilde{\psi}_0 + O(D^{-1/3}),$$

where

$$\tilde{\psi}_0 = \left[y - \tilde{\alpha}^2 \int_0^y w_{c1,\tau} \, dy \right] \Big/ w_{c0}'(x),$$

$$\frac{\partial}{\partial \tau}(\nabla_1^2 \tilde{\psi}_0) = \frac{(\tilde{\alpha}S \sin \tau + w_{c0})}{\tilde{\alpha}^2} w_{c1,y}.$$

Furthermore, the boundary layer equations (4.48) are intrinsically unsteady, and not even in the axial velocity can the steady and unsteady components of the flow be uncoupled. Smith (1975, § 6) was able to make some progress in the case $S \ll 1$, for then the leading steady term is the large-D steady solution, which interacts with the unsteadiness in a modified Stokes layer for which he derived the equations, but did not give a solution. For large S, however, even he made no progress, because the 'steady-streaming boundary layer' is predicted to be much thinner than the Stokes layer, so that Lyne's sequence of embedded boundary layers cannot describe the flow.

In summary, the situation for oscillatory flows in curved tubes, even when $\delta \ll 1$, is unsatisfactory, because although the purely oscillatory case is well understood, and although many cases with a mean pressure gradient have been analysed by Smith (1975) and by Blennerhassett (1976), the case of most physiological relevance is not apparently susceptible to asymptotic analysis, and has not yet received the necessary numerical treatment.

4.3 Fully developed unsteady flow starting from rest

Even if a solution were available for the case of sinusoidally oscillating flow in a curved tube with non-zero mean, at values of the parameters α, D, R_s pertinent to the aorta, it would not be directly applicable physiologically, because all the above expansions take the limit $\delta \to 0$ before any other limit, whereas in fact δ (≈ 0.2) is rather greater than any other relevant small parameter (α^{-1}, $D^{-1/3}$, $R_s^{-1/2}$, etc.). Furthermore, the pulse is not sinusoidal, and, in the intrinsically non-linear circumstances that obtain, the flow resulting from a realistic pressure gradient waveform is likely to be quite different from those already described. We have seen in § 3.2 that it is reasonable to assume that the flow in the aorta starts from rest each beat. In this section, therefore, we examine fully developed flow in a curved tube on this assumption, and hope that the results, expressed in powers of t, can be used in § 4.4 to describe the purely diffusive downstream flow to which unsteady entry flow must match. We also do not require the assumption $\delta \ll 1$, although the time for which the proposed solution is valid will be longer for smaller values of δ. The effects of slowly varying curvature can also be included in the expansion. If the curvature is non-uniform, the secondary motions cannot be represented by a stream function, and it is necessary to go back to the original governing equations, (4.1)–(4.4). Working from these equations also makes it easier to give a physical interpretation of the expansion; we therefore use them even for uniform δ. We also choose an axial velocity scale, $\hat{W}_0$, that is representative of the peak core velocity (not of viscous diffusion), so the Reynolds number, Re, is not equal to 1; we take it to be large.

4.3.1 Uniform curvature

We suppose that the flow is started at $t = 0$ in such a way that the dimensionless axial velocity on the centre line, $W_0(t)$, varies with time in a given way, such as that shown in fig. 3.3. In fact, we shall for simplicity make calculations using a crude polynomial representation of the physiological waveform, given by

$$W_0(t) = (4t/t_1)(1 - t/t_1) \quad \text{for } 0 \leqslant t \leqslant t_1$$
$$= 0 \quad \text{for } t_1 \leqslant t \leqslant t_2, \quad (4.49)$$

which has a maximum value of 1 at $t = \tfrac{1}{2}t_1$ (fig. 4.10). To correspond with the aortic waveform, t_2 would be approximately equal to $\hat{W}_0 \hat{T}/a$, where $\hat{W}_0$ is the peak axial velocity, $\hat{T}$ is the period of the

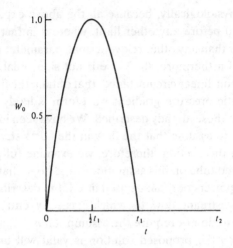

Fig. 4.10. Sketch of centre-line velocity waveform given by (4.49).

beat and a is the radius of the aorta; in a dog the values given in table 1.1 suggest that $t_2 \approx 80$, while t_1 is about a third of this value.

The sequence of events is then as follows: initially the flow everywhere except at the wall will be irrotational, so that the velocity field (for constant curvature) is

$$(u, v, w) = [0, 0, W_0(t)/h], \qquad (4.50)$$

and the corresponding pressure field is

$$p_0(r, \theta, s, t) = -s\dot{W}_0(t) - (1/2h^2)W_0^2(t), \qquad (4.51)$$

where

$$h = 1 + \delta r \cos \theta. \qquad (4.52)$$

Vorticity will immediately begin to diffuse out from the wall, and for a short time the flow can be described in the same way as in a Rayleigh layer, with the only vorticity being azimuthal and occupying a layer of thickness proportional to $(t/Re)^{1/2}$ (cf. (3.33)). In the Rayleigh layer, however, the velocity is less than W_0/h, while the pressure is still given by (4.51), and there is a pressure gradient directed from the outside of the bend towards the inside. This will drive a secondary motion in the boundary layer, from outside to inside, which will erupt back into the core at the inside wall. Initially, the axial wall shear is smaller at the outside of the bend, because W_0/h is smaller there and the boundary layer is of uniform thickness. However, the convective effect of the secondary motion will be to make the boundary layer thinner at the outside of the bend and thicker at the inside, so that the axial wall shear at the inside will decrease more rapidly than that at the outside, and after a time the shear at the outside will exceed that at the inside as in steady flow (§ 4.1). A separate effect will be the acceleration of the core flow because of the growing displacement thickness of the boundary layer. All these effects will continue throughout the cycle, although after t_1 (fig. 4.10), the forcing by $W_0(t)$ disappears, and the motion gradually diffuses away to rest. In this section we analyse the flow by representing the velocity components as power series in t, recognising that the validity of the solution will be restricted to times less than that at which higher-order terms in the expansion became as great as lower-order terms. The analysis will be developed for the case of constant curvature, and the corrections required to account for slowly varying curvature will be described afterwards.

We shall restrict attention to dimensional times short compared with the viscous diffusion time a^2/ν, because this is several times greater than the length of one cycle; in dimensionless terms this means that $t = O(1) \ll Re$. We expect that vorticity will at all times be restricted to a boundary layer of thickness proportional to $(t/Re)^{1/2}$, and therefore introduce new variables

$$\eta = (1-r)(Re/4t)^{1/2}, \qquad u' = Re^{1/2}u,$$
$$w' = w - W_0/h, \qquad p' = p - p_0,$$

the other variables remaining unchanged. The governing equations, (4.1)–(4.4), with $d\delta/ds = 0$, now become, to leading order in $Re^{-1/2}$,

$$-\frac{1}{(4t)^{1/2}}u'_\eta + v_\theta - v\frac{\delta \sin \theta}{h} = 0, \qquad (4.53)$$

$$p'_\eta = 0 \quad \text{(whence } p' \equiv 0\text{)}, \qquad (4.54)$$

$$v_t - \frac{\eta}{2t}v_\eta - \frac{u'}{(4t)^{1/2}}v_\eta + vv_\theta + \frac{\delta \sin \theta}{h}\left(\frac{2W_0 w'}{h} + w'^2\right) = -p'_\theta + \frac{1}{4t}v_{\eta\eta},$$

$$(4.55)$$

$$w_t' - \frac{\eta}{2t} w_\eta' - \frac{u'}{(4t)^{1/2}} w_\eta' + v\left(w_\theta' - w' \frac{\delta \sin \theta}{h} \right) = -\frac{1}{h} p_s' + \frac{1}{4t} w_{\eta\eta}'.$$

$$(4.56)$$

In all three equations $h = 1 + \delta \cos \theta$, and we have used the fact that $(\partial/\partial\theta)(1/h) = (\delta \sin \theta)/h^2$. The next order in $Re^{-1/2}$ would have to be included to account for the displacement effect; as in § 3.2, however, we ignore it. The boundary conditions are

$$u' = v = 0, \qquad w' = -W_0/h \quad \text{on } \eta = 0,$$

$$v, w' \to 0 \quad \text{as } \eta \to \infty.$$

The Rayleigh layer that forms immediately is described by (4.56) with $v = u' = p_s' = 0$. The solution is

$$w' = -\frac{1}{h} \frac{2}{\sqrt{\pi}} \int_\eta^\infty W_0\left(t - \frac{t\eta^2}{\mu^2} \right) e^{-\mu^2} \, d\mu, \qquad (4.57a)$$

or, for the particular form (4.49) of $W_0(t)$,

$$w' = -\frac{1}{h} \left\{ \frac{4t}{t_1} \left[\text{erfc } \eta (1 + 2\eta^2) - \frac{2}{\sqrt{\pi}} \eta \, e^{-\eta^2} \right] \right.$$

$$\left. + \frac{4t^2}{t_1^2} \left[\text{erfc } \eta (-1 - 4\eta^2 - \tfrac{4}{3}\eta^4) + \frac{2}{3\sqrt{\pi}} \eta \, e^{-\eta^2} (5 + 2\eta^2) \right] \right\}$$

$$(4.57b)$$

when $t < t_1$. As $t \to 0$, this solution and W_0 are $O(t)$, so that the secondary velocity v is $O(t^3)$ from (4.55), and the normal velocity u' must be $O(t^{7/2})$ for consistency. The effect of the secondary motion on the axial velocity profile is described by (4.56), which shows that the error in w' is $O(t^5)$, so that both terms of (4.57b) are correct. Even if we had chosen a quartic function to describe $W_0(t)$, as did Farthing (1977) in order to model the physiological waveform more closely, the quartic series derived from (4.57a) would have been valid, since the first correction is $O(t^5)$.

If we write

$$v = t^3 \sum_{n=0}^\infty t^n v_n(\eta, \theta),$$

$$u' = t^{7/2} \sum_{n=0}^\infty t^n u_n'(\eta, \theta),$$

$$w' = \frac{1}{h} t \sum_{n=0}^\infty t^n w_n'(\eta, \theta),$$

where w_0' and w_1' are independent of θ and are defined by (4.57b), and $w_2' = w_3' = 0$, we can obtain successively all the unknown functions. For example, the equation satisfied by v_0 is

$$v_{0\eta\eta} + 2\eta v_{0\eta} - 12v_0 = \frac{4\delta \sin \theta}{h^3} w_0'(w_0' + 8/t_1), \qquad (4.58)$$

which has to be solved subject to the boundary conditions $v_0 = 0$ at $\eta = 0$ and as $\eta \to \infty$. This is most easily done numerically, and the solution has been computed by Farthing (1977), who obtained

$$v_{0\eta}|_{\eta=0} = \frac{13.9}{t_1^2} \frac{\delta \sin \theta}{h^3}$$

(needed for the leading term of the secondary wall shear). Similarly, the normal velocity u_0' satisfies

$$u_{0\eta}' = 2\left(v_{0\theta} - \frac{\delta \sin \theta}{h} v_0\right),$$

and this can be integrated, subject only to $u' = 0$ at $\eta = 0$. As $\eta \to \infty$,

$$u_0' \to \frac{3.04}{t_1^2}\left(\frac{\delta \cos \theta}{h^3} + \frac{2\delta^2 \sin^2 \theta}{h^4}\right),$$

which gives rise in the core to the first displacement effect at $O(Re^{-1/2}t^{7/2})$. The equation for the first correction, w_4', to hw' is

$$w_{4\eta\eta}' + 2\eta w_{4\eta}' - 20w_4' = -2u_0'w_{0\eta}', \qquad (4.59)$$

and this must be solved subject to the same boundary conditions as (4.58). Farthing has also solved this equation, as well as those corresponding to several higher powers of t. His results for the two components of dimensionless wall shear stress are

$$\left(\frac{Re}{4t}\right)^{1/2} w_\eta'|_{\eta=0} = \left(\frac{Re}{4t}\right)^{1/2}\left\{\frac{16t}{ht_1\sqrt{\pi}}\left(1 - \frac{4t}{3t_1}\right)\right.$$

$$+ \frac{t^5}{t_1^3 h^4}\left(\delta \cos \theta + \frac{2\delta^2 \sin^2 \theta}{h}\right)$$

$$\left. \times \left(0.962 - 2.18\frac{t}{t_1}\right) + O(t^7)\right\}, \qquad (4.60)$$

$$\left(\frac{Re}{4t}\right)^{1/2} v_\eta'|_{\eta=0} = \left(\frac{Re}{4t}\right)^{1/2}\left[\frac{t^3}{t_1^2} \frac{\delta \sin \theta}{h^3}\left(13.9 - 24.4\frac{t}{t_1}\right) + O(t^5)\right].$$

$$(4.61)$$

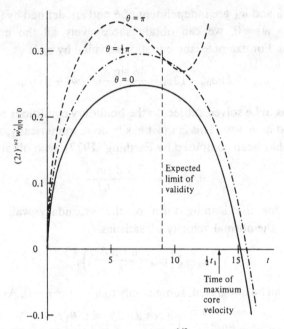

Fig. 4.11. Dimensionless wall shear ($\times Re^{-1/2}$) on the outside ($\theta = 0$), side ($\theta = \frac{1}{2}\pi$) and inside ($\theta = \pi$) walls as a function of t, calculated from (4.60) with $\delta = 0.2$, $t_1 = 27$. The theory is not expected to be accurate for $t \geqslant 9.0$.

Equation (4.60), evaluated at $\theta = 0$, $\theta = \frac{1}{2}\pi$ and $\theta = \pi$, is plotted against t in fig. 4.11, for $\delta = 0.2$, $t_1 = 27$, values appropriate to the aorta. As expected, the shear on the inner wall ($\theta = \pi$, so $h = 1 - \delta$) is initially greater than that on the outer wall ($h = 1 + \delta$), but it reaches its maximum value earlier and starts to fall rapidly; the shear on the side-wall ($\theta = \frac{1}{2}\pi$) is intermediate between the two. The shear everywhere reaches a peak long before the time of peak core velocity, $t = \frac{1}{2}t_1$. The shear on the inner wall is not predicted to reverse before that on the outer wall (despite our physical expectations) because this series solution becomes invalid before that would occur. This may be assessed by, for example, calculating the time at which $t^5\delta \cos \theta / t_1^3 h^4$ is equal to about 0.4 times $16t/t_1 h \sqrt{\pi}$ (see (4.60)) at $\theta = \pi$, when h is smallest; the critical value of t is about 9.0 ($= \frac{1}{3}t_1$). Thereafter one might expect that subsequent, neglected terms in the series become as big as those retained. This time, $t = 9.0$, which is also approximately the time at which the shear on the inner wall falls as low as that on the side-walls (fig. 4.11), corresponds in a dog's aorta to about $0.056\,\mathrm{s}$ after the beginning of systole. We may notice, however, that the series for w' and v are effectively power series in δt^3, so that they remain valid for a longer time as δ is reduced.

In addition to computing the above expansions, Farthing has performed a direct finite-difference integration of the boundary layer equations, (4.53)–(4.56). His computations confirm the series solutions exactly for sufficiently small t, and the numerical solutions also break down near $\theta = \pi$ shortly before peak systole $(t = \tfrac{1}{2}t_1)$. This breakdown is associated with a marked thickening of the boundary layer at the inside wall $(\theta = \pi)$, which is where the secondary motions begin to encroach on the previously undisturbed core flow. From that time on, the core flow will be increasingly modified and the present expansion cannot represent it. However, it gives a good description of the initial generation and modification of the secondary flow.

4.3.2 Slowly varying curvature

The above expansion can be extended to cover the case where δ varies with s over a dimensionless length-scale $1/\varepsilon \gg 1$; we can write $\delta \equiv \delta(\varepsilon s)$. The irrotational core flow is altered, because if the velocity field is given by (4.50), (4.1) is not satisfied to $O(\varepsilon)$ when h depends on εs. Therefore the core flow is given by

$$(u, v, w) = [0, 0, W_0(t)/h] + \varepsilon\delta'(\varepsilon s) W_0(t)\nabla\phi_1 + O(\varepsilon^2)$$
(4.62)

and the pressure by

$$p = p_0 + \varepsilon\delta'(\varepsilon s)p_1 + O(\varepsilon^2).$$
(4.63)

The perturbation velocity potential ϕ_1 is given by

$$\nabla^2\phi_1 = r\cos\theta/h^3, \qquad \phi_{1r} = 0 \quad \text{on } r = 1$$

where

$$\nabla^2\phi_1 = \phi_{1rr} + \frac{1}{r}\phi_{1r} + \frac{1}{r^2}\phi_{1\theta\theta} + \frac{\delta\cos\theta}{h}\phi_{1r} - \frac{\delta\sin\theta}{rh}\phi_{1\theta} + \frac{1}{h^2}\phi_{1ss},$$

together with an integral constraint representing the condition that the volume flux along the tube is uniform. The integral constraint states that

$$W_0(t) \int_0^{2\pi} \int_0^1 \left(\frac{1}{h} + \frac{\varepsilon \delta'(\varepsilon s)}{h} \phi_{1s} \right) r \, dr \, d\theta$$

should be independent of s. Now

$$h = 1 + \delta(\varepsilon s) r \cos \theta \approx 1 + \delta(0) r \cos \theta + \varepsilon s \delta'(0) r \cos \theta + O(\varepsilon^2)$$

as long as $s = O(1)$: this implies a particular choice of $s = 0$. Hence the integral constraint becomes, approximately,

$$\int_0^{2\pi} \int_0^1 \frac{\phi_{1s}}{1 + \delta_0 r \cos \theta} r \, dr \, d\theta = \int_0^{2\pi} \int_0^1 \frac{sr \cos \theta}{(1 + \delta_0 r \cos \theta)^2} r \, dr \, d\theta,$$

$$(4.64)$$

where $\delta_0 = \delta(0)$. It is clear from the formulation of this problem that it can be solved in the form

$$\phi_1 = \tfrac{1}{2} \kappa s^2 + \phi_2(r, \theta), \qquad \kappa = \text{constant}, \qquad (4.65a)$$

and (4.64) shows that

$$\kappa = (1/\delta_0)[1 - (1 - \delta_0^2)^{-1/2}] < 0. \qquad (4.66)$$

However, the explicit solution for ϕ_2 can apparently be derived only as a power series in δ_0. The first two terms in ϕ_1 are

$$\phi_1 = \tfrac{1}{8}(r^3 - 3r) \cos \theta + \delta_0 [-\tfrac{1}{4} s^2 + (2r^2 - r^4)(\tfrac{7}{64} + \tfrac{13}{96} \cos 2\theta)] + O(\delta_0^2).$$

$$(4.65b)$$

The pressure perturbation p_1, in (4.63), is

$$p_1 = -\dot{W}_0(t)\phi_1 - (W_0^2/h^2)\kappa s. \qquad (4.67)$$

We can now, in principle, calculate the $O(t)$ perturbations to the boundary layer solutions given in § 4.3.1. There will be a straightforward perturbation to the primary Rayleigh layer (4.57) because of the altered axial core velocity. However, the presence of the $O(t)$ secondary motions in the core means that there will be secondary motions in the boundary layer from the beginning, with $v = O(t)$, which even for a very small value of ε might make themselves felt before those of $O(t^3)$ driven by centrifugal forces associated with the primary Rayleigh layer. These will also cause a correction to w at $O(t^3)$, not $O(t^5)$ as in § 4.3.1. Here we restrict our attention to a brief analysis of the $O(t)$ swirl velocity, v, and the corrections of w to $O(t^3)$.

The boundary layer equations are (4.53)–(4.56) with p' not identically zero, but equal to $\varepsilon\delta'p_1$, given by (4.67). The outer boundary conditions are now

$$v \sim \varepsilon\delta'W_0\phi_{1\theta}|_{r=1}, \qquad w' = (\varepsilon\delta'W_0/h)\phi_{1s} \quad \text{as } \eta \to \infty.$$

Write

$$v = \varepsilon\delta'\Phi(\theta)\tilde{v}(t, \eta), \tag{4.68a}$$

where

$$\Phi(\theta) = \phi_{1\theta}|_{r=1}, \tag{4.69}$$

and the equation (4.55) becomes a Rayleigh layer equation for $\tilde{v}$:

$$\tilde{v}_{\eta\eta} + 2\eta\tilde{v}_\eta - 4t(\tilde{v}_t - \dot{W}_0) = O(t^3),$$

with solution

$$\tilde{v} - W_0(t) = -\frac{2}{\sqrt{\pi}} \int_\eta^\infty W_0\left(t - \frac{t\eta^2}{\mu^2}\right) e^{-\mu^2} \, d\mu + O(t^3), \tag{4.68b}$$

which is the same as (4.57a) and can be evaluated as in (4.57b). To $O(t^2)$,

$$\tilde{v} - W_0(t) = tw_0'(\eta) + t^2w_1'(\eta).$$

The corresponding correction $\tilde{u}'$ to the normal velocity u' is, from (4.53), given by

$$\frac{1}{(4t)^{1/2}}\tilde{u}_\eta' = \varepsilon\delta'\left[\Phi'(\theta) - \frac{\delta_0 \sin\theta}{h}\Phi(\theta)\right]\tilde{v}(t, \eta). \tag{4.70}$$

The first corrections to w' are of two kinds. One is needed to account for the correction to the axial velocity in the core, and can be effected by multiplying (4.57a or b) by $1 + \varepsilon\delta'\phi_{1s}$ ($= 1 + \varepsilon\delta'\kappa s$ from (4.65a)); here w' is taken to be the difference between w and the *corrected* core velocity. These constitute corrections at $O(t)$ and $O(t^2)$. The other kind of correction involves terms at $O(t^3)$ and $O(t^4)$, driven by $\tilde{v}$ via $\tilde{u}$. If this correction is written as

$(\varepsilon\delta'/h)\tilde{w}'(t, \theta, \eta)$, (4.56) gives

$$\tilde{w}'_{\eta\eta} + 2\eta\tilde{w}'_{\eta} - 4t\tilde{w}'_{t} = -\left[\Phi'(\theta) - \frac{\delta\sin\theta}{h}\Phi(\theta)\right]$$

$$\times 4t^2(w'_{0\eta} + tw'_{1\eta})\int_0^{\eta} \tilde{v}\, d\eta + O(t^5), \quad (4.71)$$

where the product of the $O(t^2)$ term in $\int \tilde{v}\, d\eta$ with the $O(t)$ term in the preceding bracket is not to be calculated, being incorporated into the $O(t^5)$ term.

Equation (4.71) has also been integrated, subject to the usual boundary conditions, by Farthing (1977), and he has found that the correction to the axial component of dimensionless wall shear stress is

$$\left(\frac{Re}{4t}\right)^{1/2}\frac{\varepsilon\delta'}{h}\left[\Phi'(\theta) - \frac{\delta\sin\theta}{h}\Phi(\theta)\right]\frac{t^3}{t_1^3}\left(1.45 - 2.43\frac{t}{t_1}\right).$$

$$(4.72a)$$

As yet this can be evaluated only to $O(\delta_0)$, from (4.69) and (4.65b), whence $1/h$ times the square bracket is equal to

$$\tfrac{1}{4}[\cos\theta - \tfrac{19}{6}\delta_0(1 - \tfrac{26}{19}\sin^2\theta) + O(\delta_0^2)]. \quad (4.72b)$$

These results show that when δ is increasing (and small) the axial wall shear initially increases on $\theta = 0$ and decreases on $\theta = \pi$. This is due to the thinning of the boundary layer at $\theta = 0$ and the thickening at $\theta = \pi$ consequent upon the normal inflow/outflow there: as $\eta \to \infty$ an integration of (4.70) yields

$$\frac{1}{(4t)^{1/2}}\tilde{u}' \sim \varepsilon\delta'\left[\Phi'(\theta) - \frac{\delta_0\sin\theta}{h}\Phi(\theta)\right]\frac{t}{t_1}\left(18.5 - 18.8\frac{t}{t_1}\right),$$

which is initially positive when δ_0 is small, δ' is positive and $\theta = 0$. Later, when $t/t_1 \approx 0.6$, the sign of the correction to the axial wall shear stress is predicted to change, from (4.72a), but the small-t expansions are almost certainly invalid by then (see § 4.3.1). Note that at $\theta = \pm\tfrac{1}{2}\pi$, where $\cos\theta = 0$, the $\sin^2\theta$ term in (4.72b) indicates an initial slight increase in wall shear when $\delta' > 0$. These effects are independent both of the uniform decrease in shear associated with the deceleration of the axial velocity in the core as δ increases (from the $\frac{1}{2}\kappa s^2$ term in (4.65a), since $\kappa < 0$) and of the deceleration/acceleration at the outer/inner wall consequent upon the change in h as δ_0 increases.

Farthing (1977) has taken the above expansions somewhat further, and has also examined the effect of torsion of the tube axis, which is not negligible in the aorta. He made calculations for a tube with a right-handed twist, from the point of view of an observer looking along the tube in the downstream direction with the instantaneous plane of curvature horizontal and with the centre of curvature to the right. He found, not unexpectedly, that a right-handed spiral motion was superimposed on the whole flow, resulting in a deceleration of the axial flow in the top half of the tube and an acceleration in the bottom half. The axial wall shear was decreased in the top half and increased in the bottom half, while the reverse was true of the azimuthal wall shear.

4.4 Entry flow with a flat entry profile

The analysis of the previous section has taken the theory of steady and unsteady fully developed flow in a curved tube about as far as possible without a major numerical effort, involving the full Navier–Stokes equations, not boundary layer approximations to them. We still lack a complete description of the flow for values of δ that are not vanishingly small, except during the early part of systole, just after the motion begins. Nevertheless, the main qualitative features of the flow, including the initiation of secondary motions and the consequent modification of the axial velocity profile, are fairly clear. This section is devoted to flow near the entrance of a curved tube, and the aim is to extend the unsteady theory of § 3.2 to take account of (possibly non-uniform) vessel curvature. Far downstream from the entrance, the flow is expected to tend to one of the fully developed flows already described. This section is concerned with steady or unsteady flow that enters a curved tube from a reservoir of uniform total pressure, with an effectively flat velocity profile; this may be applicable to the entrance to the aorta, and there are model experiments in uniformly curved tubes with which the theory can be compared. The next section examines steady flow in a tube that is initially straight (and contains Poiseuille flow) but starts curving at some given point. This example is included not because there is an immediate application to a particular artery, but because (a) it sheds light on the interaction between Poiseuille flow and non-uniformities in the tube, a problem that is examined further in the following chapter on branching tubes, and (b) it can be extended to analyse the flow in a tube that becomes straight after experiencing a bend, as in the descending aorta.

Here we suppose the flow to enter the tube from a reservoir of uniform total pressure, so that the inviscid core flow is expected to be the same irrotational flow as that already derived for a fully developed flow starting from rest. Thus in the core, using the same non-dimensionalisation as in § 4.3, and considering a tube with non-uniform curvature $\delta(\varepsilon s)$, we have from (4.62), to $O(\varepsilon)$,

$$(u_c, v_c, w_c) = \left[0, 0, \frac{W_0(t)}{h_0}\left(1 - \frac{\varepsilon s \delta_0' r \cos \theta}{h_0}\right)\right]$$

$$+ \varepsilon \delta_0' W_0(t) \nabla \phi_1, \qquad (4.73)$$

where $h_0 = 1 + \delta_0 r \cos \theta$, $\delta_0 = \delta(0)$, $\delta_0' = \delta'(0)$ and ϕ_1 is given to $O(\delta_0)$ by (4.65b). The entrance to the tube is at $s = 0$ and a suffix 'c' refers to the inviscid core. For the case of uniform curvature δ_0, this reduces to $[0, 0, W_0(t)/h_0]$. The corresponding pressure is, from (4.51) and (4.66),

$$p_c = -s\dot{W}_0(t) - \frac{W_0^2(t)}{2h_0^2}$$

$$- \varepsilon \delta_0'\left[\dot{W}_0(t)\phi_1 + \frac{W_0^2(t)s}{h_0^2}\left(\kappa - \frac{r \cos \theta}{h_0}\right)\right]. \qquad (4.74)$$

We shall confine attention to functions $W_0(t)$ that are positive for extended periods of time, so that an approximately quasi-steady flow can be realised.

The mechanism by which secondary motions are generated, and the axial flow modified, as s increases is physically very similar to the mechanism described above for s-independent flow as t increases. For the moment consider constant and uniform δ. For very small s, the primary action of viscosity is to generate a Blasius boundary layer in which the axial velocity, in (4.73), is reduced to zero. At a given point on the tube wall, this will have thickness $\tilde{\Delta}$ proportional to $[\nu s_1(s, \theta)/W_\infty(s, \theta)]^{1/2}$ where s_1 is the distance to the point in question, greater along the outside wall of the tube than the inside, and W_∞ is the axial velocity just outside the boundary layer, equal to $w_c|_{r=1}$ and greater on the inner wall. Thus the wall shear, proportional to $W_\infty/\tilde{\Delta}$, will be greater on the inner wall. However, the transverse pressure gradient (p_{cr}, $p_{c\theta}/r$), required to balance the centrifugal force in the core, also acts on the more slowly moving fluid in the Blasius boundary layer and generates a secondary velocity v there, from the outside of the bend to the inside. This has the effect of relatively thinning the boundary layer at the outside and thickering it at the inside, so that after a certain distance the shear distribution may be expected to reverse and to tend towards that in steady flow. This has been confirmed by the detailed boundary layer analysis of Singh (1974). We here reproduce the essential elements of Singh's analysis, modified so as not to require vanishingly small δ, and extended to take account of unsteady flow and non-uniform curvature. Both extensions are important in the aorta, because the flow is highly unsteady and because the ascending aorta is not very greatly curved initially, the main curvature of the arch developing after 2 or 3 cm.

The full governing equations are again (4.1)–(4.4), with Re again taken to be large. We transform them to the boundary layer form appropriate for quasi-steady flow (§ 3.2) by changing the variables as follows:

$$\eta = Re^{1/2}[(1-r)/h_0][W_0(t)/2s]^{1/2},$$

$$u = (Re^{-1/2}/h_0)[W_0(t)/2s]^{1/2}u_1, \qquad v = W_0(t)v_1, \qquad (4.75)$$

$$w = [W_0(t)/h_0]w_1, \qquad h_0 = 1 + \delta_0 \cos \theta.$$

The inclusion in these scales of the θ-dependent factor h_0 incorporates the facts that both W_∞ and $1/\tilde{\Delta}$ are larger at the outside of the bend than the inside, and mean that the leading order Blasius flow is a function only of η. This greatly simplifies the subsequent analysis (like the inclusion of $W_0(t)$ to describe quasi-steady flow). Failure to make this choice of scaling was probably the reason why Singh (1974) confined his analysis to small values of δ_0.

The governing equations now are

$$u_{1\eta} + \eta w_{1\eta} - 2sw_{1s} = 2sh_0^2\left[v_{1\theta} + \frac{\delta_0 \sin\theta}{h_0}(\eta v_{1\eta} - v_1)\right]$$

$$+ \frac{\varepsilon s\delta_0' \cos\theta}{h_0}(\eta w_{1\eta} - 2sw_{1s})$$

$$- 2\varepsilon\delta_0's^2 \sin\theta\, v_1, \qquad (4.76)$$

$$w_{1\eta\eta} + u_1 w_{1\eta} + w_1(\eta w_{1\eta} - 2sw_{1s}) - \frac{2sh_0^2}{W_0^2(t)} p_{cs}$$

$$= 2sh_0^2\left[\frac{\dot{W}_0}{W_0^2}(w_1 + \tfrac{1}{2}\eta w_{1\eta}) + \frac{\dot{w}_1}{W_0} + v_1\left(w_{1\theta} + \frac{\delta_0 \sin\theta}{h_0}\eta w_{1\eta}\right)\right]$$

$$+ \frac{\varepsilon s\delta_0' \cos\theta}{h_0} w_1(\eta w_{1\eta} - 2sw_{1s}) - 2\varepsilon s^2\delta_0' \sin\theta\, v_1 w_1, \qquad (4.77)$$

$$v_{1\eta\eta} + u_1 v_{1\eta} + w_1(\eta v_{1\eta} - 2sv_{1s}) - \frac{2sh_0^2}{W_0^2(t)} p_{c\theta}$$

$$= 2sh_0^2\left[\frac{\dot{W}_0}{W_0^2}(v_1 + \tfrac{1}{2}\eta v_{1\eta}) + \frac{\dot{v}_1}{W_0} + v_1\left(v_{1\theta} + \frac{\delta_0 \sin\theta}{h_0}\eta v_{1\eta}\right)\right.$$

$$\left. + \frac{\delta_0 \sin\theta}{h_0^3} w_1^2\right]$$

$$+ \frac{\varepsilon s\delta_0' \cos\theta}{h_0} w_1(\eta v_{1\eta} - 2sv_{1s}) + 2\varepsilon s^2\delta_0' \sin\theta\, \frac{w_1^2}{h_0^2}. \qquad (4.78)$$

In the last two equations, the pressure terms can be derived from (4.74), and include both $O(1)$ and $O(\varepsilon)$ terms. The other $O(\varepsilon)$ terms arise because, in (4.1)–(4.4), δ and h are not equal to δ_0 and h_0 respectively. The boundary conditions are that w and v should tend to w_c, v_c, from (4.73), as $\eta \to \infty$, and u, v, w are all zero on $\eta = 0$.

The scaling has been performed in such a way that, as $s \to 0$ and for $\varepsilon = 0$, the solution is obviously the Blasius solution,

$$w_1 = w_{10}(\eta) = f_0'(\eta), \qquad u_1 = u_{10}(\eta) = f_0 - \eta f_0', \qquad v_1 = 0. \tag{4.79}$$

For non-zero but small s and ε, let us write

$$w_1 = w_{10}(\eta) + \sum_{n=1}^{\infty} s^n (w_{1n} + \varepsilon \bar{w}_{1n} + \cdots),$$

$$u_1 = u_{10}(\eta) + \sum_{n=1}^{\infty} s^n (u_{1n} + \varepsilon \bar{u}_{1n} + \cdots), \qquad (4.80)$$

$$v_1 = \varepsilon \bar{v}_{10}(\eta, \theta) + \sum_{n=1}^{\infty} s^n (v_{1n} + \varepsilon \bar{v}_{1n} + \cdots),$$

where the leading terms for w_1 and u_1 are the Blasius solution. However, there is an $O(\varepsilon)$ swirl velocity to account for (a) the outer boundary condition

$$v_1 \to \varepsilon \delta_0' \Phi(\theta) \quad \text{as } \eta \to \infty,$$

where $\Phi(\theta)$ is defined by (4.68), and (b) the corresponding term in the azimuthal pressure gradient $p_{c\theta}$. Writing

$$\bar{v}_{10} = \delta_0' \Phi(\theta) g(\eta) \qquad (4.81)$$

and substituting into (4.78), we see that $g(\eta)$ satisfies

$$g'' + f_0 g' = 0, \qquad g(0) = 0, \qquad g(\infty) = 1,$$

which has the unique solution $g(\eta) = f_0'(\eta)$.

4.4.1 *Uniform curvature*

We consider first the $\varepsilon = 0$ solutions for $s > 0$. The first corrections to the axial and radial velocities, w_{11} and u_{11}, turn out to be zero in steady flow. In unsteady flow they are the same as in a straight tube or on a flat plate, and were calculated in § 3.2.2. In the notation of this section the solution is

$$w_{11} = \frac{h_0^2(\theta) \dot{W}_0(t)}{W_0^2(t)} f_{11}'(\eta),$$

$$\qquad (4.82)$$

$$u_{11} = \frac{h_0^2(\theta) \dot{W}_0(t)}{W_0^2(t)} (3f_{11} - \eta f_{11}'),$$

with $f_{11}(\eta)$ given by (3.13).

Even in steady flow, a swirl velocity sv_{11} develops at $O(s)$. This is given by

$$v_{11} = \frac{\delta_0 \sin \theta}{h_0} g_{11}(\eta), \qquad (4.83)$$

where

$$g''_{11} + f_0 g'_{11} - 2f'_0 g_{11} = 2(f'^2_0 - 1), \qquad g_{11}(0) = g_{11}(\infty) = 0.$$

This equation was solved numerically by Singh (1974), but he did not report the value of $g'_{11}(0)$ (>0), required for a knowledge of secondary shear stress. Farthing (1977) has recomputed the function, and finds $g'_{11}(0) = 1.54$. Note that (4.83) differs from Singh's $O(s)$ swirl velocity only through the factor $1/h_0$, although no assumption of small δ_0 has been made.

The first effect of curvature on the axial and radial velocities (in the case $\varepsilon = 0$ still) comes in at $O(s^2)$ in the terms w_{12}, u_{12}. These terms satisfy

$$u_{12\eta} + \eta w_{12\eta} - 4w_{12} = 2h_0 \delta_0 \cos \theta \, g_{11} + 2\delta_0^2 \sin^2 \theta \, \eta g'_{11},$$

$$\begin{aligned} w_{12\eta\eta} + f_0 w_{12\eta} + (\eta f''_0 - 4f'_0)w_{12} + f''_0 u_{12} \\ = (h_0^4 \dot{W}_0^2 / W_0^4)(-3f_{11}f''_{11} + 2f'^2_{11} - 2f'_{11} + \eta f''_{11}) \\ + (2h_0^4 \ddot{W}_0 / W_0^3)f'_{11} + 2\delta_0^2 \sin^2 \theta \, \eta f''_0 g_{11}, \end{aligned}$$

with the usual boundary conditions. Still there is no important interaction between the unsteadiness and the curvature (this comes in first in the terms $s^2 v_{12}$ and $s^3 w_{13}$ etc.), and the solution for w_{12}, u_{12} can be expressed as the sum of three terms, two of which are essentially the same as the $O(x^2)$ terms in the solution for unsteady flow over a flat plate (§ 3.2.2), and one of which accounts for curvature. We can write:

$$\begin{aligned} w_{12} &= (h_0^4 \dot{W}_0^2 / W_0^4)f'_{21}(\eta) + (h_0^4 \ddot{W}_0 / W_0^3)f'_{22}(\eta) \\ &\quad + \delta_0(h_0 \cos \theta - \delta_0 \sin^2 \theta)f'_{12}(\eta), \\ u_{12} &= (h_0^4 \dot{W}_0^2 / W_0^4)(5f_{21} - \eta f'_{21}) + (h_0^4 \ddot{W}_0 / W_0^3)(5f_{22} - \eta f'_{22}) \qquad \text{(4.84)} \\ &\quad + \delta_0(h_0 \cos \theta - \delta_0 \sin^2 \theta)\left(5f_{12} - \eta f'_{12} + 2\int_0^\eta g_{11}\, d\eta\right) \\ &\quad + 2\delta_0^2 \sin^2 \theta \, \eta g_{11}. \end{aligned}$$

The functions $f_{21}(\eta)$ and $f_{22}(\eta)$ are the same as those arising in § 3.2.2 in (3.13), and their contributions to the wall shear, $f''_{21}(0)$ and $f''_{22}(0)$, are given in table 3.1. The function $f_{12}(\eta)$ satisfies the equation

$$f_{12}''' + f_0 f_{12}'' - 4f_0' f_{12}' + 5f_0'' f_{12} = -2f_0'' \int_0^\eta g_{11}\, d\eta,$$

together with boundary conditions $f_{12}(0) = f_{12}'(0) = f_{12}'(\infty) = 0$. This function was computed by Singh (1974), and he obtained $f_{12}''(0) = 0.256$, $f_{12}(\infty) = 0.522$. The contributions to the axial and wall shear stress from all the functions f_{mk}, $\bar{f}_{mk}$, g_{mk}, $\bar{g}_{mk}$ arising in this section (e.g. $f_{mk}''(0)$, $g_{mk}'(0)$) are listed in table 4.1.

Table 4.1. *Contributions to axial (f) and azimuthal (g) wall shear stress from all functions arising in § 4.4, together with the terms in the expansion (4.80) to which they refer*

Term	w_{10}	w_{11}	w_{12}	w_{12}	w_{12}	w_{13}	w_{13}
Function $F(\eta)$	f_0'	f_{11}'	f_{21}'	f_{22}'	f_{12}'	f_{31}'	f_{32}'
$F'(0)$	0.470	1.200	0.383	−0.664	0.256	−0.839	1.008

Term	$\bar{w}_{11}$	$\bar{w}_{11}$		v_{11}	v_{12}	$\bar{v}_{10}$	$\bar{v}_{11}$
Function $F(\eta)$	$\bar{f}_{11}'$	$\bar{f}_{12}'$		g_{11}	g_{12}	g	$\bar{g}_{11}$
$F'(0)$	0.117	2.886		1.536	−1.196	0.470	1.200

The first real interaction between curvature and unsteadiness comes in the next terms of each of the series (4.80). The solutions are written below, *excluding* those terms in u_{12} and w_{12} that come purely from the $O(s^3)$ terms in the expansions representing the effect of time-dependence on the flat plate boundary layer (and which are written as 't.d.t.' below). We have

$$v_{12} = \delta_0 h_0 \sin\theta\, (\dot{W}_0 / W_0^2) g_{12}(\eta),$$

$$w_{13} = \delta_0 h_0^2 (h_0 \cos\theta - \delta_0 \sin^2\theta)(\dot{W}_0 / W_0^2) f_{31}'$$
$$+ \delta_0^2 h_0^2 \sin^2\theta\,(\dot{W}_0 / W_0^2) f_{32}' + \text{t.d.t.}, \qquad (4.85)$$

$$u_{13} = \delta_0 h_0^2 (h_0 \cos\theta - \delta_0 \sin^2\theta)\frac{\dot{W}_0}{W_0^2}\Big(7 f_{31} - \eta f_{31}' + 2\int_0^\eta g_{12}\, d\eta\Big)$$
$$+ \delta_0^2 h_0^2 \sin^2\theta\, \frac{\dot{W}_0}{W_0^2}\Big(7 f_{32} - \eta f_{32}' + 2\eta g_{12} - 4\int_0^\eta g_{12}\, d\eta\Big) + \text{t.d.t.},$$

where

$$g_{12}'' + f_0 g_{12}' - 4f_0' g_{12} = 2g_{11} f_{11}' - 3f_{11} g_{11}' + 4f_0' f_{11}' + \eta g_{11}' + 2g_{11},$$

$$f_{3k}''' + f_0 f_{3k}'' - 6f_0' f_{3k}' + 7f_0'' f_{3k} = \mathscr{F}_{3k} \quad (k = 1, 2),$$

$$\mathscr{F}_{31} = 2f_{12}' + \eta f_{12}'' - 3f_{12}'' f_{11} - 5f_{11}'' f_{12} + 6f_{11}' f_{12}'$$

$$- 2f_{11}'' \int_0^\eta g_{11} \, d\eta - 2f_0'' \int_0^\eta g_{12} \, d\eta,$$

$$\mathscr{F}_{32} = -4f_{11}' g_{11} + 4f_0'' \int_0^\eta g_{12} \, d\eta,$$

and g_{12}, f_{3k} all satisfy the usual homogeneous boundary conditions. Numerical solution of these equations yields

$$g_{12}'(0) = -1.196, \qquad f_{31}''(0) = -0.839, \qquad f_{32}''(0) = 1.008$$

$$f_{31}(\infty) = -1.665, \qquad f_{32}(\infty) = 1.589.$$

In quasi-steady flow $(s\dot{W}_0/W_0^2 \ll 1)$ the axial wall shear is proportional to

$$-w_r|_{r=1} = (Re\,W_0^3/2s)^{1/2}(1/h_0^2)$$

$$\times [f_0''(0) + s^2 \delta_0(h_0 \cos\theta - \delta_0 \sin^2\theta)f_{12}''(0) + O(s^4)]$$

$$= (Re\,W_0^3/2s)^{1/2}(1/h_0^2)$$

$$\times [0.470 + 0.256s^2 \delta_0(\cos\theta + \delta_0 \cos 2\theta) + O(s^4)].$$

$$(4.86)$$

Because of the factor $h_0^{-2} = (1 + \delta_0 \cos\theta)^{-2}$, this is initially greater on the inside wall $(\theta = \pi)$ than on the outside wall $(\theta = 0)$. However the $O(s^2)$ term is greater on the outside wall, and (assuming that the $O(s^4)$ term can be neglected) cross-over is predicted at a value of s given by

$$s = 1.92(1 - \delta_0^2)^{-1/2}. \qquad (4.87)$$

Singh (1974) also found cross-over at $s \approx 1.9$ in the small-δ_0 limit; the result (4.87), however, makes no assumptions about the smallness of δ_0 except that $\delta_0 < 1$. The above expansion is valid only for small values of s; in particular, if δ_0 is quite small we can expect validity for $s \ll \delta_0^{-1/2}$. When δ_0 is given by its maximum value in the arch of the aorta, about 0.2, the expansion is likely to be qualitatively useful only for $s \lesssim 2$, i.e. for about one diameter from the entrance; the cross-over value (4.87) of s is in that case predicted to be about 1.95, so the prediction is unlikely to be quantitatively accurate, although it would be for smaller δ_0. In the very first part of the aorta, δ_0 is markedly smaller than 0.2, so the prediction may be accurate, subject to modifications introduced by non-uniformity of curvature.

When the flow is not quasi-steady, additional terms appear in the square brackets in (4.86). Some are the same unsteady flow corrections as were computed in § 3.2.2, multiplied by functions of θ; these terms, times the h_0^{-2} outside the square bracket, are:

$$s(\dot{W}_0/W_0^2)f_{11}''(0)+s^2h_0^2[(\dot{W}_0^2/W_0^4)f_{21}''(0)+(\ddot{W}_0/W_0^3)f_{22}''(0)]+O(s^3).$$

Thus the effect of curvature is felt at $O(1)$ and at $O(s^2)$ even without a genuine interaction between curvature and unsteadiness, because of the factors h_0^{-2} and h_0^2 respectively. The first term that represents the interaction is

$$h_0^{-2}\cdot s^3\delta_0h_0^2(\dot{W}_0/W_0^2)[(h_0\cos\theta-\delta_0\sin^2\theta)f_{31}''(0)+\delta_0\sin^2\theta f_{32}''(0)]$$

$$=s^3\delta_0(\dot{W}_0/W_0^2)[-0.839h_0\cos\theta+1.847\delta_0\sin^2\theta].\qquad(4.88)$$

The square bracket here is positive at the inside wall ($\theta=\pi$) and negative at the outside wall ($\theta=0$); the value of θ at which it is zero depends on δ_0, but is always less than $\frac{1}{2}\pi$ (when $\delta_0=0.2$, for example, it is about 75°). Thus, when the flow is accelerating ($\dot{W}_0>0$), this term is of opposite sign to the $O(s^2)$ term in (4.86), and therefore tends to inhibit the cross-over of maximum shear predicted above. When the flow is decelerating, on the other hand, this interaction tends to promote cross-over at a smaller value of s.

The leading terms in the expansion for azimuthal wall shear are proportional to

$$-v_r|_{r=1}=\left(\frac{ReW_0^3}{2s}\right)^{1/2}s\frac{\delta_0\sin\theta}{h_0^2}\left[g_{11}'(0)+s\frac{h_0^2\dot{W}_0}{W_0^2}g_{12}'(0)\right].$$

$$(4.89)$$

As one would predict from the physical origin of the secondary motions, this is positive as $s\to0$ for all $\theta\neq0$, π. However, $g_{12}'(0)$ is negative, so the effect of an acceleration of the flow ($\dot{W}_0>0$) is to cause the secondary component of wall shear to fall below its steady value as s increases. Ultimate reversal would be predicted by (4.89) but validity of the equation for values of s greater than 1–2 is unlikely.

The above expansion appears to break down when $s = O(\delta_0^{-1/2})$ because the centrifugal force term which drives the azimuthal motion in (4.78) becomes $O(\delta_0^{1/2})$. Hence v_1 also becomes $O(\delta_0^{1/2})$, with the consequence that the first term on the left-hand side of the continuity equation (4.76) becomes $O(1)$, modifying the basic w_1, u_1 boundary layer. Singh (personal communication) has stated that some analytical progress is possible in this region ($S = s\delta_0^{1/2} = O(1)$) when $\delta_0 \ll 1$, but he has been unable to compute any results from the intractable non-linear equations that arise (except for a series solution in powers of S that contains the same terms as the above expansion in powers of s).

Yao & Berger (1975) investigated the steady entry flow in a uniformly curved pipe (at large values of the Dean number) by a complicated momentum integral method. However the fully developed flow to which their solution was made to tend was assumed to have parallel secondary streamlines in the core (as in the attempted solution by Barua (1963)), and this overspecifies the problem as shown in § 4.1.3 (it is equivalent to assuming $w'_c(x) \equiv 1$ in (4.24)). Thus their results show separated secondary flow, unlike the fully developed momentum integral solution of Ito (1969), which would have been a better downstream limit.

Yao & Berger's analysis leads to a prediction of entry length $\hat{s}_e$ approximately proportional to $aRe^{1/2}\delta^{-1/4}$ (compared with aRe in a straight tube; $Re = \hat{W}_0 a/\nu$). This scaling can be deduced as follows. The simplest order-of-magnitude argument suggests that the entry length should be the value of $\hat{s}$ at which the thickness of the Blasius boundary layer at the entrance, $\propto (\nu\hat{s}/\hat{W}_0)^{1/2}$, is comparable with that in fully developed, large-Dean-number flow far downstream, $\propto aD^{-1/3}$. Now $\hat{W}_0$ is related to D through the scaling of § 4.1.3, where

$$w' = (2\delta)^{1/2}(a\hat{W}_0/\nu) = O(D^{2/3}). \qquad (4.90)$$

This leads to $\hat{s}_e/a \propto \delta^{-1/2}$, which is very much smaller than Yao & Berger's estimate, and indeed is equivalent to $s = O(\delta^{-1/2})$, where

Singh's (1974) entry solution first breaks down. However, this simple prediction fails to account for the time required for the transport of secondary vorticity from the boundary layer into the core, which will require a time of $a/\hat{V}$ to develop, where $\hat{V}$ is a scale for the secondary velocity in the core. Hence a distance of $\hat{W}_0 a/\hat{V}$ will be needed. Now $\hat{V} = (\nu/a)D^{1/3}$ (see § 4.1.3), so this gives

$$\hat{s}_e/a = \hat{W}_0/\hat{V} = Re/D^{1/3} = Re^{1/2}\delta^{-1/4} \qquad (4.91)$$

from (4.90). The experiments reported in § 4.4.3 below confirm that this estimate of entry length is much nearer the truth than either $a\delta^{-1/2}$ or aRe.

4.4.2 Slowly varying curvature

We now turn to the first corrections that must be included in the above expansions in order to describe the effect of non-uniform curvature. These are the terms $\varepsilon s(\bar{u}_{11}, \bar{v}_{11}, \bar{w}_{11})$ in (4.80). We neither restrict the flow to being quasi-steady, nor do we limit the analysis to small values of δ_0. When account is taken of the solutions for $u_{10}, w_{10}, u_{11}, w_{11}, \bar{v}_{10}$, given in (4.79), (4.81) and (4.82), the equations for the present variables, from (4.76) to (4.78), are

$$\bar{u}_{11\eta} + \eta\bar{w}_{11\eta} - 2\bar{w}_{11} = \eta f_0''(\eta)(\delta_0'/h_0)[\cos\theta + 2h_0^2\delta_0\sin\theta\,\Phi(\theta)]$$

$$+ f_0'(\eta)2\delta_0'h_0^2\left[\Phi'(\theta) - \frac{\delta_0\sin\theta}{h_0}\Phi(\theta)\right],$$

$$\bar{w}_{11\eta\eta} + f_0\bar{w}_{11\eta} + (\eta f_0'' - 2f_0')\bar{w}_{11} + f_0''\bar{u}_{11} + 2\delta_0'\left(\kappa - \frac{\cos\theta}{h_0}\right)$$

$$= \eta f_0'f_0''(\delta_0'/h_0)[\cos\theta + 2h_0^2\delta_0\sin\theta\,\Phi(\theta)],$$

$$\bar{v}_{11\eta\eta} + f_0\bar{v}_{11\eta} - 2f_0'\bar{v}_{11} = \delta_0'h_0^2[\dot{W}_0(t)/W_0^2(t)]\Phi(\theta)$$

$$\times[-3f_0''f_{11} + 2(f_0' - 1) + \eta f_0''],$$

with the usual boundary conditions at $\eta = 0$, and

$$\bar{w}_{11} \to \delta_0'\left(\kappa - \frac{\cos\theta}{h_0}\right), \qquad \bar{v}_{11} \to 0 \quad \text{as } \eta \to \infty.$$

We recall that $\Phi(\theta)$ is given by (4.69) and (4.65), while κ is given by (4.66); these quantities represent the perturbation to the irrotational core flow. These equations have the following solutions:

$$
\left.
\begin{aligned}
\bar{w}_{11} &= \delta_0'\left[\frac{\cos\theta}{h_0} + 2h_0^2\Phi'(\theta)\right]\bar{f}_{11}'(\eta) + \delta_0'\left(\kappa - \frac{\cos\theta}{h_0}\right)\bar{f}_{12}'(\eta), \\
\bar{u}_{11} &= \delta_0'\left[\frac{\cos\theta}{h_0} + 2h_0^2\Phi'(\theta)\right](3\bar{f}_{11} - \eta\bar{f}_{11}') \\
&\quad + (\delta_0'/h_0)[\cos\theta + 2h_0^2\delta_0\sin\theta\,\Phi(\theta)](\eta f_0' - f_0) \\
&\quad + 2\delta_0' h_0^2\left[\Phi'(\theta) - \frac{\delta_0\sin\theta}{h_0}\Phi(\theta)\right]f_0(\eta) \\
&\quad + \delta_0'\left(\kappa - \frac{\cos\theta}{h_0}\right)(3\bar{f}_{12} - \eta\bar{f}_{12}'), \\
\bar{v}_{11} &= \delta_0' h_0^2\Phi(\theta)(\dot{W}_0(t)/W_0^2(t))\bar{g}_{11}(\eta),
\end{aligned}
\right\}
\tag{4.92}
$$

where the functions $\bar{f}_{11}$ and $\bar{f}_{12}$ satisfy

$$
\bar{f}_{1k}''' + f_0\bar{f}_{1k}'' - 2f_0'\bar{f}_{1k}' + 3f_0''\bar{f}_{1k}
\begin{aligned}
&= -f_0 f_0'' \quad (k=1) \\
&= -2 \qquad\;\; (k=2)
\end{aligned}
\tag{4.93}
$$

and $\bar{g}_{11}$ satisfies

$$
\bar{g}_{11}'' + f_0\bar{g}_{11}' - 2f_0'\bar{g}_{11} = -3f_0''f_{11} + 2(f_0' - 1) + \eta f_0''. \tag{4.94}
$$

The outer boundary conditions are

$$
\bar{f}_{11}'(\infty) = 0, \qquad \bar{f}_{12}'(\infty) = 1, \qquad \bar{g}_{11}(\infty) = 0;
$$

the solution for $\bar{g}_{11}$ is $\bar{g}_{11} \equiv f_{11}'$, from (3.13). Numerical solution of the equations for $\bar{f}_{11}$ and $\bar{f}_{12}$ yields $\bar{f}_{11}''(0) = 0.117$, $\bar{f}_{12}''(0) = 2.886$. Finding later terms in the expansion (4.80) can be reduced to the solution of ordinary differential equations in a similar way, but the labour rapidly becomes prohibitive.

These solutions result in additional contributions to the axial and azimuthal components of wall shear, in (4.86) and (4.89) respectively. Since $\Phi(\theta)$ is known only to $O(\delta_0)$ (from (4.69) and (4.65)), we present the results only to that order. The first correction to the square brackets of (4.86), multiplied by h_0^{-2}, is

$$
\varepsilon s\delta_0'[-2.71\cos\theta + \delta_0(-1.32 + 8.06\cos^2\theta) + O(\delta_0^2)]. \tag{4.95}
$$

The first term here has the same sign as that in (4.88), so, if δ_0 is small, a gradual increase in curvature has a similar effect as an acceleration of the core flow, and tends to inhibit the cross-over of maximum shear. For larger values of δ_0, (4.95) suggests that the increase in curvature would tend to increase the wall shear on both the outside and the inside walls of the bend, reducing it on the side-walls ($\theta = \pm\frac{1}{2}\pi$). However, the expansion in powers of δ_0 would be invalid if the second term became as large as the first ($\delta_0 \approx 0.34$). It no doubt gives qualitatively correct results for $\delta_0 \lesssim 0.2$, the value in the aorta.

The corrections introduced by variable curvature into the azimuthal wall shear consist of the following additions to (4.89):

$$(Re W_0^3/2s)^{1/2}[\varepsilon\delta_0'\Phi(\theta)/h_0][0.47 + 1.20 sh_0^2(\dot{W}_0/W_0^2)],$$

where

$$\Phi(\theta) = \tfrac{1}{4}\sin\theta - \tfrac{13}{48}\delta_0\sin 2\theta + O(\delta_0^2).$$

Thus if $\delta_0' > 0$ and δ_0 is small, we see that an increase in curvature enhances the leading term of (4.89), i.e. increases the azimuthal wall shear at $\theta = \frac{1}{2}\pi$, while it reduces the effect of the second term (since $g_{12}'(0) < 0$), and therefore inhibits the tendency for secondary flow reversal during an acceleration of the core.

Having developed the above expansions (and having taken them even further), Farthing (1977) constructed a composite picture of wall shear near the entrance to a rigid tube whose geometry approximated as closely as possible to that of the dog's aorta, as measured from arteriograms taken by Ettinger & Suter (1970). The tube consisted of two segments, of different diameters: the first had diameter 1.5 cm and represented the ascending aorta, while the second had diameter 1.0 cm and represented the descending thoracic aorta. The change in cross-sectional area is a real phenomenon (cf. fig. 1.4), and is related to the fact that a significant proportion of the blood flow leaves through the branches at the arch. It was therefore plausible for Farthing to assume that the cross-sectionally-averaged velocity in the aorta was the same function of time in each segment, although there is no direct evidence in support of this. The waveform $W_0(t)$ was taken to be a polynomial fit to one measured by Clark & Schultz (1973). It is more complicated than the crude approximation (4.49), but is not qualitatively different during the period of forward flow, $0 \le t \le t_1$.

Farthing further assumed that the presence of the branches has no effect on the flow near the inside wall of curvature, and its effect near the outer wall can be represented by starting a new quasi-steady boundary layer at the last flow divider. This is a gross assumption, but enabled Farthing to continue his expansions downstream of the branches. Each of the aortic segments was taken to have variable curvature, as shown in fig. 4.12(a), where the values of $\delta(s)$ are given at each cross-section; torsion was also included in the model.

Farthing used the small-s expansions of this section to give the approximately quasi-steady flow near the entrance to the aorta, and the small-t expansion of the previous section to give the downstream diffusive flow to which it must match. He found that the most convenient way of effecting the match was to use an extended version of the modified Oseen approximation of Lewis & Carrier (1949) (cf. p. 154). Some of his results are shown in fig. 4.12(a)–(d). In each part of the figure, the aortic wall is viewed from the animal's left side, as if it were untwisted and flattened, retaining the curvature of the inside wall (the outside wall is therefore somewhat stretched). Then at five points on each cross-section chosen, the computed shear stress on the far wall is represented as a vector; where the stress on the near wall is different (on account of the slight torsion) it is represented by a second vector. The four parts of fig. 4.12 represent four times during the cycle, $0.1T$, $0.2T$, $0.3T$ and $0.5T$ respectively, where T is the total period of the cycle (the time t_1 at which the entering core flow first reverses is approximately $0.6T$ on this scale). At $0.1T$, the flow is purely diffusive; the axial component of wall shear is greater towards the inside of the bend, consistent with the potential vortex core flow (4.50); a small azimuthal component of shear can be seen to have developed near the region of maximum tube curvature. At $0.2T$ the secondary motion has become much stronger, and the axial component of shear near the entrance is greater than that further downstream because of the quasi-steady influence of the leading edge. At $0.3T$, just after the peak in the entering core flow, flow reversal has occurred at the inside wall of the arch, and the secondary components of shear near

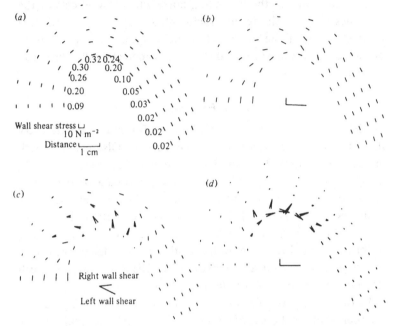

Fig. 4.12. Predictions of wall shear stress in the upper part of a dog's aorta. The aortic wall is viewed from the animal's left side, as if it were untwisted and flattened, retaining the curvature of the inside wall. Numbers by each cross-section in (a) represent the value of the local curvature ratio δ. The shear on the far wall at each point is represented by a vector; where the shear at the near wall is different (because of torsion) it is represented by a second vector. The different panels represent different times during the cycle (total period, T): (a) 0.1T, (b) 0.2T, (c) 0.3T, (d) 0.5T. (After Farthing, 1977.)

maximum curvature exceed all others. This suggests a breakdown of the small-t expansion, as foreseen above. The effect of torsion is just noticeable. At $0.5T$ the axial flow in the boundary layer has reversed almost everywhere, but the predicted shear values cannot be expected to be accurate.

These calculations represent the furthest that one can go in making analytical predictions of wall shear in the aorta. Since the curvature ratio δ is not taken to be vanishingly small, nor to be uniform ($\varepsilon \neq 0$), the present approach is likely to give more realistic quantitative information than any other. The big defects of the theory, however, are that it can be accurate only in the early part of systole, and only in the ascending aorta where the effects of the branches at the arch are not dominant. Apart from a qualitative understanding of the effects of branches (see chapter 5) future predictions will almost certainly have to be made by fully numerical means.

4.4.3 *Experiments*

Model experiments have recently been performed on steady entry flow in uniformly curved tubes by two workers. Olson (1971), with air as the working fluid, used a single hot-wire anemometer to measure axial velocity profiles and a double, pulsed wire to measure the secondary velocity field (see § 5.1 for his measurements on branched tubes). Agrawal (1975) used laser-Doppler anemometry to measure the complete velocity field, in water (see also Agrawal, Talbot & Gong, 1978). Agrawal investigated only flat entry profiles (as examined theoretically above), while Olson investigated both flat and parabolic entry profiles (cf. § 4.5).

In the case of the flat entry profile at moderately high Dean numbers, both authors confirmed that, for small s, the axial velocity profile in the core is slightly skewed towards the inside of the bend, as in the predicted potential vortex flow. This is also consistent with the skewing of the velocity profile in the upper part of the ascending aorta, as shown in fig. 1.22(b). They then reported that the core flow was eaten away by the rapid thickening of the boundary layer at the inside of the bend, and that eventually (after about 180° of bend) the axial profile had a peak near the outside of the bend and decreased approximately linearly across the core. Typical profiles are shown in figs. 4.13(b) and 4.14(b), where qualitative, but not quantitative, agreement can be seen, although the values of δ and D', where

$$D' = 2(a/R)^{1/2} \hat{W}_0 a / \nu, \qquad (4.96)$$

were comparable (in Olson's experiments (fig. 4.14) $\delta = \frac{1}{16}$ and $D' = 257$, in Agrawal's (fig. 4.13) $\delta = \frac{1}{20}$ and $D' = 251$). Note that here $D' = \sqrt{2}D^{2/3}$ when D is defined by (4.9), because D' is a Dean number based on axial velocity, while D is based on pressure gradient (see (4.90)). The discrepancy in the results shown lies in the more rapid encroachment of the boundary layer in Agrawal's experiments, since his results at a bend angle of about 35° ($s = 12.2$) are comparable with Olson's at an angle of 60° ($s = 16.8$). The two authors concur in their conclusion that the flow is virtually fully developed by the last station measured in each case ($s = 57$ for Agrawal, and $s = 50$ for Olson). These values are comparable with Yao & Berger's prediction of $Re^{1/2}\delta^{-1/4}$ from (4.91) (equal to 50 for Agrawal, 46 for Olson) and are much greater than $\delta^{-1/2}$ and much smaller than Re.

Figs. 4.13(c), (d) show Agrawal's measurement of the x-component of secondary velocity ($x = r\cos\theta$, measured across the cross-section of the tube in the plane of the bend) at the same downstream stations as the axial profiles in figs. 4.13(a) and (b). They confirm the qualitative picture of a jet-like secondary motion in the boundary layer, from outside to inside of the bend, and a slower drift back across the core; however, the details are more complicated than can be predicted by the theories.

Figs. 4.14(a), (c) and (d) are Olson's measurements for a parabolic entry profile, when the curved tube was fixed to the end of a long straight tube. The flow appears to become fully developed at about the same value of s as for a flat entry profile (fig. 4.14(a)), when the secondary motions (fig. 4.14(c)) and the contour plot of axial velocity (fig. 4.14(d)) also show good correspondence. The latter, at a value of $D' = 258$, corresponding to $D = 2460$, shows good qualitative agreement with the theoretical contour plot (at $D = 5000$) shown in fig. 4.2(c). The developing axial profile in fig. 4.14(a) shows that the peak velocity rapidly veers to the outside of the bend, in contrast to the case of a flat entry profile. This is consistent with a model in which the flow in the centre of the tube carries on undisturbed, while that near the walls is forced around, again in a secondary jet-like boundary layer. This then erupts into the core again at the inside of the bend, where a second maximum in the axial velocity profile occurs at first, subsequently dying away. The initial stages of this development are described by the theory to be presented in § 4.5.

We note finally that neither author made accurate measurements of wall shear in his models.

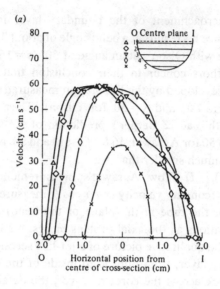

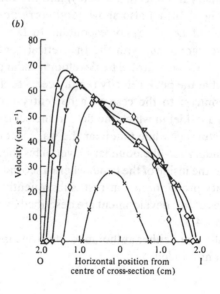

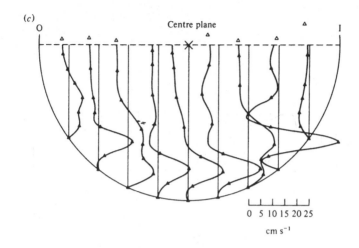

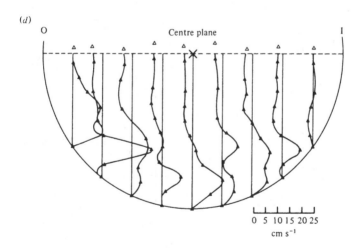

Fig. 4.13. Measurements of entry flow in a curved tube. Axial velocity profiles along horizontal traverses: (a) $\delta = \frac{1}{20}$, $D' = 251$, $s = 12.2$; (b) as (a), but with $s = 57.2$. Secondary velocity profiles along vertical traverses: (c) $\delta = \frac{1}{20}$, $D' = 251$, $s = 12.2$; (d) as (c), but with $s = 57.2$. (After Agrawal, 1975.) Dr Agrawal has pointed out that some of the wiggles in these profiles are artefacts and were not present in the more recent curves of Agrawal *et al.* (1978).

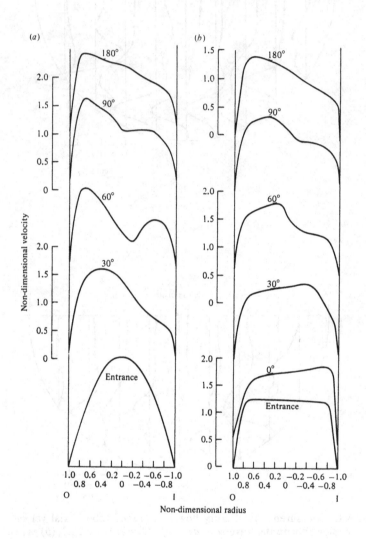

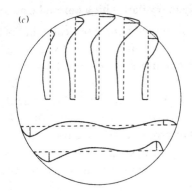

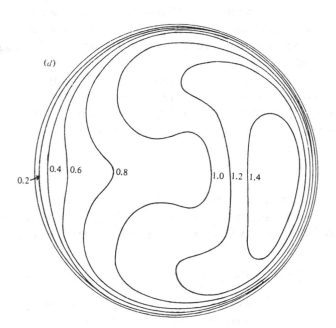

Fig. 4.14. Measured velocity patterns in a curved tube, with $D' = 258$, $\delta = \frac{1}{16}$, and 180° of bend (after Olson, 1971). (a) Axial velocity profiles in the plane of the bend; parabolic entry profile. (b) Axial velocity profiles in the plane of the bend; flat entry profile. (c) Transverse velocity profiles after 180° of bend; parabolic entry profile. (d) Contours of axial velocity after 180° of bend; parabolic entry profile.

4.5 Steady entry flow with a parabolic entry profile

The theory to be presented here is basically that of Smith (1976b), who analysed steady flow in a tube that was straight for $s < 0$ and had uniform small curvature δ_0 for $s > 0$. He took the Dean number D', given by (4.96), to be $O(1)$ (note that $D' \propto \delta_0^{1/2} Re$), and analysed the flow in the region $|s| = O(1)$. His analysis ceases to be valid if either $D' = O(Re^{5/6})$ or $s = O(\delta_0^{-1/2}) = O(D'^{-1}Re)$. Smith took axes (r, θ, s) appropriate to the curved tube with fixed curvature δ_0, which made a description of the tube wall, and of the undisturbed flow, rather cumbersome for $s < 0$. We adhere to axes appropriate to the *local* curvature (equal to 0 in $s < 0$ and δ_0 in $s > 0$ for Smith's case) so that the tube wall is $r = 1$ for all s, and extension of the theory to tubes of variable (but small) curvature in $s > 0$ becomes a simple matter. In particular, it becomes possible to analyse the flow in a tube that suffers a small bend and subsequently becomes straight again. This is expected to give results qualitatively relevant to flow in the descending aorta, where the vessel straightens out after a bend.

If we non-dimensionalise velocities with respect to $\hat{W}_0$ (the axial velocity-scale) and lengths with respect to c, the governing equations are again (4.1)–(4.4) with $Re \gg 1$, and with the $\partial/\partial t$ terms absent for steady flow. We describe the (variable) tube curvature by the equation

$$\delta = \delta_0 \Delta(s), \qquad \Delta = 0 \quad \text{for } s < 0,$$

where δ_0 is a suitable scale. We take the Reynolds number to be large, and D' defined by (4.96) to be $O(1)$, so δ_0 is formally $O(Re^{-2})$. Sufficiently far upstream the motion in the straight tube is Poiseuille flow:

$$w = W_0(r) = \tfrac{1}{4}(1 - r^2), \qquad u = v = 0, \qquad p = P_0 - s/Re,$$

where P_0 is a given constant. Since the Reynolds number is large and the curvature small, we expect the perturbation to this flow in the core to be inviscid and $O(\delta_0)$ for $s = O(1)$. Far from the tube wall, therefore, we expect the flow field to have the form

$$w = W_0(r) + \delta_0 w_1(s, r, \theta) + o(\delta_0),$$
$$u = \delta_0 u_1(s, r, \theta) + o(\delta_0),$$

$$v = \delta_0 v_1(s, r, \theta) + o(\delta_0),$$

$$p = P_0 - s/Re + \delta_0 p_1(s, r, \theta) + o(\delta_0).$$

Substitution into the governing equations gives the inviscid disturbance equations:

$$u_{1r} + u_1/r + v_{1\theta}/r + w_{1s} = 0,$$

$$W_0 u_{1s} - \Delta \cos\theta\, W_0^2 = -p_{1r},$$

$$W_0 v_{1s} + \Delta \sin\theta\, W_0^2 = -p_{1\theta}/r,$$

$$W_{0r} u_1 + W_0 w_{1s} = -p_{1s},$$

with wall boundary condition $u_1(s, 1, \theta) = 0$. The unique solution which also matches to the oncoming flow in $s < 0$ is

$$p_1 = 0, \qquad w_1 = -W_0'(r) \cos\theta \int_0^s (s - s') \Delta(s') \, ds',$$

$$u_1 = W_0(r) \cos\theta \int_0^s \Delta(s') \, ds', \qquad v_1 = -W_0(r) \sin\theta \int_0^s \Delta(s') \, ds'. \tag{4.97}$$

This means that the Poiseuille flow is not distorted until the bend, and after it the flow carries straight on, not to this order in δ_0 being diverted by the presence of the bend.

The axial velocity in the core does not satisfy the no-slip condition on the wall, so a viscous boundary layer must be interposed there. Near the wall it is appropriate to rewrite W_0 in the form

$$W_0(r) = \tfrac{1}{2}(1 - r) - \tfrac{1}{4}(1 - r)^2,$$

which is $O(b)$ if b is a scale for boundary layer thickness. Then the equation of continuity, (4.1), shows as usual that the scale for u in the boundary layer is of order b times the scale for the perturbation in w. A balance between the leading order inertial and viscous forces in either the swirl or the axial momentum equation then shows that $b = O(Re^{-1/3})$. The core flow, from (4.97), also indicates that the perturbation axial velocity is $O(\delta_0) = O(Re^{-2})$. We therefore seek a boundary layer expansion in which

$$w = \tfrac{1}{2}\zeta Re^{-1/3} - \tfrac{1}{4}\zeta^2 Re^{-2/3} + Re^{-2} w_F(s, \zeta, \theta) + O(Re^{-7/3}),$$

$$u = -Re^{-7/3} u_F(s, \zeta, \theta) + O(Re^{-8/3}),$$

$$v = Re^{-2} v_F(s, \zeta, \theta) + O(Re^{-7/3}),$$

$$p = P_0 - s/Re + Re^{-7/3} p_F(s, \zeta, \theta) + O(Re^{-8/3}),$$

where $\zeta = Re^{1/3}(1-r)$. It is crucial to the subsequent solution of the problem (a) that the pressure perturbation $Re^{-7/3}p_F$ be included, although there is no pressure perturbation to $O(Re^{-2})$ in the core, and (b) that the swirl velocity v be $O(Re^{-2})$ although its core value (4.97) is $O(Re^{-7/3})$ as $r \to 1$. Without this pressure and swirl velocity no u- or w-field of the proposed form could satisfy mass conservation: the Poiseuille flow, impinging on the outside of the bend, must raise the pressure there and force fluid around towards the inside of the bend, as in fully developed flow.

The boundary layer equations governing the motion are

$$\left. \begin{aligned} u_{F\zeta} + v_{F\theta} + w_{Fs} &= 0, \\ 0 &= p_{F\zeta}, \\ \tfrac{1}{2}\zeta v_{Fs} &= -p_{F\theta} + v_{F\zeta\zeta}, \\ \tfrac{1}{2}u_F + \tfrac{1}{2}\zeta w_{Fs} &= -p_{Fs} + w_{F\zeta\zeta}, \end{aligned} \right\} \tag{4.98}$$

and the outer boundary conditions, as $\zeta \to \infty$, are

$$v_F \to 0, \qquad w_F \to \tfrac{1}{2}D'' \cos\theta \int_0^s (s-s')\Delta(s')\,ds', \tag{4.99}$$

where $D'' = \delta_0 Re^2$ ($= \tfrac{1}{4}D'^2$ from (4.96)). Following Smith (1976b), we solve these equations by the use of Fourier transforms. We write

$$(w_F, u_F, p_F) = (W, U, P)\cos\theta, \qquad v_F = V\sin\theta,$$

where U, V, W are functions of ζ and s but P depends only on s, and define generalised Fourier transforms of the form

$$\tilde{W}(\omega, \zeta) = \int_{-\infty}^{\infty} W(s, \zeta)\,e^{-i\omega s}\,ds. \tag{4.100}$$

Eliminating the radial velocity U, we obtain the following pair of equations for $\tilde{V}$, $\tilde{W}$, $\tilde{P}$:

$$\left. \begin{aligned} \tilde{W}_{\zeta\zeta} + \tfrac{1}{2}\tilde{V} &= \tfrac{1}{2}i\omega\zeta\tilde{W}_\zeta, \\ \tilde{V}_{\zeta\zeta} + \tilde{P}(\omega) &= \tfrac{1}{2}i\omega\zeta\tilde{V}. \end{aligned} \right\} \tag{4.101}$$

The homogeneous parts of these equations can be reduced to Airy's equation by the substitution

$$t = (0 + \tfrac{1}{2}i\omega)^{1/3}\zeta,$$

where the notation implies that $(0 + \tfrac{1}{2}i\omega)^{1/3}$ has a branch cut extending from i0+ along the positive imaginary ω-axis. The solution of

(4.101) satisfying the conditions that $\tilde{V}$ vanishes at $\zeta = 0$ and as $\zeta \to \infty$ is

$$\tilde{V}(\omega, \zeta) = -(0 + \tfrac{1}{2}i\omega)^{-2/3} \tilde{P}(\omega) L(t), \Big\}$$

where

$$L(t) = Ai(t) \int_0^t \frac{dq}{Ai^2(q)} \int_\infty^q Ai(\xi) \, d\xi \Big\}$$

(4.102)

and Ai is the Airy function. Then, from (4.101), the solution for $\tilde{W}$ that is zero at $\zeta = 0$ and bounded as $\zeta \to \infty$ is given by

$$\tilde{W}_t = \tilde{B}(\omega) Ai(t) + \tfrac{1}{2}\tilde{P}(\omega)(0 + \tfrac{1}{2}i\omega)^{-5/3} M(t), \qquad (4.103)$$

where

$$M(t) = L'(t) + Ai(t)/3Ai^2(0).$$

The unknown functions $\tilde{B}(\omega)$ and $\tilde{P}(\omega)$ are determined from two further conditions. One is the outer boundary condition on $\tilde{W}$, derived from (4.99). This gives

$$\tfrac{1}{3}\tilde{B} + \frac{\tfrac{1}{2}\tilde{P}}{(0 + \tfrac{1}{2}i\omega)^{5/3}} \frac{1}{9Ai^2(0)} = \frac{\tfrac{1}{2}D''}{(i\omega)^2} \tilde{\Delta}(\omega), \qquad (4.104)$$

where $\tilde{\Delta}(\omega)$ is the Fourier transform of $\Delta(s)$. The other condition comes from the axial momentum equation, the last of (4.98), which was differentiated in order to arrive at (4.100). Setting $\zeta = 0$ in the transform of this equation, we obtain

$$\tilde{W}_{\zeta\zeta}|_{\zeta=0} = i\omega\tilde{P},$$

i.e.

$$\tilde{B}Ai'(0)(0 + \tfrac{1}{2}i\omega)^{2/3} = i\omega\tilde{P} - (\tilde{P}/i\omega)(1 + Ai'(0)/3Ai^2(0)).$$

(4.105)

Eliminating $\tilde{B}$ from (4.104) and (4.105), we finally obtain for the pressure transform $\tilde{P}$:

$$\tilde{P} = [2^{-5/3}\gamma D''(0 + i\omega)^{-1/3}/(\omega^2 + 1)]\tilde{\Delta}(\omega), \qquad (4.106a)$$

where

$$\gamma = -3Ai'(0) = 0.7765.$$

This Fourier transform, and those describing the other quantities of interest – the perturbation axial and azimuthal wall shear-rates,

$V_\zeta(0, s)$, $W_\zeta(0, s)$ – can be inverted when a particular form of $\Delta(s)$ is chosen. The shear-rate transforms are

$$\tilde{V}_\zeta(0, \omega) = \frac{2^{-4/3}\gamma D''}{3Ai(0)} \frac{(0+i\omega)^{-2/3}}{\omega^2+1} \tilde{\Delta}(\omega), \qquad (4.106b)$$

$$\tilde{W}_\zeta(0, \omega) = 3Ai(0)2^{-4/3}D''(0+i\omega)^{-5/3}\tilde{\Delta}(\omega)[1 - \gamma/9Ai^2(0)(\omega^2+1)]. \qquad (4.106c)$$

Before these are inverted, two general points should be made. The first is that both V and W (and hence v_F and w_F) tend to their limits algebraically as $\zeta \to \infty$. In fact, they are each proportional to $1/\zeta$, which means that 'slip velocities' of $O(Re^{-7/3})$ are generated at the edge of the inviscid core, which will drive further inviscid motions in the core at the next order in the large-Re expansion. The fact that the decay of v_F and w_F is algebraic does not in this case mean that the whole theory breaks down, as it does in conventional boundary layer theory, because there is a perfectly self-consistent solution of the next-order inviscid core problem, in which the tangential velocities are singular at $r = 1$, but the normal velocity remains $O(1)$, and the pressure is also bounded (Smith, 1976c). The second general point is that (4.106) has a pole at $\omega = -i$ in the lower half ω-plane. This means that although there is no $O(Re^{-2})$ perturbation in the core upstream of the curve ($s < 0$), there is a significant perturbation in the boundary layer, which has to adjust itself ahead of the disturbance.

Smith (1976b) took the case (a) of uniform curvature in $s > 0$, so that

$$\Delta(s) = H(s) \quad \text{and} \quad \tilde{\Delta}(\omega) = 1/i\omega, \qquad (4.107a)$$

where $H(s)$ is the Heaviside step function. Other cases which it might be of interest to examine are (b) a single sharp bend in an otherwise straight tube:

$$\Delta(s) = l\delta(s), \qquad \tilde{\Delta}(\omega) = l; \qquad (4.107b)$$

(c) a region of uniform curvature between two straight segments at an angle:

$$\Delta(s) = H(s) - H(s-l), \qquad \tilde{\Delta}(\omega) = (1-e^{-i\omega l})/l; \qquad (4.107c)$$

(*d*) a kink between two straight, parallel segments:

$$\Delta(s) = H(s) - 2H(s-l) + H(s-2l), \qquad \tilde{\Delta}(\omega) = (1-e^{-i\omega l})^2/i\omega.$$

$$(4.107d)$$

We first consider the inversion of (4.106*a*, *b*, *c*) for $s < 0$, which requires completion of the inversion contour in the lower half-plane. The result is

$$\frac{p_F(s)}{2^{-5/3}\gamma D'' \cos\theta} = \frac{v_{F\zeta}(0,s)3Ai(0)}{2^{-4/3}\gamma D'' \sin\theta} = -\frac{w_{F\zeta}(0,s)3Ai(0)}{2^{-4/3}\gamma D'' \cos\theta} = \tfrac{1}{2}e^s\,\tilde{\Delta}(-i),$$

$$(4.108)$$

which is directly proportional to e^s in every case, and shows precisely how the upstream adjustment is made. On the wall approaching the outside of the bend ($\theta = 0$) the pressure rises but the axial wall shear falls, while on the wall approaching the inside of the bend ($\theta = \pi$) the opposite happens. The secondary flow is from the outside to the inside. The difference between this and what happens in a uniformly curved tube lies in the *fall* of the axial wall shear at the outside. This results from the fact that the core flow is undisturbed to this order, and the boundary layer therefore remains of uniform thickness: the boundary layer flow is effectively two-dimensional in the '*s*–θ plane'. The removal of fluid at $\theta = 0$ by the secondary motions causes the axial velocity to slow down by mass conservation; the reverse is true at $\theta = \pi$. This sort of upstream response will also be seen in the case of unsymmetrically branched tubes (§ 5.2).

We now turn to the downstream solutions (for $s > 0$). In case (*a*), in which $\tilde{\Delta} = 1/i\omega$, we invert (4.106*a*) to give

$$\frac{p_F(s)}{2^{-5/3}\gamma D'' \cos\theta} = \frac{1}{2\pi}\int_{-\infty}^{\infty} \frac{e^{i\omega s}}{(0+i\omega)^{4/3}}\left[1 + \frac{(i\omega)^2}{1+\omega^2}\right]d\omega.$$

The first term is evaluated using Lighthill's (1958) table 1, to give

$$[s^{1/3}/(\tfrac{1}{3}!)] = (3\sqrt{3}/2\pi)\Gamma(\tfrac{2}{3})s^{1/3}.$$

For the second term, the contour of integration is deformed to go along both sides of the branch cut along the positive imaginary axis, and around the pole at $\omega = \mathrm{i}$. The final result is

$$p'_F = \frac{p_F(s)}{2^{-5/3}\gamma D'' \cos\theta} = \frac{3\sqrt{3}}{2\pi}\Gamma(\tfrac{2}{3})s^{1/3} - \tfrac{1}{4}e^{-s} - \frac{\sqrt{3}}{2\pi}\int_0^\infty \frac{e^{-\xi s}\,\xi^{2/3}}{1-\xi^2}\,d\xi,$$

(4.109)

where the integral assumes its principal value. This is the result obtained by Smith (1976b). Similarly, the following results are obtained for the correction to the axial wall shear and for the azimuthal wall shear:

$$\tau = \frac{w_{F\zeta}|_{\zeta=0}\,6Ai(0)}{2^{-4/3}\gamma D'' \cos\theta}$$

$$= \frac{9\sqrt{3}}{10\pi}\Gamma(\tfrac{1}{3})C_1 s^{5/3} + \frac{e^{-s}}{2} - \frac{\sqrt{3}}{\pi}\int_0^\infty \frac{\xi^{-2/3}\,e^{-\xi s}}{1-\xi^2}\,d\xi, \quad (4.110)$$

$$\tau_v = \frac{v_{F\zeta}|_{\zeta=0}\,6Ai(0)}{2^{-4/3}\gamma D'' \sin\theta}$$

$$= \frac{3\sqrt{3}}{2\pi}\Gamma(\tfrac{1}{3})s^{2/3} + \frac{e^{-s}}{2} - \frac{\sqrt{3}}{\pi}\int_0^\infty \frac{\xi^{1/3}\,e^{-\xi s}}{1-\xi^2}\,d\xi, \quad (4.111)$$

where $C_1 = 9Ai^2(0)/\gamma - 1 > 0$. In each case the only term to grow with distance downstream is the first, which represents the asymptotic expansion to the solution for large s. We note that the axial shear stress on $\theta = 0$ increases with s for sufficiently large s, whereas upstream of $s = 0$ it decreases. Hence, as with the flat entry profile, there must be a cross-over for some $s > 0$ (in fact at $s \approx 1.51$). Smith performed the integrals in (4.109)–(4.111) numerically, and fig. 4.15 shows the variation with s of the quantities evaluated, as well as two more quantities, the axial and azimuthal 'slip velocities' at the edge of the boundary layer, which drive the next-order core flow.

The downstream development of the flow in cases (b)–(d), where the bend is of finite length, can be calculated in a similar way. We restrict attention to the large-s asymptotic expansions. In case (b), where $\Delta(s) = l\delta(s)$, both the pressure and the azimuthal shear stress (and hence the azimuthal velocity for all ζ) fall with s, while the axial

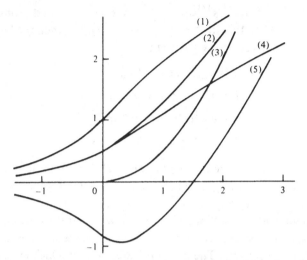

Fig. 4.15. Solution curves for the wall shear stresses and the slip velocities, axially and azimuthally. (1) $2p_F'$; (2) $v_F(s, \infty, \theta)/\sin \theta$; (3) $w_F(s, \infty, \theta)/\cos \theta$; (4) τ_v; (5) τ. (After Smith, 1976b.)

shear grows like $s^{2/3}$:

$$p_F' \sim l\frac{\sqrt{3}}{2\pi}\Gamma(\tfrac{2}{3})s^{-2/3},$$

$$\tau_v \sim l\frac{\sqrt{3}}{\pi}\Gamma(\tfrac{1}{3})s^{-1/3}, \qquad \tau \sim l\frac{3\sqrt{3}}{2\pi}\Gamma(\tfrac{1}{3})C_1 s^{2/3}. \tag{4.112}$$

Case (c), where the bend is of finite length and, as in case (b), turns through a given angle, has exactly the same large-s expansion as case (b), as one would expect. In case (d), where the tube straightens out again after a kink, even the perturbation to the axial shear falls as $s \to \infty$:

$$\tau \sim l^2(\sqrt{3}/\pi)\Gamma(\tfrac{1}{3})C_1 s^{-1/3}. \tag{4.113}$$

In this case alone, therefore, the presence of the bend causes no significant disturbance as $s \to \infty$; this is because the effect of a bend one way is exactly cancelled out by the equal bend in the opposite sense.

The above expansion, for small δ_0 and $s = O(1)$, breaks down when $s = O(\delta_0^{-1/2})$, as can be seen either from the core flow solution (4.97) or from the fact that the boundary layer thickness is proportional to $Re^{1/3}s^{1/3} \propto \delta^{1/6}s^{1/3}$ as $s \to \infty$ (derived from the inversion of $\tilde{V}$ or $\tilde{W}$ with $\zeta \neq 0$). If we let $S = \delta_0^{1/2}s$, then the large-s expansions derived above suggest that

$$w = w_L(S, r, \theta) + O(\delta_0),$$

$$u = \delta_0^{1/2}u_L(S, r, \theta) + O(\delta_0^{3/2}),$$

$$v = \delta_0^{1/2}v_L(S, r, \theta) + O(\delta_0^{3/2}),$$

$$p = \bar{p}_L(S) + \delta_0 p_L(S, r, \theta) + O(\delta_0^2),$$

where the separate identity of the original Poiseuille flow is lost. Substitution into the governing equations (4.1)–(4.4) shows that most terms are retained, with the exception of those describing longitudinal diffusion. Smith (1976b) showed that the leading term in the small-S expansion is exactly the same as the leading term in the large-s expansion. The problem parallels that which arises in a straight tube (§ 3.2), but numerical solution of the curved-tube equations has not been performed; presumably the fully developed solution for a given Dean number D' would result as $S \to \infty$.

In the case of a tube that straightens out again after a short bend, the large-s problem is much simpler than in the case of constant curvature, both because the centrifugal driving force is absent, and because, at the entrance to the straight portion, the flow still represents only a small perturbation to Poiseuille flow (as long as $l = O(1)$), as can be seen from (4.97). Thus it will everywhere be only a small perturbation to Poiseuille flow. In cases (b) and (c), therefore, we can write

$$w = W_0(r) + \delta_0^{1/2}lw_M(S, r)\cos\theta + o(\delta_0^{1/2}),$$

$$u = \delta_0 lu_M(S, r)\cos\theta + o(\delta_0),$$

$$v = \delta_0 lv_M(S, r)\sin\theta + o(\delta_0),$$

$$p = P_0 - SD''^{-1/2} + \delta_0^{3/2}p_M(S, r)\cos\theta + o(\delta_0^{3/2}),$$

while in case (d), we have

$$w = W_0(r) + \delta_0 l^2 w_M(S, r)\cos\theta + o(\delta_0),$$

$$u = \delta_0^{3/2}l^2 u_M(S, r)\cos\theta + o(\delta_0^{3/2}),$$

$$v = \delta_0^{3/2}l^2 v_M(S, r)\sin\theta + o(\delta_0^{3/2}),$$

$$p = P_0 - SD''^{-1/2} + \delta_0^2 p_M(S, r)\cos\theta + o(\delta_0^2).$$

The governing equations reduce in both cases to the linear forms

$$
\left.
\begin{aligned}
&u_{Mr} + \frac{u_M}{r} + \frac{v_M}{r} + w_{MS} = 0, \\[2mm]
&W_0(r)u_{MS} = -p_{Mr} + D''^{-1/2}\left(\nabla_2^2 u_M - \frac{u_M}{r^2} - \frac{2v_M}{r^2}\right), \\[2mm]
&W_0(r)v_{MS} = \frac{1}{r}p_M + D''^{-1/2}\left(\nabla_2^2 v_M - \frac{v_M}{r^2} - \frac{2u_M}{r^2}\right), \\[2mm]
&W_0'(r)u_M + W_0(r)w_{MS} = D''^{-1/2}\nabla_2^2 w_M,
\end{aligned}
\right\} \qquad (4.114)
$$

where $\nabla_2^2 \equiv \partial^2/\partial r^2 + (1/r)\partial/\partial r - 1/r^2$. The boundary conditions are $u_M = v_M = w_M = 0$ on $r = 1$, $u_M, v_M, w_M, p_M \to 0$ as $S \to \infty$, and all quantities match to the large-s expansions derived above as $S \to 0$. The three cases differ only in this upstream condition as $S \to 0$: in cases (b) and (c), w_M is proportional to S in the core, and u_M and v_M are independent of S there, while u_M is proportional to $S^{1/3}$ in the boundary layer although v_M and w_M retain the same S-dependence; in case (d), all powers of S are reduced by 1. Analytical progress can be made in cases (b) and (c) by eliminating p_M from the middle two of (4.114) and by using the Laplace transform in S directly; in case (d), the initial conditions require that the variables be written as S^{-1} times new variables before the Laplace transform is applied. Because of the complexity of the initial conditions, inversion of the Laplace transform is likely to be complicated, and it may in the long run be simpler to solve the problem by a direct finite-difference integration, marching forwards in S from the initial profiles. Neither method has yet been applied.

Finally, we note that the whole linear theory for $s = O(1)$ ceases to be valid when $D'' = O(Re^{5/3})$, i.e. $\delta_0 = O(Re^{-1/3})$, because the perturbation velocity is as large as the original Poiseuille velocity in the boundary layer, so that the boundary layer equations become non-linear and must be solved numerically. Only far upstream can one anticipate a small departure from Poiseuille flow, and there the problem reduces effectively to that which has already been solved, except that the magnitude of the perturbation, for a given s, is unknown. For example, Smith (1976b) shows that the pressure far upstream must be of the form

$$p = Re^{-2/3}\beta \, e^s \cos(\theta + \varepsilon),$$

where β and ε are unknown constants that only a numerical solution can determine.

CHAPTER 5

FLOW PATTERNS AND WALL SHEAR
STRESS IN ARTERIES
III BRANCHED TUBES AND FLOW INSTABILITY

5.1 Flow in symmetric bifurcations

Most branchings in the cardiovascular system are asymmetric, the only major exception in man being the bifurcation where the aorta divides to form the iliac arteries. This is in contrast to the bifurcating airways of the lung, for which the assumption of symmetry is more appropriate, and which have been the subject of extensive research (Pedley, 1977). Furthermore, the precise definition of an asymmetric bifurcation requires the specification of several more parameters than that of a symmetric one (e.g. the ratios of the flow-rates in, and the diameters of, the two daughter tubes, as well as the different angles of branching). There is, therefore, considerably more fluid mechanical information available on the subject of symmetric bifurcations, and this chapter begins with a survey of it (taken largely from the review by Pedley, 1977). It should be said at the start, however, that the problem is still very complicated, and most of the data have been obtained experimentally not theoretically, with steady rather than unsteady flow. Clearly much work remains to be done.

In all the investigations described in this section and the next, the geometry of the bifurcations is taken to be fully three-dimensional. There has been relatively extensive theoretical and experimental work on two-dimensional bifurcations, but since that geometry rules out all secondary motions it is unlikely to have much relevance to the cardiovascular system.

5.1.1 *Model experiments*

A full fluid dynamical description of the flow through a single junction requires the complete specification of the geometry of the junction, as well as of the flow-rate and velocity profile upstream. Thus, in addition to the diameters of the tubes, we need to know the

angles of branching, the sharpness of the flow divider, the radii of
curvature of the tube walls at the junction, the way the parent tube
changes shape as it approaches the bifurcation, and whether the
tubes are themselves curved or straight. The experiments that have
been done do not cover all possible values of the dimensionless
parameters that define such a geometry. Indeed, the most detailed
experiments have been done with a single geometry, chosen to be
representative of bifurcations in the bronchial airways. Such a
'typical' junction is depicted in fig. 5.1, and has the following
properties. (a) The diameter ratio (d_2/d_1) is 0.78, corresponding to
an area ratio (both daughters to parent) of 1.2. (b) The angle of
branching is 70° (qualitative experiments have been performed with
different values of this and the previous parameter). (c) The flow
divider is sharp. (d) The radius of curvature of the outer wall of a
junction (R, fig. 5.1) lies between 1 and 30 times the radius ($\frac{1}{2}d_1$) of
the parent tube. (e) As it approaches the junction, the parent tube
first becomes elliptical without change of cross-sectional area,
before changing both shape and area as the daughter tubes emerge.
Note that the cross-sectional area at the junction is not uniquely
defined, since it depends on whether the plane of measurement is
perpendicular to the axis of the parent tube or that of a daughter
tube. (f) Daughter tubes are initially curved and of constant area,
becoming straight when the branching angle has been achieved.
(The data in (c) to (f) above are all taken from Horsfield *et al.*,
(1971); such detailed data are not available in arteries, even
for the aortic bifurcation.) (g) Finally, we assume that the flow in
the parent tube, far upstream of the bifurcation, is axisymmetric
and is either Poiseuille flow or partially developed laminar
entrance flow; the Reynolds number is taken to be large, of
the order of 500. (Of course, the parent tube of one junction is
in general the daughter of another, so the flow in it is unlikely to
be symmetric.)

It is not difficult to construct a qualitative picture of the flow
pattern to be expected in the daughter tube (Pedley, Schroter &
Sudlow, 1971). First, the flow is split into two streams, so that a new
boundary layer is formed on the inside wall, with maximum axial
velocity just outside it (fig. 5.2). Secondly, the flow into each
daughter tube turns a corner, so that secondary motions are set up

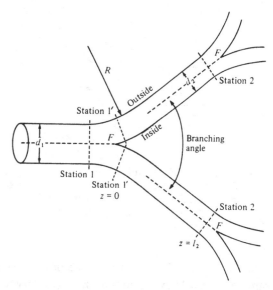

Fig. 5.1. Diagram of a 'typical' symmetric bifurcation, with a second generation downstream. F, flow dividers. Stations 1, 1' and 2 are the stations referred to in the text.

as in a uniform curved tube: the faster-moving fluid moves towards the inside wall of the junction (the outside of the bend), tending to keep the point of maximum axial velocity close to that wall, and the slower-moving fluid near the walls moves towards the outside. In addition, depending on the sharpness of the corner in the outer wall (i.e. on $2R/d_1$; see fig. 5.1), there may be a region of separated flow.

Two types of experiment have been carried out in models of single bifurcations: flow visualisation (using smoke as a tracer in air flow, or dye in water flow) and measurement of velocity profiles (using hot-wire anemometry or similar techniques). All flow-visualisation studies have confirmed the above qualitative picture, that fluid particles move downstream in the daughter tubes in two helical paths, for various branching angles, area ratios, and Reynolds numbers (in the approximate range 100–1400). Schroter & Sudlow (1969) used a 'typical' bifurcation, as described above, except that $2R/d_1$ was small (=1), the parent tube did not become elliptical before the junction, and the daughter tubes were not

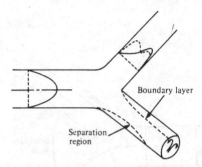

Fig. 5.2. Qualitative picture of flow downstream of a single junction with Poiseuille flow in the parent tube. Direction of secondary motions, new boundary layer and separation region are indicated in the lower branch; velocity profiles in the plane of the junction (continuous curve) and the normal plane (broken curve) are indicated in the upper branch.

initially curved (all other flow-visualisation studies have suffered from the same defects). They found that the secondary flows were strong enough to complete one cycle within three diameters of the junction. This indicates that the secondary velocities could be as high as 50% of the average axial velocities. However, the velocity measurements of Olson (1971) show that the maximum value is about 30% when $2R/d_1 = 7$, and is lower still for more gradually curved tubes (at the given branching angle of 70°). This suggests that the unnaturally sharp curvature of the outer wall is responsible for such large secondary velocities. Schroter & Sudlow also observed separation and reversed flow at the outer wall of the junction, although this does not imply a region of dead water, because the secondary motions sweep new fluid into the separation bubble from the sides (Brech & Bellhouse, 1973). Zeller, Talukder & Lorenz (1970) further confirmed that if the flow-rates in the two daughter tubes are unequal, separation readily occurs at the outer wall of the tube with the smaller flow-rate. Olson (1971) did not observe separation in his more gradually curved models, despite having the same area expansion of 1.2 as Schroter & Sudlow. Experiments on a bifurcation with a blunt flow divider suggest that even there the influence of the bifurcation on flow in the parent tube is insignificant more than one diameter upstream of the flow divider (Pacome, 1975; Pedley, Schroter & Sudlow, 1977).

The best measurements of velocity downstream of a bifurcation have been made by Olson (1971); they remain largely unpublished although a few have been reported by Pedley *et al.* (1977) and by Pedley (1977). Olson's measurements are the most reliable for the following reasons. (*a*) His models were more carefully made than any others, and conform closely to the 'ideal' bifurcation described above, including the gradual change in shape of the parent tube, and the initial curvature of the daughter tubes (with $2R/d_1$ equal to 7 or more). (*b*) He made careful measurements *both* with conventional hot-wires, which can be interpreted as measurements of axial velocity, w, as long as the component of transverse velocity normal to the wire is much less than w (in most cases this was the case), *and* with a specially designed probe consisting of two wires, one of which is given a pulse of heat and the other of which records the temperature downstream. Rotation of the probe about the pulsed wire until the response of the receiver wire is greatest gives both the magnitude and the direction of the velocity component in the plane perpendicular to the wire. Repeating the measurement at the same point in a perpendicular plane gives the complete velocity vector. Using this procedure at many points, Olson was able to map out profiles of the transverse velocity components u and v, as well as those of w. (*c*) Finally, Olson made measurements at more points in a given cross-section and at more cross-sections than other workers. However, because the experimental procedure was so laborious, he reported measurements of secondary velocities at only two Reynolds numbers, in one model bifurcation, and with a flat entry profile a few diameters upstream.

Typical velocities measured by Olson in the daughter tubes of the bifurcation are presented in figs. 5.3–5. Fig. 5.3 shows the development of the axial velocity profile in the plane of the bifurcation from a station 1.66 daughter-tube diameters upstream of the flow divider to many diameters downstream, at a daughter-tube Reynolds number Re_2 of 530. As expected, the maximum velocity occurs close to the inner wall of the bifurcation, although, near the flow divider, there is evidence of a slight skewing of the profile away from the inner wall, outside the boundary layer. This presumably reflects the inviscid development of the flat profile upstream as it rounds the bend (cf. § 4.4 above). Similar results were found at

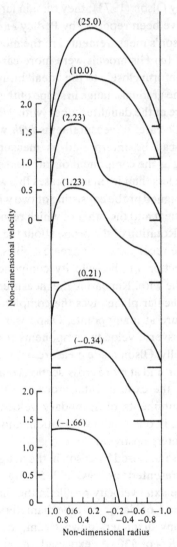

Fig. 5.3. Velocity profiles in the plane of the junction for inspiratory flow through a single bifurcation; flat entry profile, $Re = 530$. Flow divider on *left*. Figures in brackets are numbers of tube diameters from flow divider. (After Olson, 1971.)

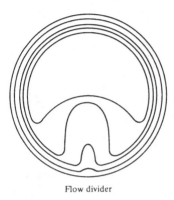

Flow divider

Fig. 5.4. Contour plot of axial velocity in the daughter tube at 2.2 diameters from the junction. Inspiratory flow; flat entry profile; $Re = 660$. Contours are at steps of 0.4 non-dimensional velocity (they have been slightly smoothed in reproduction). (After Olson, 1971.)

other values of Re_2 (in the range 300 to 1500). Olson found no separation from the outside wall. He also remarked that a parabolic entry profile made little difference to the results, at least after two or three diameters.

In fig. 5.4 is plotted a contour map of longitudinal velocity, measured at 2.2 diameters downstream of the flow divider and at $Re_2 = 660$. The main features are (a) that the region of high shear on the inside wall (the boundary layer) extends more than half-way round the tube, and (b) that the contours have a winged appearance, demonstrating that a velocity profile in the plane perpendicular to the bifurcation will be M-shaped, with a significant dip in the middle (cf. fig. 5.2).

In fig. 5.5 we reproduce Olson's measurements of the profiles of the secondary velocity components at a somewhat higher Reynolds number ($Re_2 = 935$). They are shown at four stations, three downstream and one $0.34d_2$ upstream of the flow divider. At this upstream station, the profile is plotted first in a plane perpendicular to the parent tube axis, and then in a plane perpendicular to the daughter tube axis. The difference is striking. Whereas in the first case the secondary motion is seen to be primarily towards the outside of the *junction*, in response to the change of tube shape (except on the minor axis where the effect of the imminent flow

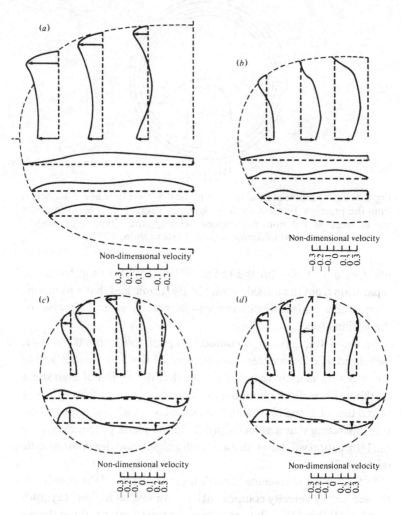

Fig. 5.5. Velocity profiles of the secondary flow at different positions upstream and downstream of the flow divider, which is on the *right* of the diagram; flat entry profile; $Re = 935$. (a) $0.34d$ upstream of flow divider (d = diameter of daughter tube); plane of profiles is perpendicular to axis of parent tube; (b) $0.34d$ upstream; plane of profiles is perpendicular to axis of daughter tube; (c) $0.21d$ downstream; (d) $2.23d$ downstream; (e) $5.0d$ downstream. (After Olson, 1971.)

(e)

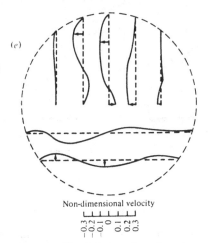

Non-dimensional velocity

Fig. 5.5. (continued)

divider is strong), in the second case it corresponds to a strong sideways motion towards the outside of the *bend* (the flow divider). It is this aspect of the secondary flow that presents itself first to the daughter tubes, and it can be seen to persist downstream of the junction. At 0.21 diameters it is still in evidence, with a boundary layer developing at the walls in which the secondary flow back to the outside wall is to be seen, as expected from the flow-visualisation studies. Further downstream, however, at 2.23 diameters, the situation has become very confusing again; here the secondary motion is almost all directed towards the outer wall of the junction, and there is no obvious way to explain this. Further downstream still, the effect is less strong. It is as if the flow pattern varied sinusoidally with distance downstream, as has been reported for turbulent flow in pipe bends of large angle (Rowe, 1970). However, no such variation is obvious in the axial profiles.

The picture of the secondary motions in the elliptical transition region was confirmed by Olson from measurements in a tube that gradually became elliptical (with constant area), but that had no bifurcation at the other end. An example of the measurements, at a (parent tube) Reynolds number of 1620, is shown in fig. 5.6(a). The similarity between this and fig. 5.5(a) is obvious. For comparison, the corresponding plot with a parabolic entry profile is shown in fig. 5.6(b); the secondary motions are far less strong.

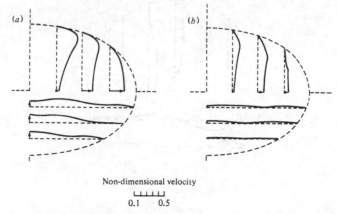

Non-dimensional velocity

0.1   0.5

Fig. 5.6. Velocity profiles of the secondary flow in a tube that is initially circular and becomes increasingly elliptical with constant area; $Re = 1620$, (a) flat entry profile, (b) parabolic entry profile. (After Olson, 1971.)

A number of authors have made less accurate measurements in less carefully made bifurcations. Among the most important (because of the use subsequently made of them – see below) were those by Schroter & Sudlow (1969). They made hot-wire measurements of (longitudinal) velocity profiles in the plane of the junction and the plane perpendicular to it, usually with a parabolic inlet profile. They too found peak velocities near the inside wall, and an M-shaped profile in the plane perpendicular to the junction, and qualitatively their results differed from Olson's only in the absence of the skew outside the boundary layer on the inside wall (because the inlet profile was not flat) and in the presence of a minimum in the profile in the plane of the junction, with a second maximum near the outside wall. This they took to be associated with the separation demonstrated by their flow-visualisation experiments, since the hot-wire anemometer gives a positive signal whatever the direction of flow. Qualitatively similar results have been obtained by Schreck & Mockros (1970), Berger, Calvet & Jacquemin (1972) and Brech & Bellhouse (1973). Berger *et al.*'s measurements were made in the first generation of a branched tube network containing several generations, with the later generations still attached. It is not clear, however, what effect this might have had on the measured profiles.

A limited number of experiments have been performed in networks containing several generations of geometrically similar symmetric bifurcations, in which the length-to-diameter ratios of intermediate tubes were taken to be typical of the lung (i.e. about 3.5). Schroter & Sudlow (1969) measured profiles in a two-generation model, with one of the second-generation bifurcations being in the plane of the first and the other perpendicular to it. All the bifurcations were geometrically similar. They measured velocity profiles in the same two planes as before, at several distances from the flow divider, and at three Reynolds numbers, in each case. Berger *et al.* used a two-generation model, with all bifurcations in the same plane, and with a different area ratio at each generation, as given by the first two generations of a symmetric model of the lung proposed by Weibel (1963). Pacome extended this model to five generations, also using Weibel's data, and also keeping the first three generations coplanar, but for geometrical reasons the last two were not coplanar.

All these authors used their velocity-profile measurements in an estimate of pressure drop (see below), and the only profiles presented are in the daughters of the second bifurcation. When this is coplanar with the first, but not otherwise, the velocity fields are symmetric about the plane of the junctions, as expected. The slopes of the velocity profiles near the wall are again high at least half-way round the tube from the flow divider (note, however, that extrapolation of a measured velocity profile to zero is a notoriously inaccurate way of obtaining the shear-rate at the wall). However, the main impression is of a very complicated flow, varying significantly with distance downstream because of the redistribution of axial momentum by the secondary motions. Such disturbed (but still laminar) flow is to be anticipated almost everywhere in the lung and in much of the cardiovascular system.

Pressure drop
A quantity of fluid dynamical significance (and of great physiological importance, especially in the study of respiration) is the pressure loss associated with flow through a bifurcation. Direct measurement of the pressure in a single bifurcation does not give useful results, because the pressure differences are very small, the pressure varies

as much within the cross-section of the tube as it does across a junction (Jaffrin & Hennessey, 1972), and the contribution to the pressure drop of kinetic energy changes is large. One should therefore estimate the pressure drop by using the measured velocity profiles, and check the estimates by measuring the pressure drop across a model with several generations. Furthermore, we cannot use a momentum equation to estimate the pressure drop, because the walls of the tubes are not parallel, and so the flow divider, for example, exerts a longitudinal component of force on the flow in the parent tube (and, in any case, wall shear stress is extremely difficult to measure accurately). It is therefore necessary to use an energy equation.

We refer to a junction and its daughter tubes as a single unit (between stations 1 and 2 in fig. 5.1). The energy equation for the fluid within the unit can then be written as

$$(\hat{p}_1 + \tfrac{1}{2}\rho\widehat{q_1^2}) - (\hat{p}_2 + \tfrac{1}{2}\rho\widehat{q_2^2}) = D/Q, \qquad (5.1)$$

where Q is the volume flow-rate through the unit, D is the total rate of viscous energy dissipation in the unit, p is pressure, q is total fluid velocity, and suffixes 1 and 2 refer to stations 1 and 2. The symbol $\hat{\ }$ refers to the average of a quantity, weighted by the longitudinal velocity component w, across the cross-section of the tube: thus

$$\hat{p} = \frac{1}{Q} \int pw \, \mathrm{d}A, \qquad (5.2)$$

etc. (Pedley *et al.* 1977). Note that if either p or w is uniform across the tube, then $\hat{p}$ is equal to $\bar{p}$, the conventional average pressure. Also, if we write

$$\widehat{q^2} = \beta\bar{w}^2, \qquad (5.3)$$

where $\bar{w}$ is the average velocity, we note that $\beta = 1$ for parallel flow with a flat profile, and $\beta = 2$ for Poiseuille flow. For complicated three-dimensional flows such as ours, the only rational meaning that can be given to the words 'pressure drop' is that derived from (5.1) and (5.2).

In applying (5.1) to a bifurcation, two more simplifying assumptions are made. (*a*) We neglect the viscous dissipation at the junction itself, i.e. between stations 1 and 1', where the flow is thought to be largely determined by non-viscous mechanisms; this is reasonable since most of the dissipation takes place in the boundary layer in the daughter tube, where the velocity gradient is high. (b) We neglect the contributions to $\hat{q}^2$ and to D of the transverse components of velocity, u and v; this is very crude, leading to a possible underestimate of at least 10% in D (when u and v reach 30% of $\bar{w}$), but no more accurate, yet still manageable, alternative has been suggested.

Pedley, Schroter & Sudlow (1970a) used rather crude interpolation procedures to estimate the dissipation per unit length (dD/dz) at each station in each daughter tube, and at each Reynolds number for which profile measurements were made. The results were expressed as the ratio, Y, of the actual dissipation-rate to that for Poiseuille flow at the same Reynolds number in the same tube. D was calculated from dD/dz, and its ratio, Z, to the corresponding Poiseuille-flow value was reported. The results showed considerable scatter, but two general conclusions could be drawn: Y (i.e. dD/dz) decreased on average with increased distance, z, from the flow divider; both Y and Z decreased (on average) as Re decreased. Because most of the energy dissipation is expected to take place in the boundary layer growing on the flow divider, these authors proposed a model in which Y and Z would depend on Re, z, and the tube length, l, and diameter, d, in the same way as for entry flow in a straight tube. According to this model, Y and Z in any tube would be given by

$$Y = \tfrac{1}{2}\gamma(Re\,d/z)^{1/2}, \qquad Z = \gamma(Re\,d/l)^{1/2}, \qquad (5.4)$$

where the dimensionless constant γ is independent of Re, z, l, and d. These equations can be used to deduce values of γ from the values of Y and Z derived from the measured velocity profiles. Pedley et al. (1970a) found no systematic dependence of the values of γ obtained in this way on either Re or z/d (l/d was constant in the experiments), and therefore regarded the entry flow model as confirmed. The only systematic variation was that the values of γ obtained from second-generation tubes were uniformly smaller than those obtained from the first, by 25% on average. This indicates that greater complexity in the flow modifies the boundary layer on the flow divider. Nevertheless, this error was within two standard deviations, and the authors chose to use their overall mean value of $\gamma = 0.33$ for predictions of energy dissipation in the lung. Those predictions were in good qualitative agreement with physiological experiment (Pedley, Schroter & Sudlow, 1970*b*).

The values of β (see (5.3)) obtained from the measured profiles at a distance of 6 cm from the flow divider (close to station 2, fig. 5.1) showed *no* systematic variation with Re or with generation number, and the average value of 1.7 was used in all predictions.

The above results concerning dissipation suggest that in any system of symmetric branched tubes, with the same branching angle and area ratio, and with Reynolds numbers in the same range, D will be proportional to $Q^{5/2}$, and the loss of total head, averaged as in (5.1) and (5.2), will be proportional to $Q^{3/2}$. Berger *et al.* (1972) obtained conflicting results by a similar method. They found that a good fit to their data is given by $D \propto Q^2$, so that the loss of total head is directly proportional to Q. Such a linear relation is inherently very unlikely, although the constant of proportionality is much greater than in Poiseuille flow, and as yet is has not proved possible to explain the discrepancy (Pedley, 1977). However, it is interesting to note that a student of Calvet's, Pacome (1975), has extended the measurements of Berger *et al.* to five generations, but he measured pressure drops as well as velocity profiles (the lateral variations in pressure are small compared with the overall pressure drop in a network of this size). His most important result was that, when the kinetic energy terms ($\frac{1}{2}\beta\rho\bar{w}^2$) in (5.1) are correctly accounted for, the total head loss (D/Q) is accurately proportional to $Q^{3/2}$. This vindicates the model of Pedley *et al.* (1970*a*), even if doubts remain about the estimation of dissipation.

Berger *et al.* also disagreed with Pedley *et al.* in the value to be taken for the kinetic energy factor β. They found a value lying between 1.44 and 1.55 in generation 1, and between 1.09 and 1.29 in generation 2, and being in general lowest for the highest Reynolds numbers: the parent-tube Reynolds number took several values between 420 and 2800, at the highest of which the flow was turbulent. Douglass (1973) on the other hand found values of β around 2.0 in generation 1, and around 1.5 in generation 2, for turbulent flow at $Re > 10^4$, with no systematic dependence on Re.

All these discrepancies suggest that further experiments are still required. Finally we may note that a few measurements have been made on unsteady laminar flow in model branched tubes. Jaffrin & Hennessey (1972) measured the variation of static pressure at several sites in a two-generation model when the flow-rate was varied sinusoidally. The peak values of Re and the Womersley parameter in the parent tube were taken to be 1560 and 2.7 respectively in one case, and 1230 and 1.9 in another. There was a small phase shift between pressure drop and flow-rate at the higher frequency, but no amplitude response. The measurements confirmed that transverse pressure variations in a daughter tube are as great as the longitudinal variations in two generations. The pressure drop for expiratory flow was greater than that for inspiration. Brech & Bellhouse (1973) measured velocity profiles and shear stresses in pulsatile flow (with non-zero mean) through a single bifurcation, with parent-tube Re equal to 750 and 1500, and α equal to 22. However, they reported that the flow was quasi-steady, which at such a large value of α is extremely difficult to believe and casts some doubt on the accuracy of their measurements.

5.1.2 Theory

Attention is still restricted to steady laminar flow at high Reynolds number. The geometrical configuration of even a single bifurcation is so complicated that a complete analytical theory is out of the question, and we must await the development of suitable numerical methods before we can investigate all the details of the flow. (The flow in two-dimensional bifurcations has been analysed numerically, but since that inevitably excludes all the interesting phenomena, and gives no clue to the structure of the three-dimensional flow even in an idealised example (Smith, 1977a), it is not worth describing the results. One hopes that the methods can soon be extended to three-dimensional bifurcations.) Theoretical insight into the development of the secondary motions and the distortion of the axial velocity profile will be possible only with the help of crude simplifying assumptions, which must always be assessed critically. One such is to neglect the effect of viscosity on the gross changes in the core flow that take place in the region of the junction itself, on the grounds that viscosity has no time to influence the flow pattern except in the boundary layers. This is likely to be reasonable for a short distance beyond the junction, as long as separation does not occur, and will break down where secondary motions cause fluid from near the wall to penetrate into the core from the outside of the bend. The development of the boundary layers, both on the flow divider in the daughter tubes, and upstream of the junction in the parent, may be analysed separately.

Olson's observations on the flow in the elliptic transition region suggest that the non-viscous core problem can also be divided into two stages. First, one should compute the secondary motions (and the distortion of the axial velocity profile) caused by streamline curvature in the increasingly elliptic parent tube upstream of the flow divider. This neglects any blocking effect that the presence of the flow divider may have on the core flow just upstream, in the hope (based on the flow-visualisation studies reviewed above), that this is confined to a small region, especially when the flow divider is sharp. Secondly, one should calculate the effect that the initial curvature of the daughter tube will have on the already distorted flow entering it.

A tentative start has been made to the first stage by Sobey (1976a), who has calculated the distortion of a weakly sheared axisymmetric flow in a tube that is circular (radius a_0) for $z < 0$, and that has slowly varying (but not necessarily small) ellipticity downstream ($z > 0$). He solved the problem using classical secondary flow methods (reviewed by Horlock & Lakshminarayana, 1973), which involve the following stages:

(i) Calculate the potential flow in the slowly varying tube as a power series in ε (equal to a_0/l, where l is the length-scale for longitudinal variations). The leading term is a uniform stream with velocity equal to $\hat{W}_0/ab$, where $\hat{W}_0$ is the average velocity in $z < 0$, and $a_0a(z)$, $a_0b(z)$ are the major and minor axes of the elliptical cross-section. The next term in the velocity potential, of order ε, can easily be calculated, and is determined uniquely by the condition that volume flux is uniform along the tube. The error term is $O(\varepsilon^3)$.

(ii) Let the transverse vorticity in the upstream flow be $O(\delta\hat{W}_0/a_0)$, where $\delta = o(\varepsilon)$; then the streamlines of the $O(\varepsilon)$ potential flow will be a first approximation to the actual streamlines.

Assuming that vortex lines are carried along the potential flow streamlines one can compute a first approximation to the disturbance vorticity field. This includes longitudinal components, which are associated with secondary velocity components, of $O(\hat{W}_0\varepsilon\delta)$, as well as an $O(\hat{W}_0\delta)$ perturbation to the axial velocity.

(iii) These result in further distortions to the axial velocity profile, of $O(\hat{W}_0\,\delta^2)$, which can also be computed without too much difficulty.

This approximation procedure is self-consistent for values of z less than $O(a_0/\varepsilon\delta)$. Its greatest weakness is that it cannot treat large, $O(\hat{W}_0/a_0)$, upstream vorticity, and therefore any results can have at most qualitative relevance to real pipe flows, except perhaps those (as in some of Olson's experiments) with almost flat velocity profiles and thin boundary layers at the inlet.

Sobey's results can be summarised quite simply. When the entry profile is perfectly flat, the secondary velocity profiles are also flat, and these motions are stronger further away from the centre of the tube. The contours of constant axial velocity are ellipses that remain similar to the tube itself. When there is a slight shear, the secondary velocities are greater near the wall of the tube, and the contours of constant axial velocity are distorted so that the axial velocity is increased near the ends of the major axis, and decreased near the ends of the minor axis. The predicted secondary profiles agree qualitatively with those of fig. 5.6(a), except, of course, in the boundary layers on the walls.

An extremely crude estimate of the flow in the (straight) daughter tubes of a symmetrical bifurcation has been made by Scherer (1972), who ignored any perturbation to the flow in the parent tube upstream of the flow divider. He also assumed that the flow in the daughter tubes did not change with distance along them, and then calculated the simplest possible flow pattern consistent (i) with this assumption, (ii) with the non-viscous equations of motion, and (iii) with the fact that the flow must be symmetric about the plane containing the axes of all tubes in the bifurcation. The flow entering from the parent tube was taken into account by arranging that the axial and transverse velocity components on the axis of the daughter tube at its entrance were the same as would occur at that point if the parent tube continued uninterrupted. He worked out the details for a flat entry profile, of velocity $\hat{W}_0$, and his results are in rough qualitative agreement with Schroter & Sudlow's profiles, including the M-shaped profile in the plane normal to that of the junction.

Before further progress can be made on the core flow problem a theory will have to be developed to analyse the rapid inviscid distortion, over a short length of tube, of a flow with large transverse vorticity. It is difficult to see how any perturbation theory (such as Sobey's) can suffice.

Progress has also been made with the boundary layer analysis, though this is necessarily restricted to an idealised situation in which the core flow is not grossly distorted by secondary vorticity. Smith (1977a) examined the effect of dividing the flow into two without making it simultaneously turn a corner, by considering steady flow in an infinitely long, straight pipe of circular cross-section with radius a_0 (whose axis is the line $x = y = 0$) which is split by a semi-infinite plane $x = 0$ in $z > 0$. Poiseuille flow

$$w = W_0(x, y) \equiv \tfrac{1}{4}(1 - x^2 - y^2), \qquad p = p_0 - z, \qquad (5.5)$$

is present as $z \to -\infty$; here x, y, z are made dimensionless with respect to the tube radius a_0, and the velocity and pressure are non-dimensionalised with respect to $\hat{W}_0$ (equal to 4 times the centre-line velocity) and to $\hat{W}_0 \mu / a_0$ (equal to a_0 times the upstream pressure gradient) respectively. Smith analyses the flow on a long, boundary layer length-scale, $z = O(Re)$, where $Re = \hat{W}_0 a_0 / \nu$, but focusses attention on small values of $Z = z/Re$; at large values of Z the flow will tend towards the form of Poiseuille flow appropriate to tubes of semi-circular cross-section. A Blasius-type boundary layer is set up on the splitter plate, corresponding to the y-dependent outer velocity $w = \tfrac{1}{4}(1 - y^2)$. This generates a y-dependent normal velocity u which causes both a displacement of the inviscid core flow and a transverse velocity v inside the boundary layer and in the core. The net effect of these secondary motions is to transport fluid from the outer corner of the tube ($x = 0$, $y = 1$; see fig. 5.7) to the point of symmetry on the wall ($x = 1$, $y = 0$). Such a flow, however, cannot satisfy the equation of continuity unless there is an acceleration of the axial flow in the core near the point of symmetry, and this requires the presence of an additional axial pressure gradient of $O(Z^{-1/2})$. The associated pressure is itself responsible for a

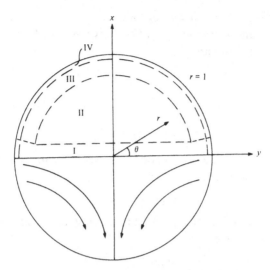

Fig. 5.7. Diagram of the four principal regions of the flow, with coordinate system and sketch of secondary flow direction in the inviscid core (region II). The flow divider occupies the plane $x = 0$ ($\theta = 0, \pi$).

vigorous secondary motion in a boundary layer near the outer wall $(r = (x^2 + y^2)^{1/2} = 1)$, which turns out to have a double structure, with an outer, inviscid region of thickness proportional to $Z^{1/4}$, and an inner viscous region of thickness proportional to $Z^{3/8}$ to reduce the tangential velocity to zero. We outline Smith's solution in the four principal regions of the flow, shown schematically in fig. 5.7.

The expectation of a Blasius-type boundary layer indicates that the variables should be rescaled according to

$$(u, v) = Re^{-1}(U, V), \qquad w = w, \qquad p = ReP$$

for $Z = O(1)$. The governing equations then are

$$\left.\begin{array}{c} U_x + V_y + w_Z = 0, \\ Uw_x + Vw_y + ww_Z = -P_Z + \nabla^2 w, \\ U\Omega_x + V\Omega_y + w\Omega_Z - \Omega w_Z - w_x U_Z + w_y V_Z = \nabla^2\Omega, \end{array}\right\} \quad (5.6)$$

where $\nabla^2 = \partial^2/\partial x^2 + \partial^2/\partial y^2$, Ω is the axial component of vorticity,

$$\Omega = -U_y + V_x,$$

and the pressure $P(Z)$ is independent of x and y.

The Blasius boundary layer, region I
Here the z- and x-components of velocity are given by the standard Blasius theory, and are

$$w = \tfrac{1}{4}(1 - y^2)f_0'(\tilde{\eta}) + O(Z^{1/2}), \left.\begin{matrix} \\ \\ \end{matrix}\right\}$$
$$U = Z^{-1/2}[\tfrac{1}{8}(1 - y^2)]^{1/2}(\tilde{\eta}f_0' - f_0) + O(1),$$

(5.7)

where

$$\tilde{\eta} \doteq xZ^{-1/2}[\tfrac{1}{8}(1 - y^2)]^{1/2}$$

and $f_0(\tilde{\eta})$ is the Blasius function. Associated with this flow is a transverse component of velocity,

$$V = v_0(\tilde{\eta}, x) + O(Z^{1/2}),$$

which can readily be calculated. This does not tend to zero as $\tilde{\eta} \to \infty$, but is given instead by

$$v_0 \sim -3\beta_0 y\tilde{\eta}/(1 - y^2) + O(1),$$

(5.8)

where

$$\beta_0 = \lim_{\tilde{\eta} \to \infty} (f_0 - \tilde{\eta}) > 0.$$

Thus there must also be a secondary core flow which is directed in towards the centre of the plate. The pressure perturbation mentioned above does not affect the flow region I to this order; it turns out that a self-consistent structure to the flow elsewhere can be formulated only if

$$dP/dZ = P_1 Z^{-1/2} + O(Z^{-1/4}),$$

(5.9)

where P_1 is a constant.

The core, region II
The Z-dependence of the above solutions as $Z \to 0+$ suggests that the perturbation to the oncoming flow in the core takes the form

$$w = W_0(x, y) + Z^{1/2}W_1, \qquad U = Z^{-1/2}U_1, \qquad V = Z^{-1/2}V_1.$$

Substitution into the governing equations (5.6) shows that U_1, V_1 and W_1 satisfy linearised inviscid disturbance equations that have a solution of the form

$$(U_1, V_1) = \frac{1}{W_0}\left(\frac{\partial\phi}{\partial x}, \frac{\partial\phi}{\partial y}\right), \qquad W_1 = -\frac{8}{1 - r^2}\left(\nabla^2\phi + \frac{2r}{1 - r^2}\frac{\partial\phi}{\partial r}\right),$$

(5.10)

where

$$(1 - r^2)\nabla^2\phi + 4r\,\partial\phi/\partial r = P_1(1 - r^2) \qquad (5.11)$$

and

$$\phi \sim \gamma x(1 - y^2)^{3/2} \quad \text{as } x \to 0. \qquad (5.12)$$

Here $\gamma = \beta_0/8\sqrt{2}$ and (r, θ) are polar coordinates in the transverse plane (fig. 5.7). The condition (5.12) determines ϕ uniquely; the only boundary condition that is satisfied on $r = 1$ is $\partial\phi/\partial r = 0$, which follows from (5.11).

For a given value of P_1, the problem for ϕ can, in principle, be solved; a method for doing so is given by Smith. This inviscid solution, of course, does not satisfy the no-slip condition on $r = 1$: if $A_0(\theta)$ stands for the unknown value of ϕ at $r = 1$, then (5.11) shows that

$$\phi = A_0(\theta) + \tfrac{1}{2}(1 - r^2)[A_0''(\theta) - P_1] + O[(1 - r)^3]$$

as $r \to 1$. The corresponding velocity components are

$$w \sim \tfrac{1}{2}(1 - r) - 4Z^{1/2}A_0''(\theta)/(1 - r) + \cdots$$
$$U^* \sim -2Z^{-1/2}(A_0'' - P_1) + \cdots, \qquad V^* \sim [2Z^{-1/2}/(1 - r)]A_0'(\theta),$$
$$(5.13)$$

where we have introduced velocity components in the r- and θ-direction such that

$$U^* = (Z^{-1/2}/W_0)\partial\phi/\partial r, \qquad V^* = (Z^{-1/2}/W_0)(1/r)\partial\phi/\partial\theta.$$

We note that this solution breaks down when the second term in w becomes comparable with the first, which occurs when $1 - r = O(Z^{1/4})$. A boundary layer must therefore be interposed at this distance from the wall. ·

For future reference it is useful to write down the forms taken by the function $A_0(\theta)$ at small values of θ and of $\alpha = \tfrac{1}{2}\pi - \theta$. Series solutions of the problem for ϕ lead to

$$A_0(\theta) \sim \tfrac{1}{2}P_1\theta^2 + \tfrac{8}{5}\gamma\theta^{5/2} - \tfrac{1}{6}P_1\pi\theta^3 + O(\theta^{7/2}) \qquad (5.14)$$

and

$$A_0(\theta) \sim A_0(\tfrac{1}{2}\pi) - \kappa\alpha^2 + O(\alpha^4), \qquad (5.15)$$

where κ is a constant depending on γ and P_1, and

$$A_0(\tfrac{1}{2}\pi) = 0.241\,P_1 + 0.721\,\gamma.$$

The inertial and viscous side-wall layers, regions III and IV
Here the appropriate normal coordinate is $\zeta = (1-r)Z^{-1/4}$, and
(5.13) shows that the corresponding scaling for the velocity
components is

$$w = Z^{1/4}w_1(\zeta, \theta), \qquad U^* = Z^{-1/2}u_1(\zeta, \theta), \qquad V^* = Z^{-3/4}v_1(\zeta, \theta).$$

Symmetry considerations mean that only the range $0 < \theta < \tfrac{1}{2}\pi$ need
be considered. Substituting into (5.6) and retaining only the leading
term as $Z \to 0+$, we find that the governing equations in region III
are inviscid, and represent a non-linear interaction between fluid
inertia and the pressure gradients in the axial and azimuthal directions (represented by P_1 and $\tfrac{1}{2}A_0'(\theta)$). The no-slip condition cannot
be satisfied at $\zeta = 0$, the only condition that can be imposed there
being that of zero normal velocity u_1. Thus as $\zeta \to 0$ the following
representation of the flow is anticipated:

$$w_1 \sim C(\theta), \qquad u_1 \sim \zeta D(\theta), \qquad v_1 \sim E(\theta),$$

where C, D, E are functions of θ to be found. Putting $\zeta = 0$ in the
governing equations shows that they satisfy

$$\left.\begin{aligned}
\tfrac{1}{4}C &= D - E', \\
\tfrac{1}{4}C^2 + EC' &= -P_1, \\
\tfrac{3}{4}CE - EE' &= \tfrac{1}{2}A_0'.
\end{aligned}\right\} \tag{5.16}$$

The viscous terms in the equations of motion become as large as
the inertia terms when $(1-r) = O(Z^{3/8})$; a viscous sublayer on this
scale will enable the no-slip condition to be satisfied. The appropriate equations in this sublayer (region IV) are obtained by setting

$$w = Z^{1/4}\tilde{w}(\eta, \theta), \qquad U^* = Z^{-3/8}\tilde{u}(\eta, \theta), \qquad V^* = Z^{-3/4}\tilde{v}(\eta, \theta),$$

where $\eta = (1-r)Z^{-3/8}$; the outer boundary conditions are

$$\tilde{w} \to C(\theta), \qquad \tilde{u} \sim \eta D(\theta), \qquad \tilde{v} \sim E(\theta) \quad \text{as } \eta \to \infty.$$

Note that the above scaling predicts values of the wall shear that are
much greater than those in the oncoming Poiseuille flow, being
$O(Z^{-1/8})$ as $Z \to 0+$.

Smith (1977a) has not obtained a full numerical solution to the
non-linear partial differential equations governing the flow in
regions III and IV. However, he has answered the most important

questions, which are whether there can be a solution of the proposed form, and if so how the unknown induced pressure gradient term can be determined. He makes it very plausible that a solution for the viscous layer (region IV) can be found for any negative value of P_1. What determines P_1 is the condition that the swirl velocity component V^* should be zero at each end of the boundary layers in regions III and IV (i.e. at $\theta = 0$ and $\theta = \frac{1}{2}\pi$). He does not *prove* that there cannot be a very rapid change of secondary flow direction in a small region near the corner $\theta = 0$, $r = 1$, but if a solution can be found for which this is merely a region of passive adjustment, then it is reasonable to expect it to be correct. Thus the function $E(\theta)$ derived from (5.16) is taken to vanish at both $\theta = 0$ and $\theta = \frac{1}{2}\pi$.

The nature of the solutions of (5.16) near $\theta = 0$ can be obtained with the help of (5.14). It is necessary that P_1 be negative (i.e. the induced pressure gradient is favourable, leading to the expected increase in pressure drop over that in undisturbed Poiseuille flow), and the following relations emerge:

$$C(\theta) \to C_0, \qquad D(\theta) \to D_0, \qquad E(\theta) \sim E_1(\theta) \quad \text{as } \theta \to 0+,$$

where

$$C_0 = 2|P_1|^{1/2}, \qquad D_0 = \tfrac{1}{4}(5 - \sqrt{17})|P_1|^{1/2}, \qquad E_1 = \tfrac{1}{4}(3 - \sqrt{17})|P_1|^{1/2}.$$
$$(5.17)$$

The choice of sign of the square root is discussed below. An expansion in integer powers of $\theta^{1/2}$, which is valid as long as there is no singular behaviour in the corner, shows that $C \equiv C_0$. This is confirmed by an expansion in powers of $\alpha = \frac{1}{2}\pi - \theta$, which with the help of (5.15), gives

$$C \equiv \bar{C}_0, \qquad D \to \bar{D}_0, \qquad E \sim \alpha\bar{E}_1 \quad \text{as } \alpha \to 0+,$$

where

$$\bar{C}_0 = 2|P_1|^{1/2}, \qquad \bar{D}_0 = \tfrac{1}{4}|P_1|^{1/2}[5 - (9 - 16\kappa/P_1)^{1/2}],$$
$$\bar{E}_1 = \tfrac{1}{4}|P_1|^{1/2}[-3 + (9 - 16\kappa/P_1)^{1/2}].$$

The last of equations (5.15), with $C \equiv 2|P_1|^{1/2}$, leads to the following non-linear ordinary differential equation for $E(\theta)$:

$$(\tfrac{3}{2}|P_1|^{1/2} - E')E = \tfrac{1}{2}A_0'.$$

This can be solved in $\theta > 0$, subject to the condition that $E \sim E_1\theta$ as $\theta \to 0+$, for all values of P_1, since when P_1 is fixed, $A_0(\theta)$ is

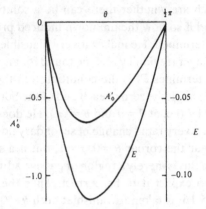

Fig. 5.8. Solution curves for the swirling velocities $A_0'(\theta)$, $E(\theta)$ at the outer edges of layers III and IV respectively ($P_1 = -3.35$). (After Smith, 1977a.)

determined from the solution of (5.11) and (5.12). The condition $E(\frac{1}{2}\pi) = 0$ then serves to determine P_1; Smith's numerical solution eventually yields the unique value $P_1 = -3.35$. The corresponding solutions for $E(\theta)$ and $A_0'(\theta)$, representing the swirl velocity at the inner and the outer edges of region III respectively, are shown in fig. 5.8.

These curves suggest that the swirl velocity is always in the same sense in region III, and this is confirmed by detailed solutions for small θ and for small α. However, solution of the equations for the viscous layer IV at both small θ and small α shows that the swirl velocity very near the wall is in the opposite sense to that in region III and the core. This indicates that there is a region of closed secondary streamlines near the wall, represented schematically in fig. 5.9; it should also be remembered that the magnitude of the swirl velocity in regions III and IV is very large compared with that in the core, being $O(Z^{-3/4})$. The region III and IV solutions also give the clue as to why the negative root $-\sqrt{17}$ was chosen in (5.17). Choice of the positive root would have implied an extra change of sign of swirl velocity in region III, in addition to the change of sign in region IV, and there seems to be no physical reason for this. However, there is not yet a mathematical or numerical demonstration that the positive root does not correspond to a self-consistent solution, and the possibility of non-uniqueness cannot be ruled out.

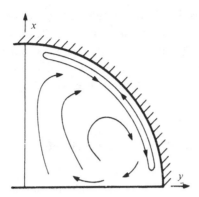

Fig. 5.9. Representative diagram of the induced secondary motion in one-half of a daughter tube. The flow velocities within the vortices adjoining the curved wall are much greater than within the core. (After Smith, 1977a.)

Another feature of the present solution worth noting is that the predicted secondary motion in the core is in the opposite sense to that which would be generated by curvature of the daughter tubes away from the junction. The experimental observations described in § 5.1.1 therefore demonstrate that the curvature present in the model bifurcations is a much more potent source of secondary motion than the branching *per se*, as analysed here.

Final validation of the flow structure outlined above will depend on a solution being obtained for the transition region, corresponding to the scale $z = O(1)$, between the upstream Poiseuille flow and the downstream boundary layer flow investigated here. Smith (1977a) has not attempted such a solution, but the flow 'far' upstream $(z \to -\infty)$ will presumably be a small perturbation to Poiseuille flow, and can therefore be described by an analysis such as that used both in § 4.5 for flow entering a bend (Smith, 1976b) and in § 5.2.2 below for flow upstream of a weak side-branch, as represented by (5.21) (Smith, 1976c). Although such an analysis is linearised, it does indicate a tendency towards three-dimensional flow separation at a distance $O(a_0 \log Re)$ upstream of a non-axisymmetric deformation (such as branching) in a straight tube. Smith (1977b) has analysed upstream separation in two-dimensional channels and in axisymmetrically distorted tubes, but not yet in the non-axisymmetric conditions of interest here. His axisymmetric results, however, are likely to be relevant to the flow upstream of a symmetric arterial stenosis, of either natural or artificial origin. An important conclusion is that the core flow at $O(a_0)$ upstream of a severe obstruction will be quite different from the Poiseuille flow assumed to be present far upstream.

5.2 Flow in asymmetric bifurcations

5.2.1 *Model experiments*

Flow visualisations and velocity profile measurements downstream of various asymmetric bifurcations have been made by Talukder (1975) and by Talukder & Nerem (1978) who have considerably extended the earlier results of Zeller *et al.* (1970). There are three possible causes of asymmetry in the flow through a single bifurcation (fig. 5.10), even if the velocity profile far upstream in the parent tube is axisymmetric. These are: different areas of the daughter tubes $(A_1 \neq A_2)$, different angles of branching $(\theta_1 \neq \theta_2)$, and different flow-rates $(Q_1 \neq Q_2)$. Talukder's (1975) experiments compared an asymmetric model $(\theta_1 = 15°, \theta_2 = 30°, A_1/A_2 = 2.12, A_1 + A_2 = 1.2\ A_0)$ with two symmetric models $(\theta_1 = \theta_2 = 15°$ and $30°$ respectively; $A_1 + A_2 = 1.2\ A_0)$, while his more recent work has examined right-angle branches off a straight tube $(\theta_1 = 0°, \theta_2 = 90°, A_1 = A_2 = A_0)$. In each case he has considered steady flow with various flow-rate ratios (Q_1/Q_2) and parent-tube Reynolds numbers, and has also made limited observations on unsteady flow. Talukder's work is intended to be applied to predicting the flow patterns and wall shear stress in large arteries, in particular to the junctions between the aorta and vein grafts which are used surgically to by-pass diseased coronary arteries. Some flow visualisations have also been made on flow in right-angle junctions by Dr Elspeth Brighton and Dr C. D. Bertram, with intended application to the avian respiratory system. In all cases the flow far upstream of the branch was fully developed, i.e. was Poiseuille flow when steady.

   The main features of steady flow, as observed by flow visualisation, can be described qualitatively as follows. Except when one of the branching angles is zero, the streams of fluid into each branch have to change their direction, as in a symmetric bifurcation. Thus

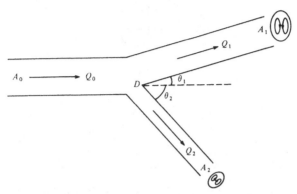

Fig. 5.10. Sketch of an asymmetric bifurcation, showing some of the parameters that must be specified.

secondary motions are generated in each stream by the same mechanism as in a curved tube, and the peak velocity is maintained near the outside bend of each stream. Even when one of the branching angles is zero, as in the case of a right-angled branch off an otherwise undisturbed parent tube, the motion near the walls of the parent tube has a complicated three-dimensional structure, and secondary motions are observed. For example, if the flow-rate down the side-branch in a right-angle junction is comparable with that in the straight branch, fluid particles that are initially near the far wall of the parent tube are carried transversely, in the boundary layer still, towards the side-branch. However, they may travel some way down the straight branch before turning back, eventually to enter the side-branch (fig. 5.11). Thus it is clear that the dividing stream surface can intersect the tube walls in a very complicated curve, which would defy detailed quantitative description. Only in the plane of symmetry is it possible to say that the dividing stream surface intersects the walls somewhere near the 'flow divider' (D in fig. 5.10), the exact position depending on the flow-rate ratio.

Another almost universal feature of the flow, absent in many symmetric bifurcations, is flow separation. This occurs *either* because a fluid stream has to negotiate a sharp corner, as at S_1 in fig. 5.12(a), *or* because there is sideways flow requiring a relatively strong transverse pressure gradient away from a boundary, which causes a pressure rise on the boundary and hence an adverse

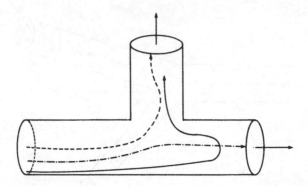

Fig. 5.11. Sketch of streamlines in steady flow in a T-junction, when the flow-rates in the two daughter tubes are comparable. Continuous curve, streamline near wall, remaining close to it; broken curve, streamline near centre line of parent tube; dot–dash curve, streamline between the two.

pressure gradient for flow along the boundary (S_2 in fig. 5.12(a), S_3 in fig. 5.12(b)). In all cases studied, the separation bubbles are bounded downstream, where the flow re-attaches, and away from the plane of symmetry, where the top and bottom boundary layers are continuous from parent tube to daughters. If regions of low wall shear are predisposed to the development of atherosclerosis (§ 1.2), then such flow-separation zones are likely to be dangerous, and it is important to have some idea of where they might occur in the circulation. In particular, surgical techniques such as vein grafting should be designed to avoid post-operative flow separation, especially when it is associated with dead-water regions, as in these steady flows. Observation of pulsatile flow in such models showed vigorous oscillatory movements in the dead-water regions, resulting in a more rapid turnover of fluid there; this would presumably be beneficial *in vivo*.

Talukder & Nerem (1978) have made velocity-profile measurements only in steady flow and only in the plane of the bifurcations. The best measurements were those made with a hot-film anemometer in the right-angled model, and are reproduced in figs. 5.13 to 5.15. The measurements are made in the side-tube at various parent and daughter-tube flow-rates, and all show a consistent pattern. A separated region is clearly present at one diameter from the junction, but by 2.5 diameters re-attachment has occurred, and,

(a)   (b)

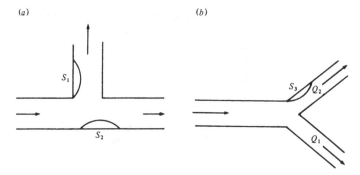

Fig. 5.12. Regions where separation is observed (a) in a T-junction, (b) in a symmetric junction with unequal flow rates ($Q_1 > Q_2$).

at least at the lower side-tube Reynolds number of 100, the velocity profile is remarkably close to its ultimate parabolic form. Very similar mean velocity profiles have been measured in vein grafts interposed (at 90°) between the iliac arteries of experimental dogs, indicating the practical relevance of these experiments (Rittgers *et al.*, 1976). It is interesting to note that the dead-water region appears to persist much further downstream in vein grafts that have been implanted for four months than in those implanted just before the measurements were made. This appeared to be associated with a chronic increase in internal diameter of about 50%, rather than with distortion of the junction wall by atheroma or other abnormalities.

Talukder (1975) also used a hot-film probe to measure wall shear stress at three sites in the daughter tube of a bifurcation with symmetric geometry and with symmetric or asymmetric flow-rates. In addition, he measured wall shear in the parent tube at a site about seven diameters upstream from the bifurcation, confirming in steady flow that the ratio between the actual wall shear and that predicted for Poiseuille flow at the same Reynolds number is very close to 1, but increases very gradually with Reynolds number. His measurements downstream of the bifurcation in steady flow, when the flow-rate through the branch where measurements were being made was only 30% of the total, showed a marked reduction below the Poiseuille value, especially at higher Reynolds numbers (500–1000 rather than 100–200). This is interpreted as evidence of a

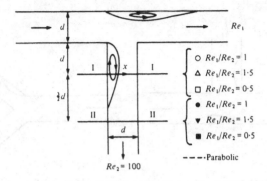

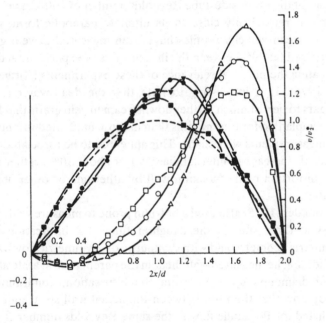

Fig. 5.13. Axial velocity profiles measured in the side-branch of a T-junction by Talukder & Nerem (1978). Re_1, Re_2 are the Reynolds numbers in the daughter tubes; v_2^* is the average velocity in the branch. Details given in diagram.

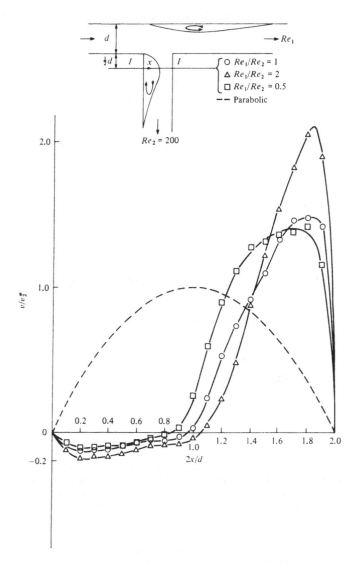

Fig. 5.14. Axial velocity profiles measured in the side-branch of a T-junction by Talukder & Nerem (1978). Re_1, Re_2 are the Reynolds numbers in the daughter tubes; v_2^* is the average velocity in the branch. Details given in diagram.

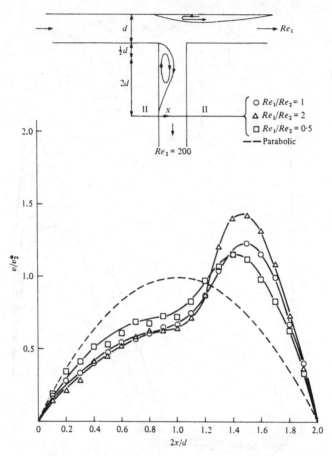

Fig. 5.15. Axial velocity profiles measured in the side-branch of a T-junction by N. Talukder & R. M. Nerem (personal communication). Re_1, Re_2 are the Reynolds numbers in the daughter tubes: v_2^* is the average velocity in the branch. Details given in diagram.

relatively stagnant region. Talukder's measurements in unsteady flow show large-amplitude fluctuations in wall shear even in the 'stagnant' region (fig. 5.16). Note the peaks on the probe output associated with reversed shear; the probe response cannot be accurate near shear reversal (§ 3.1 and appendix), and whether it is accurate at other times depends on the probe characteristics (not described by Talukder (1975)).

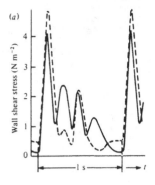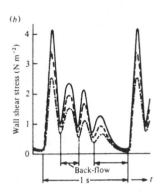

Fig. 5.16. Measurements of unsteady wall shear stress for asymmetric flow in a symmetric bifurcation; point of measurement is on the outer wall of the daughter tube containing the lower flow-rate, just downstream of the bifurcation (a region of separation in steady flow). (a) $\overline{Re}_0 = 500$; continuous curve, $\overline{Re}_2/\overline{Re}_1 = 0.43$; broken curve, $\overline{Re}_2/\overline{Re}_1 = 0.5$. (b) $\overline{Re}_2/\overline{Re}_1 = 0.43$; continuous curve, $\overline{Re}_0 = 500$; broken curve, $\overline{Re}_0 = 200$; dot–dash curve, $\overline{Re}_0 = 100$. $\overline{Re}_0$, $\overline{Re}_1$, $\overline{Re}_2$ are the mean Reynolds numbers in the parent tube and two daughter tubes, respectively. (After Talukder, 1975.)

Wall shear measurements have also been made, in a rigid cast of a canine descending aorta and its main branches, by Lutz *et al.* (1977) who used the electrochemical technique (§ 3.1). They made measurements in steady and unsteady flow at various sites on the ventral wall of the model aorta (approximately in the plane of symmetry) both upstream and downstream of the coeliac and superior mesenteric branches, but including sites actually in the bifurcation (fig. 5.17). Measurements were also made in the right renal branch, and at three positions round the circumference of the femoral artery. Typical results for steady flow are shown in fig. 5.17, and make it clear that the shear is low on the outside wall of a junction (inside of a bend) as at positions S_1 and S_3 in fig. 5.12, but is extremely high just downstream of a flow divider. This is to be expected both because new boundary layers are formed there and because high-velocity fluid is brought near to the flow divider by the splitting of the upstream flow. The effect is particularly pronounced here because the flow dividers protrude into the main stream (upper part of fig. 5.17(a)) and do not remain flush with the parent-tube wall (in this respect they resemble symmetric bifurcations); this

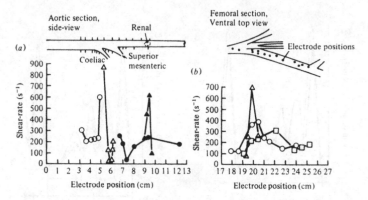

Fig. 5.17. Shear stress distribution pattern on ventral side of model artery in (a) aortic and (b) femoral sections. Electrode positions are indicated in the schematic drawing of the artery at the top of the figure. (a) Open circles, coeliac (flow-rate, 1.32 cm^3 s^{-1}); open triangles, superior mesenteric (flow-rate, 2.57 cm^3 s^{-1}); filled circles, abdominal aortic (flow-rate, 8.33 cm^3 s^{-1}); filled triangles, renal (the right renal branch, which runs perpendicular to the plane of the figure; flow-rate, 1 cm^3 s^{-1}). (b) In the femoral section, electrodes were placed on the inner wall (open triangles), centre line (open circles) and outer wall (open squares); flow-rate, 2.5 cm^3 s^{-1}. (After Lutz et al., 1977.)

protrusion probably also causes local separation, which is why the shear falls so rapidly downstream of the coeliac and superior mesenteric flow dividers. At higher flow-rates (greater than 12.5 cm^3 s^{-1}) turbulent fluctuations in wall shear were observed, consistent with the presence of separated flow at large Reynolds number. The measurements in the femoral artery (fig. 5.17(b)) suggest that relatively high shear is maintained half-way round the circumference of the tube, just downstream of the flow divider, as in curved tubes (chapter 4). Lutz et al.'s results in unsteady flow confirm those of Talukder (1975), showing large-amplitude oscillations, especially in regions where the steady flow is separated. The main qualitative lessons to be learnt from the work of Lutz et al. are (a) that small irregularities in geometry do normally exist and have a marked effect on local wall shear, and (b) that changes in the relative flow-rates in different branches can completely alter the distribution of wall shear in the parent vessel (e.g. occlusion of the coeliac branch appears to abolish the separation downstream of the superior mesenteric branch). These conclusions confirm the need for a detailed knowledge of local shear stress distribution before its effect on mass transport or arterial disease can be properly assessed.

5.2.2 *Effect of a weak branch on flow in the parent tube*

In this subsection and the next we describe the limited theoretical progress that has been made in analysing the flow through an asymmetric bifurcation in which a daughter tube branches at right angles off an otherwise straight parent. The theory summarised here enables the effect of the branch on the flow far upstream or downstream to be determined, while the next subsection presents a preliminary attempt at a detailed theory for the flow in the neighbourhood of a very small side-branch. The latter was intended as a model for flow into the intercostal arteries, but is unlikely to be quantitatively applicable because in order to allow progress the analysis was restricted to low-Reynolds-number flow in the side-branch.

Here we take a long straight parent tube of radius a_0, and model the branch as an extended, $O(a_0)$, region of the wall over which the normal velocity is a given, steady function of position and is small compared with a typical axial velocity in the parent tube. If the axial and azimuthal velocities across the mouth of the branch are of the same order of magnitude as the normal velocity, they do not influence the solution to the order retained, and can therefore be set equal to zero. However, because this theory neglects these details of the motion across the branch, it cannot describe the local flow accurately; it is expected to give correct results sufficiently far from the branch. The analysis to be presented was given by Smith (1976c), who also applied it to non-symmetric deformation of the parent tube wall (with no branch), and to unsteady side-tube velocities (or unsteady deformations), but not to unsteady parent-tube flows.

We adopt cylindrical polar coordinates $(a_0 r, \theta, a_0 s)$ in the parent tube, with corresponding velocity components $\hat{W}_0(u, v, w)$ and pressure $\mu \hat{W}_0 p / a_0$, so that the governing parameter in the Navier–Stokes equations is the Reynolds number, $Re = \hat{W}_0 a_0 / \nu$, which we take to be large. The mouth of the branch, B, is defined as the region $r = 1$, $0 < s < l$, $-\alpha(s) < \theta < \alpha(s)$, which determines the zero of the

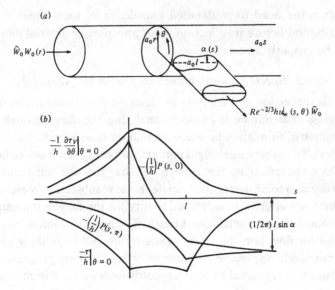

Fig. 5.18. (a) Coordinate system for flow in a side-branch. (b) Results of linearised theory, as described in text. (After Smith, 1976c.)

s-coordinate, and is taken to be symmetric about $\theta = 0$ (see fig. 5.18(a)). The boundary conditions on the velocity components are: on $r = 1$, $u = v = w = 0$ except in B, where $u = Re^{-2/3}hu_w(s, \theta)$ and h is a scale factor so that u_w is $O(1)$ for all h; as $s \to -\infty$, $w \sim W_0(r) \equiv \frac{1}{4}(1 - r^2)$, $u \sim v \to 0$, $p \to p_0 - s$, where p_0 is a constant.

The upstream condition of Poiseuille flow is the same as that imposed in the curved-tube problem of § 4.5, so it should come as no surprise that the upstream adjustment of the flow, as it approaches the asymmetric disturbance, takes the same form as in that case, i.e. the core remains undisturbed, but there is a marked interaction in the boundary layer. Indeed, it is this consideration that suggested the $Re^{-2/3}$ scaling chosen above for the normal velocity into the branch. The theory proceeds for $Re \to \infty$ while s and h remain $O(1)$. As in the curved-tube case, the scaling suggests that there are $O(Re^{-1/3})$ perturbations to the Poiseuille flow in the core, but the inviscid equations of motion are satisfied to this order by a solution that is identically zero. The core perturbation is therefore only $O(Re^{-2/3})$, and we do not calculate it (although for a detailed examination of the flow near the branch we would have to analyse perturbations of this order). In the boundary layer, however, perturbations of $O(Re^{-1/3})$ cannot be ruled out, and we write

$$u = -Re^{-2/3}U(s, \zeta, \theta) + O(Re^{-1}),$$

$$v = Re^{-1/3}V(s, \zeta, \theta) + O(Re^{-2/3}),$$

$$w = Re^{-1/3}W(s, \zeta, \theta) + O(Re^{-2/3}),$$

$$p = Re^{1/3}P(s, \theta) + O(1),$$

$$\text{(5.18)}$$

where

$$\zeta = (1 - r)Re^{1/3}$$

is the boundary layer coordinate, and P is independent of ζ from the radial momentum equation. The other governing equations are

$$\left. \begin{aligned} U_\zeta + V_\theta + W_s &= 0, \\ UV_\zeta + VV_\theta + WV_s &= -P_\theta + V_{\zeta\zeta}, \\ UW_\zeta + VW_\theta + WW_s &= -P_s + W_{\zeta\zeta}, \end{aligned} \right\} \qquad \text{(5.19)}$$

with boundary conditions:

$$\left. \begin{aligned} \text{on } \zeta = 0, \quad V = W = 0, \quad U &= -hu_w(s, \theta) \text{ on } B, \\ &= 0 \text{ elsewhere}, \\ \text{as } \zeta \to \infty, \quad U_\zeta, V, (W - \tfrac{1}{2}\zeta) &\to 0, \\ \text{as } s \to -\infty, \quad U, V, P, (W - \tfrac{1}{2}\zeta) &\to 0. \end{aligned} \right\} \qquad \text{(5.20)}$$

The use of the no-slip condition on B implies that v and w are expected to be $O(Re^{-2/3})$ there, which as explained above means that local details of the flow cannot be analysed to this order. There is no reason, however, apart from that of great labour, why the present analysis cannot be extended to the next order in $Re^{-1/3}$ and then the local details could be calculated if the distribution of v and w on B were known. In a real problem, of course, these distributions would not be known, and finding them would require a simultaneous solution of the problem in the side-tube (see § 5.2.3 below for a low-Reynolds-number example).

It is perhaps surprising that the relatively large pressure perturbation induced in the boundary layer should have negligible effect on the core flow. This is because the jet-like flow in the boundary layer is adequate to balance the induced pressure gradient. In an unbounded, external flow, or in a two-dimensional channel, on the

other hand, such pressure perturbations inevitably result in displacement of the primary flow (Stewartson, 1974; Smith 1976*d*, *e*).

Equations (5.19) are non-linear and, in general, require numerical solution (which has not been performed). However, the effect of
the disturbance on the flow far upstream can be assessed by
calculating the eigensolutions of the problem obtained by linearis-
ing about the oncoming Poiseuille flow. If we seek a solution of the
form (as $s \to -\infty$)

$$
\left.
\begin{aligned}
W &= \tfrac{1}{2}\zeta + \mathrm{e}^{\kappa s} \cos m(\theta - \varepsilon)f(\zeta), \\
U &= \mathrm{e}^{\kappa s} \cos m(\theta - \varepsilon)g(\zeta), \\
V &= \mathrm{e}^{\kappa s} \sin m(\theta - \varepsilon)k(\zeta), \\
P &= b\,\mathrm{e}^{\kappa s} \cos m(\theta - \varepsilon),
\end{aligned}
\right\}
\tag{5.21}
$$

where b, m, κ, ε are unknown constants (m is an integer), the
equations reduce to

$$
\kappa f + g' + mk = 0,
$$
$$
\tfrac{1}{2}\zeta \kappa k = mb + k'',
$$
$$
\tfrac{1}{2}\zeta \kappa f + \tfrac{1}{2}g = -\kappa b + f''.
$$

The functions f, g and k are to be zero at $\zeta = 0$ and f, g', k must be
zero as $\zeta \to \infty$. The problem is now virtually the same as that already
solved in § 4.5, and solution in the same way leads to the conclusions
that $\kappa = m$, that $g \equiv 0$ and that

$$
f = -k = [bm^{1/3}/(\tfrac{1}{2})^{2/3}]L(t),
$$

where $t = (\tfrac{1}{2}m)^{1/3}\zeta$ and $L(t)$ is defined by (4.102). The constants b,
m, ε remain undetermined. In general the solution will be the sum
over all positive integers m of terms like (5.21), with $\kappa = m$ and b, ε
depending on m; the values of b and ε in each term can be fixed only
after solution for all values of s. The term to persist furthest
upstream will be that with $m = 1$, and this will describe the maximum upstream response in any asymmetrically disturbed pipe flow
(cf. p. 226).

In the absence of a numerical solution to the non-linear equations
(5.19), detailed calculations can be made only for problems that can
be linearised for all s of $O(1)$, i.e. for which $h \ll 1$. The problem once

more becomes virtually the same as that defined by (4.98). We therefore write

$$W = \tfrac{1}{2}\zeta + h\bar{W}, \qquad (U, V, P) = h(\bar{U}, \bar{V}, \bar{P}),$$

retain only terms of $O(h)$ in (5.19), take Fourier transforms in s (defined by (4.100) and again denoted by $\tilde{P}(\omega, \theta)$ etc.), solve the resulting inhomogeneous Airy equations as in (4.102–3), and apply the boundary conditions derived from (5.20). The conditions at infinity show that $\tilde{B}(\omega, \theta)$, a term exactly analogous to the function $\bar{B}(\omega)$ introduced in (4.103), is given by

$$\tilde{B} = [\tilde{P}_{\theta\theta}/6Ai^2(0)](0 + \tfrac{1}{2}i\omega)^{-5/3}.$$

The boundary condition at $\zeta = 0$ then gives

$$(1/i\omega)\tilde{P}_{\theta\theta} + i\omega\tilde{P} = \tfrac{1}{2}\tilde{u}_w(\omega, \theta),$$

where $\tilde{u}_w$ is the Fourier transform of the normal velocity into the branch. If we now Fourier analyse in the θ-direction also, writing

$$[\tilde{u}_w, \tilde{P}](\omega, \theta) = \sum_{m=0}^{\infty} [\tilde{u}_{wm}(\omega), \tilde{P}_m(\omega)] \cos m\theta$$

since the branch (and by inference the flow into it) is symmetric about $\theta = 0$, we obtain

$$\tilde{P}_m = \tfrac{1}{2}\tilde{u}_{wm}/(i\omega - m^2/i\omega). \qquad (5.22)$$

From this result all quantities of interest are derived.

Smith (1976c) worked out the detailed results for a simple case, in which the mouth B of the branch had rectangular cross-section ($0 < s < l, -\alpha < \theta < \alpha$; α constant) and the component of velocity through it parallel to the axis of the branch (supposed to be at right angles to the axis of the parent tube) was uniform, i.e. $u_w = \cos \theta$ in B. Hence

$$\tilde{u}_{wm}(\omega) = (A_m/i\omega)(1 - e^{-i\omega l}),$$

where

$$\pi A_0 = \sin \alpha, \qquad 2\pi A_1 = 2\alpha + \sin 2\alpha,$$

$$(m^2 - 1)\pi A_m = 2(m \sin m\alpha \cos \alpha - \cos m\alpha \sin \alpha) \text{ for } m \geq 2.$$

Inverting (5.22) we have

$$P(s, \theta) = h\left[\tfrac{1}{2}A_0 sH(s) - \tfrac{1}{4} \sum_{m=1}^{\infty} \frac{A_m}{m} e^{-m|s|} \cos m\theta\right] \qquad (5.23)$$

minus a similar term with $s - l$ for s. The corresponding expressions for the perturbation to the axial wall shear, τ, and for the azimuthal wall shear, τ_v, are

$$\frac{\tau}{h} = \frac{(\frac{1}{2})^{2/3}}{6Ai(0)} \sum_{m=1}^{\infty} e^{ms} m^{-1/3} A_m \cos m\theta \quad \text{for } s < 0$$

$$= \frac{-(\frac{1}{2})^{2/3}}{3Ai(0)} \sum_{m=1}^{\infty} \frac{A_m \cos m\theta}{m^{1/3}} \left[\frac{e^{-ms}}{4} - \frac{3\sqrt{3}}{2\pi} \Gamma(\tfrac{2}{3})(ms)^{1/3} \right.$$

$$\left. + \frac{\sqrt{3}}{2\pi} \int_0^{\infty} \frac{\xi^{2/3} e^{-ms\xi}}{1 - \xi^2} d\xi \right] \quad \text{for } s > 0 \qquad (5.24)$$

and

$$\frac{\tau_v}{h} = \frac{-(\frac{1}{2})^{2/3}}{6Ai(0)} \sum_{m=1}^{\infty} e^{ms} m^{-1/3} A_m \sin m\theta \quad \text{for } s < 0$$

$$= \frac{-(\frac{1}{2})^{2/3}}{3Ai(0)} \sum_{m=1}^{\infty} \frac{A_m \sin m\theta}{m^{1/3}} \left[\frac{e^{-ms}}{4} \right.$$

$$\left. + \frac{\sqrt{3}}{2\pi} \int_0^{\infty} \frac{\xi^{-1/3} e^{-ms\xi}}{1 - \xi^2} d\xi \right] \quad \text{for } s > 0, \qquad (5.25)$$

again with similar expressions, containing $s - l$ for s, to be subtracted. The negatives of these quantities are plotted against s in fig. 5.18(b); the figure is taken from Smith's paper, and the negative sign comes from the fact that he was considering injection into the parent tube, which may be relevant to flow in veins. The curve of $-(1/h)P(s, \pi)$ against s shows that the net pressure rise (relative to the linearly falling Poiseuille pressure drop) is independent of θ. The curves of $-(1/h)P(s, 0)$, and of $-\tau/h$ and $-(1/h)\partial\tau_v/\partial\theta$ evaluated on $\theta = 0$, show that, on $\theta = 0$, the pressure falls and the axial shear rises ahead of the branch, while over the mouth of the branch itself the pressure rises sharply; downstream, the pressure perturbation continues to rise slowly to its asymptotic value, while the axial shear perturbation falls again to zero. On $\theta = \pi$ there is a pressure rise and a shear fall in the neighbourhood of the branch; it is these features that, when the outflow is stronger, lead to separation (S_2 in fig. 5.12). It is possible that such three-dimensional separation upstream of the branch could be described in a regular way by the non-linear equations (5.19) even when the outflow is too strong for the theory to be valid in the neighbourhood of the branch (as in the two-dimensional case (Smith, 1976d)). The azimuthal shear curve indicates that the secondary motion in the boundary layer is everywhere directed towards $\theta = 0$ (where the branch is), consistent with the excess pressure everywhere at $\theta = \pi$. As in the case of flow into a curve that subsequently straightens out without net change of direction, the solution breaks down when $s = O(Re)$, but the perturbation is by then very small anyway.

Another example that is relevant to the effect on the parent tube of a very small branch is that in which $u_w = \delta(\theta)\delta(s)$. In this case, $\tilde{u}_{wm} = 1/\pi$ for all m, and

$$\tilde{P}_m = \frac{1/2\pi}{i\omega - m^2/i\omega}.$$ (5.23a)

Inverting the various Fourier transforms gives:

$$\frac{P}{h} = \frac{1}{2\pi}\left[H(s) + \sum_{m=1}^{\infty} \tfrac{1}{2} e^{-m|s|} \cos m\theta \ \text{sgn}\ s\right] \quad \text{for all } s;$$ (5.26)

$$\frac{\tau}{h} = \frac{1}{2\pi \cdot 3Ai(0)} \sum_{m=1}^{\infty} (\tfrac{1}{2}m)^{2/3} e^{ms} \cos m\theta \quad \text{for } s < 0,$$

$$= \frac{1}{\pi \cdot 3Ai(0)} \sum_{m=1}^{\infty} (\tfrac{1}{2}m)^{2/3} \cos m\theta \left[\frac{\sqrt{3}}{2\pi}\Gamma(\tfrac{2}{3})(ms)^{-2/3} + \frac{e^{-ms}}{4}\right.$$

$$\left. + \frac{\sqrt{3}}{2\pi}\int_0^{\infty} \frac{e^{-ms\xi}\,\xi^{5/3}}{1-\xi^2}\,d\xi\right] \quad \text{for } s > 0;$$ (5.27)

$$\frac{\tau_v}{h} = \frac{-1}{2\pi \cdot 3Ai(0)} \sum_{m=1}^{\infty} (\tfrac{1}{2}m)^{2/3} e^{ms} \sin m\theta \quad \text{for } s < 0,$$

$$= \frac{1}{\pi \cdot 3Ai(0)} \sum_{m=1}^{\infty} (\tfrac{1}{2}m)^{2/3} \sin m\theta$$

$$\times \left[\frac{e^{-ms}}{4} + \frac{\sqrt{3}}{2\pi}\int_0^{\infty} \frac{e^{-ms\xi}\,\xi^{2/3}}{1-\xi^2}\,d\xi\right] \quad \text{for } s > 0.$$ (5.28)

The interesting features of this solution are the step rise in pressure everywhere within the boundary layer across $s = 0$, corresponding to the more gradual rise of fig. 5.18(b), and the leading term in the axial wall shear expansion for $s > 0$. The first term in the square brackets of (5.27) is, in fact, independent of m, so the axial shear has a term proportional to $s^{-2/3}[\delta(\theta) - \frac{1}{2}]$, which means that, in practice, the presence of a branch has a strong influence on the axial wall shear in a narrow band exactly downstream of the mouth of the branch. This is reminiscent of the 'corridor' downstream of a three-dimensional hump in a two-dimensional boundary layer, predicted both by Smith, Sykes & Brighton (1977), for the case where the horizontal length-scale of the hump is much greater than the boundary layer thickness, and by Brighton (1977) for the case where it is comparable with the boundary layer thickness. The wake of the side-hole will be examined in more detail in the next subsection. This wake may indeed be relevant to the observations of Cornhill & Roach (1976), reported in § 1.2.6, that the development of atheroma around the mouths of the intercostal arteries in choles-terol-fed rabbits varies from one pair of intercostals to the next. The present theory suggests that, if a second branch is directly in the wake of the first, the wall shear in the neighbourhood of the second will be modified.

The present theory, especially this last example, clearly exhibits singular behaviour near $s = 0$, and more terms in the Navier–Stokes equations must be taken into account in order to smooth them out and describe the detailed motion near the mouth of the branch. The next subsection represents an approach to the complete solution in a particular case.

5.2.3 Flow into a very small, very weak branch

Here we examine the details of the flow into a small side-branch in which the flow is so slow that it is inertia-free and satisfies the Stokes equations. This is clearly a far cry from large aortic branches, but it is the only case in which an attempt has been made to analyse the complete flow, including the effect of the velocities across the mouth of the tube as well as the normal velocity into it. The perturbation to the basic flow is so weak that the pressure perturbation in the $O(Re^{-1/3})$ boundary layer, which was central to the solutions of the last subsection, is $o(Re^{1/3})$ and therefore negligible to that order of magnitude. Thus the boundedness of the flow is irrelevant, and we can suppose the side-hole to be embedded in a plane wall. This

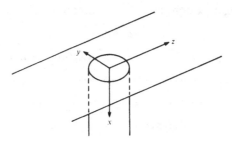

Fig. 5.19. Coordinate system for flow in a small side-branch.

problem has been formulated, and its solution outlined, by Sobey (1976b).

The geometry is shown in fig. 5.19. The side-tube is a cylinder of radius $b = \delta a_0$ ($\delta \ll 1$), coming off the parent tube (radius a_0) at right angles; the Reynolds number $Re = \hat{W}_0 a_0 / \nu$ of the parent-tube flow is still taken to be large. Let the flow-rate in the side-tube be $\hat{Q}$, and define $q = \hat{Q} a_0 / \hat{W}_0 b^3$. Using coordinates $b(x, y, z)$, where x is measured down the side-tube axis, z is the longitudinal coordinate in the parent tube, and y is at right angles to both, we define velocity components as follows (as usual, ˆ denotes a dimensional quantity).

In $x < 0$ (the parent tube)

$$(\hat{u}, \hat{v}) = \hat{W}_0 \delta q (u, v),$$

$$\hat{w} = \hat{W}_0 [\tilde{W}(-\delta x) + \delta q w],$$

where (u, v, w) depend on (x, y, z), and

$$\tilde{W}(\xi) = \tfrac{1}{4}(2\xi - \xi^2)$$

is the basic Poiseuille flow in the parent tube. In $x > 0$ (the side-tube)

$$(\hat{v}, \hat{w}) = \hat{W}_0 \delta q (v', w'),$$

$$\hat{u} = \hat{W}_0 \delta q [(2/\pi)(1 - y^2 - z^2) + u'],$$

where u' is the perturbation to the Poiseuille flow that develops for large values of x. We define pressures

$$P_0 + \rho \hat{W}_0^2 Re^{-1}(-\delta z + qp) \quad \text{in } x < 0,$$

$$\rho \hat{W}_0^2 Re^{-1} q [-(8/\pi)x + p'] \quad \text{in } x > 0.$$

The governing equations in $x < 0$ are therefore

$$u_x + v_y + w_z = 0,$$

$$\nabla^2 u - p_x = \delta Re\tilde{W}(-\delta x)u_z + \delta^2 Req(uu_x + vu_y + wu_z),$$

$$\nabla^2 v - p_y = \delta Re\tilde{W}(-\delta x)v_z + \delta^2 Req(uv_x + vv_y + wv_z), \quad (5.29)$$

$$\nabla^2 w - p_z = \delta Re[\tilde{W}(-\delta x)w_z - \delta\tilde{W}'(-\delta x)u]$$

$$+ \delta^2 Req(uw_x + vw_y + ww_z),$$

while in $x > 0$ they are

$$\nabla^2 \mathbf{u}' - \nabla p' = O(\delta^2 Req), \quad (5.30)$$

where $\mathbf{u}' = (u', v', w')$. The boundary conditions are

$$u = v = w = 0 \quad \text{on } x = 0, \quad y^2 + z^2 > 1,$$

$$u, v, w \to 0 \quad \text{as } (x^2 + y^2 + z^2) \to \infty, \quad x \le 0,$$

$$u' = v' = w' = 0 \quad \text{on } y^2 + z^2 = 1, \quad x > 0,$$

$$u', v', w' \to 0 \quad \text{as } x \to \infty,$$

together with the condition that all velocity and stress components must be continuous at the mouth of the hole. That is,

$$\{u, v, w, p - (\delta/q)z, v_x, -[\tilde{W}'(0)/q] + w_x\}$$

$$= \{u' + (2/\pi)(1 - y^2 - z^2), v', w', p', v'_x, w'_x\} \quad (5.31)$$

on $x = 0$, $y^2 + z^2 < 1$.

Since $\tilde{W}(-\delta x) = O(\delta)$ when $x = O(1)$, the right-hand side of (5.29) is $O(\delta^2 Re)$ for $x = O(1)$, and to make progress we suppose that $\delta Re^{1/2} \ll 1$, while q remains $O(1)$: note that q is the ratio of a typical shear-rate in the side-tube to that in the parent, and that $\delta^2 Req$ is the Reynolds number of the side-tube flow. In the region $|\mathbf{x}| = O(1)$ the equations reduce, to leading order, to Stokes's equations

$$\nabla^2(\mathbf{u}, \mathbf{u}') = \nabla(p, p'), \quad \nabla \cdot (\mathbf{u}, \mathbf{u}') = 0$$

and the primary velocity field in the parent tube enters the problem only through its contribution to the shear stress boundary condition in (5.31). The Stokes flow problem has a single dimensionless parameter in it, $\tilde{W}'(0)/q$; this equals $\frac{1}{2}q$ for the case of Poiseuille flow in the parent tube (the same theory can be set out for other distributions of parent-tube flow). For a given value of this parameter, the Stokes flow problem can, in principle, be solved as follows: (a) guess distributions of u, v, w over the mouth of the tube, say

$$u = U(y, z), \qquad v = V(y, z), \qquad w = W(y, z); \qquad (5.32)$$

(b) calculate the two Stokes flows, in $x < 0$ and $x > 0$, corresponding to these velocity distributions; (c) compute the pressure and shear stresses at $x = 0-$ and $x = 0+$, and equate them. This should result in three simultaneous integral equations for U, V, W that must be solved iteratively. In fact, unless there is a simple analytical form for the solutions in both $x < 0$ and $x > 0$, it will probably be simpler to solve the finite-difference form of Stokes's equations by direct numerical integration. Sobey (1977) found this to be true in the two-dimensional version of this problem (despite the existence of a simple solution in $x < 0$), but he has not yet embarked on the more difficult three-dimensional numerical problem. The solution in the present case is therefore still incomplete.

Nevertheless, it is useful to examine the nature of the solution in $x < 0$, in order to see how the wall shear in the parent tube is affected by the presence of the side-branch. Sobey (1976b) pointed out that the Stokes flow problem is a familiar one in elasticity theory (Sokolnikoff, 1956), and that the solution in $x < 0$ corresponding to the velocity distribution (5.32) at the mouth of the side-branch (B) can be constructed as the sum of three integrals over B. The integrand in each is the velocity distribution at B times the velocity field due to a particular kind of singularity at a particular point (y_0, z_0) of B. The three singularities are

(a) $v = w = 0,$ $u = \delta(y - y_0)\delta(z - z_0)$ on $x = 0,$

(b) $u = w = 0,$ $v = \delta(y - y_0)\delta(z - z_0)$ on $x = 0,$

(c) $u = v = 0,$ $w = \delta(y - y_0)\delta(z - z_0)$ on $x = 0.$

Using Fourier transforms over the (y, z)-plane, Sobey showed that the velocity and pressure fields in each case are:

(a) $u = -3x^3/2\pi\rho_0^5,$ $v = -3x^2(y - y_0)/2\pi\rho_0^5,$

$\qquad w = -3x^2(z - z_0)/2\pi\rho_0^5, \qquad p = -(3/2\pi)(2x^2/\rho_0^5 - 2/3\rho_0^3),$

$$(5.33a)$$

where $\rho_0^2 = x^2 + (y - y_0)^2 + (z - z_0)^2;$

(b) $u = -3x^2(y - y_0)/2\pi\rho_0^5,$ $v = -3x(y - y_0)^2/2\pi\rho_0^5,$

$w = -3x(y - y_0)(z - z_0)/2\pi\rho_0^5,$ $p = -3x(y - y_0)/\pi\rho_0^5;$

$$(5.33b)$$

(c) $u = -3x^2(z - z_0)/2\pi\rho_0^5,$ $v = -3x(y - y_0)(z - z_0)/2\pi\rho_0^5,$

$w = -3x(z - z_0)^2/2\pi\rho_0^5,$ $p = -3x(z - z_0)/\pi\rho_0^5.$

$$(5.33c)$$

The shear on the wall $x = 0$, a vector in the (y, z)-plane, is proportional to $\tau = [-v_x, -w_x]_{x=0}$, and is therefore given by

$$\tau = \frac{3}{2\pi} \iint_B [V(y_0, z_0)(y - y_0) + W(y_0, z_0)(z - z_0)]$$

$$\times [y - y_0, z - z_0]\rho_0^{-5} \, dy_0 \, dz_0 \qquad (5.34)$$

for $(y, z) \notin B$. This result will be true for any shape B of the mouth of the side-branch.

The most striking aspect of the result (5.34) is that it is independent of the distribution of normal velocity U into the side-branch; only the velocities *across* the mouth affect the wall shear in the region of Stokes flow. This means that any theory, such as that of Smith (1976c), which was presented in § 5.2.2 and which represents a branch by its distribution of normal velocity alone, cannot give a correct prediction of wall shear near the branch (if inertia is locally negligible). Of course, the flow-rate into the side-tube will affect the shear distribution because, through the parameter q, it enters the boundary conditions (5.31) that determine the functions $V(y_0, z_0)$, $W(y_0, z_0)$.

For large values of $\rho = (y^2 + z^2)^{1/2}$, the wall shear (5.34) can be expanded as follows:

$$\tau = (3[y, z]/2\pi\rho^5)\{yK_V + zK_W - K_{V1} - K_{W2}$$

$$+ (5/\rho^2)[y^2K_{V1} + yz(K_{V2} + K_{W1}) + z^2K_{W2}]\}$$

$$- (3/2\pi\rho^5)[yK_{V1} + zK_{W1}, yK_{V2} + zK_{W2}] + O(\rho^{-5}),$$

where

$$K_{V,W} = \iint_B [V(y_0, z_0),\ W(y_0, z_0)]\, dy_0\, dz_0,$$

$$K_{V1,2} = \iint V(y_0, z_0)[y_0,\ z_0]\, dy_0\, dz_0, \tag{5.35}$$

and similarly for $K_{W1,2}$. In cases with symmetry about the plane $y = 0$, V will be an odd function of y_0, and W an even function of y_0, so that $K_V = K_{V2} = K_{W1} = 0$. In this case the leading term in the far field depends only on $W(y_0, z_0)$, since

$$\tau = (3/2\pi)[y,\ z](zK_W/\rho^5) + O(\rho^{-4}), \tag{5.36}$$

and the distribution $V(y_0, z_0)$ of lateral velocity across the mouth of the hole is important only at the next order (through K_{V1}).

The Stokes flow solution in $x < 0$ breaks down where the neglected terms in (5.29) become $O(1)$. This occurs where

$$r = (x^2 + y^2 + z^2)^{1/2} = O(\delta^{-1}Re^{-1/2}) \gg 1,$$

which is equivalent to a dimensional distance $\hat{r}$ from the origin, where $\hat{r}/a_0 = O(Re^{-1/2}) \ll 1$. Thus the region where the Stokes flow breaks down is still very near the hole on the length-scale of the parent-tube radius, but is far from it on the length-scale of the side-tube radius. At this distance from the origin, the velocity perturbations (u, v, w) are $O(r^{-2})$ from (5.33), which is $O(\delta^2 Re) \ll 1$. Thus in this region we rescale the variables according to

$$\tilde{\mathbf{x}} = \delta Re^{1/2}\mathbf{x} = Re^{1/2}\hat{\mathbf{x}}/a_0, \quad \tilde{\mathbf{u}} = \delta^{-2}Re^{-1}\mathbf{u}, \quad \tilde{p} = \delta^{-3}Re^{-3/2}p,$$

and, noting that $\tilde{W}(-\delta x) = \tilde{W}(-Re^{-1/2}\tilde{x})$, we obtain from (5.29) (for $x < 0$ only)

$$\tilde{u}_{\tilde{x}} + \tilde{v}_{\tilde{y}} + \tilde{w}_{\tilde{z}} = 0,$$

$$\left.\begin{aligned}
\nabla^2\tilde{u} - \tilde{p}_{\tilde{x}} &= [-\tfrac{1}{2}\tilde{x} - \tfrac{1}{4}Re^{-1/2}\tilde{x}^2]\tilde{u}_{\tilde{z}} + O(\delta^3 Re^{3/2}q), \\
\nabla^2\tilde{v} - \tilde{p}_{\tilde{y}} &= [-\tfrac{1}{2}\tilde{x} - \tfrac{1}{4}Re^{-1/2}\tilde{x}^2]\tilde{v}_{\tilde{z}} + O(\delta^3 Re^{3/2}q), \\
\nabla^2\tilde{w} - \tilde{p}_{\tilde{z}} &= [-\tfrac{1}{2}\tilde{x} - \tfrac{1}{4}Re^{-1/2}\tilde{x}^2]\tilde{w}_{\tilde{z}} - \tfrac{1}{2}[1 - Re^{-1/2}\tilde{x}]\tilde{u} \\
&\quad + O(\delta^3 Re^{3/2}q).
\end{aligned}\right\} \tag{5.37}$$

With $q = O(1)$ the non-linear terms remain negligible, and with $Re \gg 1$ so does the second term in each square bracket. The boundary conditions are that $\tilde{\mathbf{u}} \to 0$ as $|\tilde{\mathbf{x}}| \to \infty$ and that $\tilde{\mathbf{u}} = 0$ on $\tilde{x} = 0$ except at the mouth of the side-branch. In the present scaling this is equivalent to

$$\tilde{\mathbf{u}} = (1, K_V, K_W)\delta(\tilde{y})\delta(\tilde{z}) \quad \text{on } \tilde{x} = 0, \tag{5.38}$$

where K_V, K_W are defined by (5.35) and $K_V = 0$ for a symmetrical flow. Since the problem is still linear its solution should incorporate the far-field of the Stokes flow, constructed from (5.33), as $|\tilde{\mathbf{x}}| \to 0$.

We may note that if near the wall but far from the hole the boundary layer approximation can be made, so that $\partial/\partial\tilde{y}$ and $\partial/\partial\tilde{z}$ are small compared with $\partial/\partial\tilde{x}$, the problem reduces to that considered at the end of the previous subsection, apart from the presence of K_V and K_W in (5.38). However the scaling of § 5.2.2 suggests that, if K_V and K_W are $O(1)$, they have no effect (to leading order) on the boundary layer solution. This would mean that the solution of the present problem should tend to that given by (5.26)–(5.28) as $(\tilde{y}^2 + \tilde{z}^2)^{1/2} \to \infty$, with the exception of those features, such as the step rise in pressure, that are a direct consequence of the confined geometry of the pipe and do not appear on the present much smaller length-scale for which the wall is represented as an unbounded plane. A useful check on both solutions would be to verify this conclusion.

The problem defined by the $O(1)$ terms of (5.37) and (5.38) can also be solved using Fourier transforms in $\tilde{y}$ and $\tilde{z}$. We follow Sobey (1976b) and examine only the case $K_V = 0$. Defining

$$\bar{f}(\tilde{x}, l, m) = \int_{-\infty}^{\infty} \int_{-\infty}^{\infty} e^{-im\tilde{y}-il\tilde{z}} f(\tilde{x}, \tilde{y}, \tilde{z}) \, d\tilde{y} \, d\tilde{z},$$

we obtain

$$(D^2 + \tfrac{1}{2}il\tilde{x})D^2\bar{u} = 0,$$

where

$$D^2 = d^2/d\tilde{x}^2 - k^2 \quad \text{and} \quad k^2 = l^2 + m^2,$$

which must be solved subject to

$$\bar{u}(0, l, m) = 1, \qquad \bar{u}(\tilde{x}, l, m) \quad \text{bounded as } \tilde{x} \to -\infty.$$

This has solution

$$\bar{u} = [1 - \alpha S(0)] e^{-k\xi} + \alpha S(\xi), \tag{5.39}$$

where $\xi = -\tilde{x}$, α is an arbitrary constant, and $S(\xi)$ is defined by

$$S(\xi) = \int_\xi^\infty \frac{\sinh k(\xi - \xi')}{k} Ai\,[\zeta(\xi')]\,d\xi',$$

where

$$\zeta(\xi) = (0 + \tfrac{1}{2}il)^{1/3}(\xi + k^2/\tfrac{1}{2}il). \tag{5.40}$$

The third of equations (5.37), together with the boundary condition (5.38) with $K_V = 0$, yields the following expression for $\bar{v}$ in terms of $\bar{p}$:

$$\bar{v}(\xi) = \frac{im\pi}{(0 + \tfrac{1}{2}il)^{1/3}} \left\{ \frac{Bi(\gamma)}{Ai(\gamma)} Ai[\zeta(\xi)] \int_0^\infty Ai[\zeta(\xi')]\bar{p}(\xi')\,d\xi' \right.$$

$$- Ai[\zeta(\xi)] \int_0^\xi Bi[\zeta(\xi')]\bar{p}(\xi')\,d\xi'$$

$$\left. - Bi[\zeta(\xi)] \int_\xi^\infty Ai[\zeta(\xi')]\bar{p}(\xi')\,d\xi' \right\},$$

where the (l, m)-dependence of $\bar{v}$ and $\bar{p}$ is omitted, and

$$\gamma = \zeta(0) = (0 + \tfrac{1}{2}il)^{-2/3}k^2. \tag{5.41}$$

The continuity equation can now be used to derive $\bar{w}$, and α is then determined by the condition that $\bar{w}$ satisfies the boundary condition (5.38), as follows:

$$\alpha = [k + ilK_W]/[S'(0) + kS(0)]. \tag{5.42}$$

Finally, substitution of $\bar{w}$ and $\bar{u}$ into the last of equations (5.37) gives the transform of the pressure:

$$\bar{p}(\xi) = (-\tfrac{1}{2}il/k^2)\{[1 - \alpha S(0)](1 + k\xi)\,e^{-k\xi}$$

$$+ \alpha[S(\xi) - \xi S'(\xi)]\} + \alpha Ai'[\zeta(\xi)](0 + \tfrac{1}{2}il)^{1/3}/k^2. \tag{5.43}$$

The particular quantities of interest are the pressure on the wall, proportional to $p(\xi = 0)$, and the perturbations to the shear on the wall, proportional to $[v_\xi, w_\xi]_{\xi=0}$. Their transforms are given by

$$\left.\begin{aligned}
\bar{p}|_{\xi=0} &= [\alpha Ai'(\gamma)(0 + \tfrac{1}{2}il)^{1/3} - \tfrac{1}{2}il]/k^2, \\
\bar{\tau}_v &= \bar{v}_\xi|_{\xi=0} = -im\bar{P}/Ai(\gamma), \\
\bar{\tau} &= \bar{w}_\xi|_{\xi=0} = (1/il)[k^2 - \alpha Ai(\gamma) - im\bar{\tau}_v],
\end{aligned}\right\} \tag{5.44}$$

where

$$\bar{P}(l, m) = \int_0^\infty Ai[\zeta(\xi')]\bar{p}(\xi')\,d\xi'.$$

Substitution of (5.43) into this integral, followed by further manipulation, leads to the following expression for $\bar{P}$ in terms of integrals whose asymptotic expansions (at least) can be deduced:

$$\bar{P} = Ai(\gamma) - (\alpha/k^2)Ai^2(\gamma) - (2K_W/k^2)(0 + \tfrac{1}{2}il)^{4/3}Ai'(\gamma)$$
$$- (il/4k^2)\{I_+(\gamma) + I_-(\gamma) + (ilK_W/k)[I_+(\gamma) - I_-(\gamma)] + 2\alpha I\}.$$
$$(5.45)$$

In this equation

$$I_\pm(\xi) = \int_0^\infty e^{\pm k\xi'}Ai[(0 + \tfrac{1}{2}il)^{1/3}\xi' + \xi]\,d\xi'$$

$$= (0 + \tfrac{1}{2}il)^{-1/3}\exp\left[\mp(0 + \tfrac{1}{2}il)^{-1/3}k\xi\right]$$

$$\times \int_\xi^{\infty \exp[i(\pi/6)\,\mathrm{sgn}\,l]} \exp\left[\pm k(0 + \tfrac{1}{2}il)^{-1/3}\zeta\right]Ai(\zeta)\,d\zeta \quad (5.46)$$

and

$$I = \int_0^\infty S(\xi)Ai[\zeta(\xi)]\,d\xi$$

$$= J_+ + J_-, \quad (5.47)$$

where

$$J_\pm = \frac{-(0 + \tfrac{1}{2}il)^{-1/3}}{2k}\int_\gamma^{\infty \exp[i(\pi/6)\,\mathrm{sgn}\,l]} I_\pm(\zeta)Ai(\zeta)\,d\zeta. \quad (5.48)$$

Note too that α (see (5.42)) is equal to $(k + ilK_W)/I_-(\gamma)$.

Numerical inversion of the above Fourier transforms has not yet been performed, and the only analysis concerns the asymptotic behaviour as

$$\rho = (\bar{y}^2 + \bar{z}^2)^{1/2} \to \infty \quad \text{and} \quad \rho \to 0.$$

The corresponding two-dimensional problem has been fully investigated by Sobey (1976b, 1977), and in addition to his numerical results he showed that the solution does match with the Stokes flow as $|\bar{z}| \to 0$ and with the appropriate small-$|\bar{z}|$ expansion of the boundary layer solution (Smith, 1974; Brighton, 1977) as

$|\tilde{z}| \to \infty$. The results given on p. 276 indicate that in the present, three-dimensional problem one should expect a different structure to the flow in the wake of the hole from that elsewhere. Thus the asymptotic form of the solution at large distances from the hole will depend on whether or not $|\tilde{y}| \to \infty$ as $\rho \to \infty$.

Wall pressure and shear far from the hole: $\rho \to \infty$, $|\tilde{y}| \to \infty$

In order to obtain the leading terms describing the flow when both $|\tilde{z}|$ and $|\tilde{y}|$ are large, we must take the leading terms of (5.44) as both l and m (and hence also k) tend to zero together. In this case $|\gamma| \ll 1$, and asymptotic analysis of the integrals in (5.46) and (5.48) shows that $I_{\pm}(\gamma) = O(|l|^{-1/3})$, $I = O(|l|^{-1})$; the leading term in $\bar{P}$ turns out to be that involving $[I_{+}(\gamma) + I_{-}(\gamma)]$. In the limit, the three quantities given in (5.44) become

$$\left.\begin{array}{l} \bar{p}|_{\xi=0} \sim -il/2k^2, \\[2mm] [\bar{\tau}_v, \bar{\tau}] \sim [-l, m][m(0+\tfrac{1}{2}il)^{-1/3}/6Ai(0)k^2], \end{array}\right\} \quad (5.49)$$

and inverting them gives

$$p|_{\xi=0} \sim \sin \Theta / 4\pi\rho, \qquad (5.50)$$

$$\left.\begin{array}{l} \tau_v \sim \dfrac{-2^{1/3}\Gamma(\tfrac{2}{3})}{18\pi Ai(0)\rho^{5/3}} \cos \left(\tfrac{5}{3}\Theta + \tfrac{1}{3}\pi\right) \operatorname{sgn} \tilde{y}, \\[4mm] \tau \sim \dfrac{\Gamma(\tfrac{2}{3})}{6\pi Ai(0)} \left[\dfrac{\sqrt{3}}{|2\tilde{z}|^{2/3}} H(\tilde{z})\delta(\tilde{y}) - \dfrac{2^{1/3} \sin \left(\tfrac{5}{3}\Theta + \tfrac{1}{3}\pi\right)}{3\rho^{5/3}} \right], \end{array}\right\} \quad (5.51)$$

where

$$\Theta = \tan^{-1} (\tilde{z}/|\tilde{y}|).$$

An important check on the present analysis is to verify that the above results for large ρ (made dimensionless with length-scale $\delta Re^{1/2}a_0$) match on to those of the previous subsection (where lengths are scaled with respect to a_0). The check is most easily made by comparing the present Fourier transforms (5.49) with those, such as (5.23a), that determine the flow (on length-scale a_0) due to a point sink in the wall of a tube. They can be seen to be exactly the same, with l in (5.49) corresponding to ω in (5.23a), apart from a factor π which arises because the θ-spectrum in § 5.2.2 is discrete, while the $\tilde{y}$-spectrum here is continuous. The functional forms of τ

and τ_v in (5.51) are somewhat different from the corresponding quantities in (5.27) and (5.28), again because of the inversion in terms of $\tilde{y}$ representing the wall as an unbounded plane, rather than θ representing the bounded geometry of the parent tube.

Despite these differences, however, the main qualitative features of the results are the same. The pressure distribution in (5.50) is the same as the limit of (5.26) as θ and $|s| \to 0$, apart from the different scaling factor of this subsection and the step rise in pressure represented by the term $H(s)$ in (5.26). This rise is once more a consequence of the inversion in terms of θ, instead of $\tilde{y}$. The azimuthal shear is also the same: if we let $|\tilde{y}/\tilde{z}| \to 0$ in (5.51) we see that $\tau_v \to 0$ upstream ($\Theta = -\frac{1}{2}\pi$) but not downstream ($\Theta = +\frac{1}{2}\pi$) where it decays as $|\tilde{z}|^{-5/3}$ (cf. (5.28)). The downstream decay of τ, on the other hand, is as $|\tilde{z}|^{-2/3}$ in a wake region whose width is zero on this length-scale, but as $|\tilde{z}|^{-5/3}$ outside it, which again agrees with (5.27). The only difference here is that the upstream decay of τ is also as $|\tilde{z}|^{-5/3}$, not purely exponential (i.e. zero) as predicted in (5.27); this too must be a consequence of the unbounded geometry assumed here (since the Fourier transform of τ is the same in each case) and would not be maintained on a length-scale a_0. More detail about the angular dependence of the wall pressure and shear can be seen from contour plots; single contours for each quantity have been derived from (5.50) and (5.51), and are shown in fig. 5.20. Further information would require numerical inversion of the Fourier transforms (5.44) for $O(1)$ values of $\tilde{y}$ and $\tilde{z}$, which has not yet been performed (see Brighton (1977) for similar results concerning the flow of a two-dimensional boundary layer over a small three-dimensional hump on the wall). Note from fig. 5.20(b) that the longitudinal component of wall shear is enhanced in regions surrounding the $\tilde{z}$-axis both upstream and downstream of the hole (as one would predict from a two-dimensional study (Sobey, 1977)), but is diminished in the region $-36° < \Theta < 72°$. Note too that according to these asymptotic results there are, in the wake of the hole, discontinuities both in the $\tilde{y}$-derivative of the axial component of wall shear (in addition to the singular term in (5.51)) and in the azimuthal shear component itself. No such discontinuities are predicted upstream. A closer examination of the wake region is required to understand this singular behaviour.

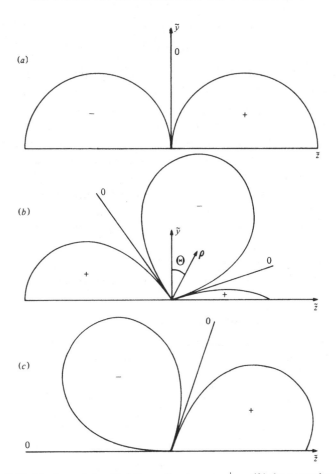

Fig. 5.20. Contour plots of (a) the wall pressure, $p|_{\xi=0}$, (b) the perturbation, τ, to the longitudinal component of wall shear, and (c) the azimuthal component of wall shear, τ_v, all derived from the asymptotic results for large ρ; see (5.50) and (5.51). Regions of the wall in which the plotted quantities are positive and negative are marked accordingly.

Wall pressure and shear far from the hole: $\rho \to \infty$, $|\tilde{y}| \ll |\tilde{z}|$

In order to examine the wake we must let m become much larger than l as $l \to 0$. The asymptotic expansions (5.49) remain unchanged as m is increased until $m = O(l^{1/3})$, when $\gamma = O(1)$ and the first two terms of (5.45) become as large as the term involving $[I_+(\gamma) + I_-(\gamma)]$. At this stage the three quantities in (5.44) take the form of powers of l times functions of $(m/l^{1/3})$; the respective powers of l

are

$$\bar{p}|_{\xi=0} \sim l^{1/3}, \qquad \bar{\tau}_v \sim l^{1/3}, \qquad \bar{\tau} \sim l^{-1/3}.$$

Upon inversion, the additional terms in the three physical quantities themselves can be seen to take the form of powers of $\tilde{z}$ times functions of a similarity variable $\eta = \tilde{y}/\tilde{z}^{1/3}$, as predicted by Jackson (1973): the power of $\tilde{z}$ corresponding to l^{ν} is $\tilde{z}^{-(4/3+\nu)}$. However, the details of this wake region for $\eta = O(1)$ are complicated and have not been examined in any further detail. Instead, we proceed to examine the innermost part of the wake, keeping $\tilde{y} = O(1)$ as $\tilde{z} \to \infty$; this requires that in the transformed quantities we keep m (and hence k) equal to $O(1)$ as $l \to 0$. We therefore require the asymptotic expansions of the integrals arising in $\bar{P}$, from (5.45), as $|\gamma|$, from (5.41), tends to infinity.

These expansions are derived by deforming the contour of integration in the integrals of both (5.46) and (5.48) to lie along the line

$$\arg \zeta = -\tfrac{1}{3}\pi \operatorname{sgn} l = \arg \gamma,$$

and then to use the method of stationary phase. The eventual results, when $\arg \xi = -\tfrac{1}{3}\pi \operatorname{sgn} l$ and $|\xi| \to \infty$, are

$$\left.\begin{aligned}
I_{-}(\xi) &\sim \frac{\exp\{i \operatorname{sgn} l[\tfrac{1}{4}\pi + |\xi|^{3/2}(\tfrac{2}{3}+\delta') - |\gamma|^{1/2}|\xi|]\}}{2\sqrt{\pi}(0+\tfrac{1}{2}il)^{1/3}(1+\delta')|\xi|^{3/4}}, \\[2mm]
I_{+}(\xi) &\sim \frac{\exp\{i \operatorname{sgn} l[|\xi|^{3/2}F(\delta') + |\gamma|^{1/2}|\xi|]\}}{2(0+\tfrac{1}{2}il)^{1/3}(2-\delta')^{1/2}} \\[2mm]
&\quad \times \operatorname{erfc}\left[e^{-i \operatorname{sgn} l\,\pi/4}\frac{|\xi|^{3/4}(1-\delta')}{(2-\delta')^{1/2}}\right],
\end{aligned}\right\} \quad (5.52)$$

where

$$\delta' = |\gamma/\xi|^{1/2} \leqslant 1 \quad \text{and} \quad F(\delta') = \tfrac{2}{3} - \delta' - (1-\delta')^2/(2-\delta').$$

This means that, as $|\gamma| \to \infty$,

$$I_{+}(\gamma) \sim \frac{\exp\{i \operatorname{sgn} l\,\tfrac{2}{3}|\gamma|^{3/2}\}}{2(0+\tfrac{1}{2}il)^{1/3}}\left[1 + \frac{e^{i \operatorname{sgn} l\,\pi/4}}{6\sqrt{\pi}|\gamma|^{3/4}}\right], \qquad (5.53)$$

which is $O(|l|^{-1/3} \exp \ldots)$, while

$$I_{-}(\gamma) = O(|\gamma|^{-3/4}|l|^{-1/3} \exp \ldots) = O(|l|^{1/6} \exp \ldots)$$

as $|l| \to 0$ with $|k| = O(1)$. The asymptotic forms for $J_{\pm}$ can be evaluated in a similar way, and we finally obtain

$$I \sim -\frac{\exp\{\mathrm{i}\,\mathrm{sgn}\,l(\tfrac{1}{4}\pi + \tfrac{4}{3}|\gamma|^{3/2})\}}{16\sqrt{\pi k}\,(0+\tfrac{1}{2}\mathrm{i}l)^{2/3}|\gamma|^{3/2}}\left(1 - \frac{e^{\mathrm{i}\,\mathrm{sgn}\,l\,\pi/4}}{2\sqrt{\pi}}\right), \qquad (5.54)$$

which is $O(|l|^{1/3}\exp\ldots)$ as $|l| \to 0$, so that the term $2\alpha I$ in $\bar{P}$ (see (5.45)) is the same order of magnitude as $I_-(\gamma)$, and smaller than $I_+(\gamma)$. Substitution of (5.52)–(5.54) into (5.44) then gives the following expansions for the Fourier transforms of the perturbation pressure and shear components at the wall:

$$\left. \begin{aligned}
\bar{p}|_{\xi=0} &\sim -2k - (\mathrm{i}l/4k^2)(\tfrac{25}{6} + 8k^2 K_W), \\
\bar{\tau}_v &\sim \mathrm{i}m[1 + (\mathrm{i}l/k)K_W + (0+\tfrac{1}{2}\mathrm{i}l)^{1/2}/2k^{3/2}], \\
\bar{\tau} &\sim \mathrm{i}l - (2k^2 - m^2)K_W/k + \tfrac{1}{4}\sqrt{\pi}k^{1/2}/(0+\tfrac{1}{2}\mathrm{i}l)^{1/2}.
\end{aligned} \right\} \quad (5.55)$$

If we ignore singularities at the origin these expressions can be inverted to give

$$\left. \begin{aligned}
p|_{\xi=0} &\sim 25\tilde{z}/48\pi\rho^2 + 1/\pi\rho^3, \\
\tau_v &\sim 3\tilde{y}\tilde{z}K_W/2\pi\rho^5 + \mathrm{sgn}\,\tilde{y}\,H(\tilde{z})/8\pi|\tilde{y}|^{1/2}|\tilde{z}|^{3/2}, \\
\tau &\sim 3\tilde{z}^2 K_W/2\pi\rho^5 - H(\tilde{z})/8\sqrt{\pi}|\tilde{y}|^{3/2}|\tilde{z}|^{1/2},
\end{aligned} \right\} \quad (5.56)$$

where the $\tilde{y}$- and $\tilde{z}$-dependence of p and of the first terms in τ_v and τ have been written out in full, but only the small-$\tilde{y}$ limits of the last terms have been given because of the complicated nature of the functions (of $\tilde{y}^2/\tilde{z}^2$) that arise. Subsequent terms in the expansion of which (5.56) gives the leading terms can be shown to be regular at $\tilde{y} = 0$ for both positive and negative values of $\tilde{z}$. Indeed, the pressure perturbation given above is everywhere regular (except at the origin) and varies in approximately the same manner as that given by (5.50).

However for $\tilde{z} > 0$ the wall shear distributions, τ_v and τ, contain terms that are singular at $\tilde{y} = 0$, although they can be written respectively as $\tilde{z}^{-5/3}$ and $\tilde{z}^{-1}$ times functions of η, as already predicted. This shows that yet a third distinct region must be considered, at the centre of the downstream wake. In this region there will be a deficit in axial wall shear (the minus sign in the last term of (5.56)) and an outwardly directed flow at the wall itself. This inner wake, or 'corridor', must have lateral length-scale comparable with the radius, b, of the side-branch, so that $y = O(1)$ ($\tilde{y} = O(\delta Re^{1/2})$), and the details of the motion are presumably dependent on the exact distribution of the flow across the mouth of the side-tube. The present analysis can be used to verify this if the actual distribution of u is used instead of the delta function in (5.38), so that $\bar{u}(\xi = 0)$ is a known function of l and m, say $\bar{u}_0(l, m)$, instead of 1. The only differences in equations (5.44), (5.45) and (5.55) are that $\alpha = (k\bar{u}_0 + ilK_W)/I_-(\gamma)$, and that all terms not involving α or K_W are multiplied by $\bar{u}_0$. As an example, suppose that Poiseuille flow enters the mouth of the side-branch:

$$\tilde{u}(0, \tilde{y}, \tilde{z}) = 1 - (\tilde{y}^2 + \tilde{z}^2)/\tilde{b}^2,$$

where $\tilde{b} = \delta Re^{1/2}$ is the dimensionless radius of that branch. Then

$$\bar{u}_0 = J_2(k\tilde{b})/\pi k^2,$$

where J_2 is a Bessel function, and the singular term in, for example, $\bar{\tau}_v$ (see (5.55)) becomes instead

$$im(0 + \tfrac{1}{2}il)^{1/2}J_2(k\tilde{b})/2\pi k^{7/2}.$$

Inversion shows that, as $\tilde{z} \to \infty$, this contribution to τ_v is

$$\frac{\tilde{b}^2 \, \text{sgn} \, \tilde{y} \, H(\tilde{z})}{64\pi^2 |\tilde{z}|^{3/2} |\tilde{y}|^{1/2}} F(\tfrac{3}{4}, \tfrac{1}{4}; 3; \tilde{b}^2/\tilde{y}^2) \quad \text{for } |\tilde{y}| > \tilde{b}$$

and

$$\frac{\tilde{b}^{1/2} \tilde{y} H(\tilde{z})}{16\pi^{5/2} |\tilde{z}|^{3/2}} \frac{\Gamma(\tfrac{3}{4})}{\Gamma(\tfrac{5}{4})} F(\tfrac{3}{4}, -\tfrac{5}{4}; \tfrac{3}{2}; \tilde{y}^2/\tilde{b}^2) \quad \text{for } |\tilde{y}| < \tilde{b},$$

where F is the hypergeometric function (Gradshteyn & Ryzhik, 1965). It can be readily confirmed that these two expressions are equal at $|\tilde{y}| = \tilde{b}$, and nowhere become infinite. Note too that they tend to zero as $\tilde{b} \to 0$ for all $\tilde{y}$.

We are thus forced to conclude that however small and weak the side-branch is, there is downstream of it a narrow corridor in which the flow has a different behaviour from that in the $\tilde{y} = O(\tilde{z}^{1/3})$ wake predicted by Jackson (1973). Such corridors have been predicted downstream of somewhat larger perturbations, such as a three-dimensional hump in a two-dimensional boundary layer, when the lateral length-scale of the hump is comparable with the thickness of the boundary layer (Brighton, 1977). However, the conclusion in this case is more surprising, because one might have expected the effect of lateral diffusion (in the $\tilde{y}$-direction) to smooth out the detailed influence of the original disturbance over such a small length-scale. That it does not do so must be a consequence of the secondary motions induced by the shear flow past the three-dimensional disturbance, although more research will be needed to understand all the implications of this result. The existence of the corridor adds considerable weight to the above argument (p. 276) that passive mechanical factors may be responsible for the observations of Cornhill & Roach (1976) on the distribution of atheroma downstream of the intercostal arteries in cholesterol-fed rabbits.

Wall pressure and shear near the hole: $\rho \rightarrow 0$
This region can be examined by letting l and m (and hence k) tend to infinity together. In this limit, too, $|\gamma| \rightarrow \infty$, so that (5.55) again gives the required Fourier transforms, and (5.56) again gives the asymptotic form of the quantities of interest, although the singular terms are relevant only when $\tilde{z} > 0$ and $|\tilde{y}/\tilde{z}| \ll 1$, i.e. in the corridor. The important thing to notice about the non-singular terms in the wall shear is that they are exactly the same as those given by (5.36); that is, the small-ρ limit of the present expansion does match with the large-ρ limit of the Stokes flow solution, apart from the existence of the corridor which is, of course, absent when there is no shear flow past the mouth of the side-hole.

5.3 The instability of flow in the aorta

We saw in § 1.2 that blood flow in the aorta of men and horses (and probably other large mammals) appears normally to become turbulent at or shortly after the time of peak aortic velocity. The results shown in fig. 1.26 indicate that turbulence tends to occur if the peak Reynolds number, $\hat{R}e$, exceeds about 250α, where α is Womersley's parameter. In this section we briefly review the possible origins of aortic turbulence.

It is clear from the relative absence of turbulence during the early part of systole and during diastole that the amplitudes of the high-frequency disturbances that constitute the turbulence grow very rapidly just after peak systole, and die away again before the end of the beat. A detailed spectral analysis by Parker (1977), in terms of wave *number* (frequency divided by instantaneous ensemble-averaged velocity), confirms the observation. This suggests that the growth of the disturbances represents some form of hydrodynamic instability rather than the convection or propagation to the measurement site of fully developed disturbances from upstream. Furthermore, Nerem *et al.* (1972) and Nerem *et al.* (1974*a*) report that the turbulence is not confined to 'first-beat blood', i.e. to the blood that has been ejected from the ventricle during the beat in which it is observed. Convection of the disturbances known to be present in the left ventricle cannot account for the presence of turbulence in 'second-beat blood'. On the other hand, these authors noted a marked reduction of turbulence intensity with distance down the aorta, which is consistent with the presence of disturbances whose amplitude, albeit growing from a local instability, is proportional to the amplitude existing before growth began. Thus the amplitude in 'first-beat blood' would be expected to be greater than that farther downstream. These observations are inconsistent with an explanation that relies on wave propagation along the vessel walls to transmit the disturbances to the measurement site, since, according to the data presented in § 2.1, that would cause the amplitude to decay too rapidly: the disturbance amplitude decays like $e^{-kx/\lambda}$, where k lies between 0.7 and 1.0 (see p. 78) and an upper limit for the wavelength λ is one tube diameter. Hence the disturbances would lose at least 50% of their amplitude per diameter, a much greater loss than is observed.

Thus the questions to be answered in this section are what mechanisms of instability might be operative in the aorta, and how might they account for the critical peak Reynolds number $\hat{R}e_c$ being approximately 250α. The proportionality of $\hat{R}e_c$ and α can be explained in general terms if the instability mechanism is quasi-steady, i.e. if the unsteadiness of the flow plays only a modulating part in the mechanism (as suggested by the short periods and growth times of the observed disturbances compared with the duration of systole). This follows because if growth of disturbances is possible whenever the local Reynolds number Re exceeds a certain critical value, such as about 2300 for Poiseuille flow in a straight tube, then infinitesimal disturbances will grow to an observable amplitude only if Re exceeds its critical value for a sufficient length of time. Higher values of α mean a higher frequency and hence a shorter time for Re to exceed its critical value. Therefore, on the assumption that the growth rate of the disturbances increases with Re, it is necessary for the peak Reynolds number $\hat{Re}$ to increase in order for turbulence still to be observed as α increases. This argument is consistent with experimental observations on slowly modulated Poiseuille flow by Sarpkaya (1966).

The most obvious mechanism that can be expected to produce instability just after peak systole is that associated with the appearance of a point of inflection in the velocity profile. In steady parallel flows this is known to cause a strong instability at values of the Reynolds number well below that at which profiles without such an inflection point become unstable. This was the mechanism favoured by Nerem et al. (1972), and will be further discussed below. First, however, we consider other mechanisms, associated with vessel-wall elasticity and vessel curvature.

Flow in an elastic tube can cause instability by the mechanism that produces aerodynamic flutter in flow past any elastic wall. That mechanism is essentially the same as Kelvin–Helmholtz instability in which inviscid flow past a plane, indefinitely extensible boundary at rest generates growing waves on that boundary. An elastic boundary resists such deformations, but if the velocity of the flow is great enough there will be instability. Shayo & Ellen (1974) have analysed the stability of inviscid flow through a finite section of an elastic tube of circular cross-section, of length l and undisturbed radius a, joined at each end to a segment of rigid tube also of radius a. The tube wall is taken to be homogeneous, isotropic, linearly elastic and thin, and its deformations are described by Flügge's shell theory. This model is clearly not directly applicable to an artery *in vivo*, but the results should be qualitatively relevant. The analysis shows that instability will occur first for either the 'beam' mode (azimuthal wave number 1) or the 'shell' mode (azimuthal wave number 2) in which the cross-section of the tube becomes elliptical. The 'beam' mode, in which the whole vessel will be displaced sideways at any station, is unlikely to be relevant for a tethered aorta, so we quote results only for the 'shell' mode. Shayo & Ellen (1974) show that the critical velocity U above which instability is expected is given approximately by

$$U^2 = E\beta^3 \omega_n^2 / 36\rho A Q_n,$$

where

$$\omega_n^2 = (4 + A^2)^2 - (4 + A^2)^{-2}[112 + 100A^2 + (24 - 9/\beta^2)A^4 + A^6],$$

$$Q_n = \frac{A}{8}\left[1 - \frac{1.68A}{n\pi^2} - \frac{A^2}{12} + \frac{0.336A^3}{n\pi^2} + \frac{A^4}{230}\right],$$

$$A = n\pi a/l,$$

and n is the longitudinal wave number. E is the Young's modulus of the wall, ρ is its density, $\beta = h/a$ the thickness–radius ratio, and we have taken the Poisson's ratio to be 0.5. If we choose $E = 5 \times 10^5$ N m^{-2}, $\beta = 0.15$, $\rho = 10^3$ kg m^{-3}, to represent the aorta, we find that the most unstable mode for $a/l = O(1)$ is $n = 3$, where the critical velocity is about 3.6 m s^{-1}. For longer elastic segments (a/l smaller) or for larger Young's modulus (as in more peripheral arteries), the critical velocity is greater. Thus the predicted critical velocities are significantly greater than those actually present in the aorta, and this is therefore unlikely to be a relevant mechanism of instability. Furthermore, it is hard to see how the buckling of the tube wall in the shell mode can lead to turbulence of the sort observed, except through local flow separation off large-amplitude constrictions, which are not observed. In fact this mode of instability is more relevant to the model experiments reported in chapter 6 and to relatively thin-walled veins than to the observation of turbulence in the aorta.

Curvature of a tube appears to offer the possibility of instability in the form of Taylor–Görtler vortices on the outside wall, where the primary flow is parallel to a surface that is concave in the flow direction. If the flow is assumed to be quasi-steady, quasi-two-dimensional and quasi-parallel to the wall, then instability is predicted to occur if

$$T = (U_\infty \delta_2 / \nu)(\delta_2 / R)^{1/2} \geqslant T_c, \qquad T_c = 0.3, \qquad (5.57)$$

where U_∞ is the velocity just outside the boundary layer on the concave wall, R is the radius of curvature of the wall, ν is the kinematic viscosity of the fluid and δ_2 is the momentum thickness of the boundary layer (Schlichting, 1968, p. 506). If the boundary layer grows gradually in the x-direction, as for the Blasius layer on a flat plate, then transition to turbulence is observed to take place if $T \gtrsim 7$. These results will be relevant to quasi-steady flow near the entrance of a curved tube as long as the developing secondary motions do not affect the stability criterion. In the case of a flat plate, the momentum thickness is given by

$$\delta_2 = 0.664(\nu x/U_\infty)^{1/2},$$

where x is the distance from the leading edge (Schlichting, 1968, p. 131). Combined with (5.57), with $T_c = 7$, this shows that transition to turbulence is predicted if

$$Re \geqslant 28\,000\delta^{-2}(a/x)^3, \tag{5.58}$$

where Re is based on the tube radius, a, and $\delta = a/R$. Now the theories of §§ 3.2 and 4.4 suggest that the boundary layer thickness does not grow indefinitely with x, being limited both by curvature (in steady flow) and by the Rayleigh layer thickness (in unsteady flow) to a value roughly equal to its value at $x = 5$ cm. In the canine aorta $a = 0.75$ cm and $\delta \approx 0.2$, so (5.58) predicts turbulence if $Re > 2360$, i.e. $U_\infty \gtrsim 1.25$ m s^{-1}. This value of Re is comparable with the critical Reynolds number for flow in a straight tube, and suggests that steady flow in a tube like the aorta might become unstable by either of the two mechanisms. Furthermore, the predicted critical velocity is comparable with the peak measured velocity in the aorta. However, according to (5.58) the flow should become more unstable (i.e. turbulent at a lower Reynolds number) as the curvature δ increases, whereas experiments by White (1929) and Taylor (1929) show that the critical Reynolds number for flow in a curved tube increases markedly as δ increases, which is not consistent with the predictions. While this does not completely rule out the possibility of quasi-steady instability associated with vessel curvature, it does suggest that the neglected secondary motions may have an important stabilising influence that should be examined further.

The above arguments have been based on the assumption of quasi-steady flow. However, Seminara & Hall (1976, 1977) have demonstrated the existence of a mechanism of instability for flow with curved streamlines for which the unsteadiness of the flow is vital. Their theory (and experiment) concerned the flow between concentric circular cylinders when the outer one is at rest and the inner one (radius a) is rotated with an angular velocity that varies sinusoidally with time with zero mean: $\Omega = (U_0/a)\cos\omega t$ (the corresponding steady motion is that which leads to Taylor vortices). They found that the flow becomes unstable when the Taylor number, $2(U_0\delta_S/\nu)^2(\delta_S/a)$, based on the thickness $\delta_S = \sqrt{(\nu/\omega)}$ of the Stokes layer on the inner cylinder, exceeds a critical value (about 164) and that the instability takes the form of toroidal vortices as in steady flow, except the lateral and longitudinal length-scale of these vortices is δ_S, not the difference in cylinder radii. Seminara & Hall have pointed out that instability by this mechanism might occur near the inside wall of the arch of the aorta, in the same way that Taylor–Görtler vortices might occur on the outside wall. The only factor that would tend to prevent it is the presence of secondary motion. No direct evidence, *in vivo* or in model curved tubes, is yet available.

We revert now to the 'obvious' instability mechanism, associated with the point of inflection that is present in the velocity profile during deceleration. The first question to be asked is whether the flow can be regarded as quasi-steady: as remarked above, the high frequency and rapid growth-rate of the observed disturbances, compared with the frequency of the basic flow, suggest that it can. More formally, a time-scale associated with eddies that fill the vessel is a/U, where a is the vessel radius and U the local ensemble-average velocity. This is an upper limit for the disturbance time-scale because most eddies are smaller. This time is short compared with the duration of the decelerative phase (approximately equal to $2\pi/6\omega$, where $\omega/2\pi$ is the heart-rate in Hz) as long as $\omega a/U \ll 1$, i.e. as long as $\alpha^2/Re \ll 1$, where Re is the instantaneous Reynolds number and α is Womersley's parameter; in the canine aorta, $\omega a/U < 0.1$ if $U \geqslant 24$ cm s^{-1}. The time-scale is also much shorter than the diffusive time-scale a^2/ν as long as $Re \gg 1$, which is clearly satisfied for almost all of systole. Thus the flow in the canine aorta is quasi-steady for much of systole. In a similar problem, that of the stability of the flow in a channel that is suddenly blocked off so that the Poiseuille flow present before the blocking decays away to zero in a time proportional to a^2/ν, Hall & Parker (1976) have confirmed the instability problem to be quasi-steady by using a

WKB expansion. A similar approach could no doubt be followed in this case, although the 'slow' time-scale would have to be chosen to be $2\pi/\omega$ not a^2/ν.

Given that the problem is quasi-steady, let us consider what happens as the velocity profile evolves from its inflection-free state at peak systole. Immediately deceleration begins, an inflection point appears near the wall. According to inviscid stability theory this would be unstable for disturbances whose wavelengths were greater than a certain critical value. However, viscous theory would show that, if the inflection point were weak enough (i.e. if d^3U/dr^3 were sufficiently small) or if it were close enough to the tube wall, disturbances could not grow. At some time after peak systole, however, the flow would become unstable to disturbances of a particular wavelength which would start to grow. A little later still, the flow would be unstable to disturbances of other wavelengths, and the most rapidly growing disturbance may not be that which first became unstable. Thus in order to predict the first disturbance to become large enough to be seen one would have to study the (linear) evolution of all disturbances, while the basic flow varied slowly, and see which first reached a certain critical amplitude (cf. Gaster (1974), on the instability of spatially varying flows). However, unstable disturbances in parallel flows rapidly become non-linear, and a catastrophic transition to turbulence takes place, so the best that can be hoped for from linear theory is a prediction of when instability first occurs and when disturbances should first be observable, involving a prediction of the growth-rates and wavelengths of the most rapidly growing disturbances. (See Davis (1976) for a broader discussion of the instability of periodic flows.)

Until some recent work by Dr A. J. Sobey (personal communication; see below), such predictions had not, as far as I know, been made for time-dependent flow in a straight circular-cross-section tube, because the corresponding steady, linear instability problem had not been solved for the family of velocity profiles involved. The only previous computations I am aware of on the instability of velocity profiles with inflection points in viscous fluids are those by Obremski, Morkovin & Landahl (1969) for the Falkner–Skan boundary layer on a wedge with a slight adverse pressure gradient, by Eagles & Weissman (1975) for flow in a slightly divergent channel, and by Hall & Parker (1976) on the decaying flow in a blocked channel. The first study considered only steady quasi-parallel flows, the second was a complete study on steady flows with slow *spatial* variation, and the last, although concerned with a temporally slowly varying basic flow (calculated by Weinbaum & Parker, 1975), did not compute the evolution of unstable disturbances when the Reynolds number is large, but concentrated instead on the properties of the neutral disturbances at different times and Reynolds numbers. Hall & Parker's results (and some experiments by K. H. Parker, personal communication) confirm that the inflection point causes instability at a much lower Reynolds number than the critical value for plane Poiseuille flow. If the channel width is $2h$ and the maximum velocity of the Poiseuille flow is U_0, then the critical Reynolds number U_0h/ν before the channel is blocked is 5750 (Grosch & Salwen, 1968), while afterwards it is about 150. Instability first occurs at a time $0.023\nu/h^2$ after blocking of the flow; if ν is the kinematic viscosity of blood and h the radius of the aorta, this time is about 1.6 ms. Because Hall & Parker's flow is stopped more suddenly than that in the aorta, their estimate of the time to instability (at $Re \approx 150$) is likely to be an underestimate, although it will, of course, be an overestimate of the time to instability of a suddenly stopped flow at higher values of Re. In any case, the time is so short that there is clearly plenty of time for such an instability to grow in the aorta.

Nerem et al. (1972) made a comparison of their results with the theory of Obremski et al. (1969). Fig. 5.21 shows plots of dimensionless wave number against Reynolds number, each based on the displacement thickness δ_1 of the boundary layer under consideration. The curves are the theoretical neutral curves computed by Obremski et al. for different values of the Falkner–Skan parameter β ($= 2n/n + 1$, where the velocity outside the boundary layer is proportional to x^n). The points are experimental ones for which the boundary layer displacement thickness was calculated from the solution for the Rayleigh layer (see (3.33)) and the wave number was chosen to be that associated with disturbances at a frequency of 125 Hz, representative of measured frequencies. This figure is consistent with the idea that the point of inflection is important for instability, since all points lie outside the unstable region for $\beta = 0$,

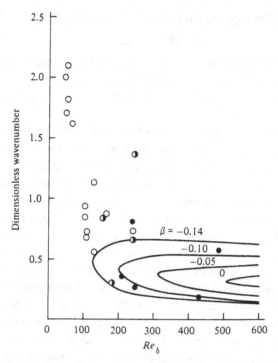

Fig. 5.21. Comparison of experimental observations with Falkner–Skan neutral stability curves from Obremski *et al.* (1969): undisturbed flow, open circles; disturbed flow, half-filled circles; highly disturbed flow, full circles. (After Nerem *et al.*, 1972.)

while many of the disturbed or turbulent points lie inside the unstable region for β between -0.10 and -0.14. More quantitative conclusions could not be made, since further progress requires computations of the sort recommended above.

Nerem *et al.* also gave an extremely crude theoretical argument for the proportionality between the critical peak Reynolds number $\hat{R}e_c$ and α. Steady boundary layer instability theory, with or without an adverse pressure gradient, shows that boundary layers become unstable if the Reynolds number based on boundary layer displacement thickness δ_1 exceeds a critical value:

$$U\delta_1/\nu > Re_{\delta c}.$$

But in unsteady periodic flows $\delta_1 = k(\nu/\omega)^{1/2}$ for some constant k, so instability is predicted if $U(\omega\nu)^{-1/2}$ exceeds $Re_{\delta c}$, i.e. if

$$Re > \alpha Re_{\delta c}/k.$$

This is the required result. Note that for a sinusoidal oscillation, $k \approx 1$, while for a Falkner–Skan flow with $\beta = -0.10$, $Re_{\delta c} \approx 200$ (fig. 5.21). Hence $Re_{\delta c}/k \approx 200$, which is not far from the observed value of about 250.

Some more detailed computations on flow in a straight circular-cross-section tube have now been made by Dr A. J. Sobey at Imperial College, London (personal communication). He chose an undisturbed velocity profile consisting of a mean Poiseuille flow plus a component corresponding to a single sinusoidal oscillation in pressure gradient of frequency ω (c.f. (2.34) or (3.1)). He restricted attention to quasi-steady disturbances (i.e. $Re \gg \alpha^2$ and $Re \gg 1$, see p. 296), and performed numerical integration of the Orr–Sommerfeld equations for various values of the three remaining dimensionless parameters: α, γ (the ratio of the pressure-gradient amplitude to its mean), and T $(=(\overline{Re}/\alpha^2)\omega\hat{t})$ which represents the time within the cycle. At the time of writing not many results were available, but one example that has been calculated is that for which $\alpha = 10$ and γ is chosen so that the centre-line velocity just comes to rest each beat. Then instability is first predicted very shortly after the time of peak centre-line velocity (when deceleration begins and an inflection point first appears), and at that time the Reynolds number based on boundary layer displacement thickness, Re_δ, is approximately 175. Considering that some time must elapse, and the boundary layer become thicker, before the instability is observed, this too is consistent with the in-vivo value of 250 quoted by Nerem et al. (1972).

CHAPTER 6

FLOW IN COLLAPSIBLE TUBES

6.1 Physiological and experimental background

6.1.1 *Physiological phenomena*

The main emphasis of the previous four chapters has been on the explanation and prediction of the normal distributions of pressure, velocity and wall shear stress in large arteries. These are the areas in which theoretical fluid mechanics has already made a considerable contribution. In this chapter we turn to an area that has had far less extensive study, although it is medically important and throws up fluid mechanical problems quite as challenging as those examined above. The phenomena we wish to describe are those that occur when the flow of fluid through an elastic tube causes it to collapse, i.e. to suffer a large reduction in cross-sectional area. We saw in § 1.1 that blood vessels (and rubber tubes) are very distensible, and therefore readily experience collapse, when the transmural pressure $\hat{p}_{tm}$ (the difference between the internal and external pressure) is close to zero. This has the consequence that veins above the level of the heart, with the exception of those in the skull, are normally collapsed. The mean flow-rate through a raised venous bed (e.g. in the arm) is, of course, unaltered because the impedance of the whole circulatory bed is primarily determined by the viscoelastic properties of the arteries and by the resistance of the microcirculation; venous calibre is unimportant. Hence the mean blood velocity in the veins is likely to increase when they collapse.

Collapse also occurs in abdominal veins just before they enter the thorax, since abdominal pressure (in a horizontal subject) is normally slightly above atmospheric (about $0.26 \, \text{kN m}^{-2}$) and intrathoracic pressure is variable, but normally subatmospheric (by $0.5 \, \text{kN m}^{-2}$ or more), except during forced expiration. Thus the blood passes through a chamber of high pressure into a chamber of low pressure and the veins tend to collapse just outside the chest. A similar collapse occurs in veins entering the top of the chest, but since the pressure outside them is atmospheric, the pressure in the chest must be slightly lower before collapse occurs. This collapse has the important consequence that venous return to the heart, and hence the cardiac output, is largely unaffected by the actual value of the pressure in the chest, as long as that is subatmospheric, since it is determined by the end-capillary pressure and by the extent of the collapse, i.e. by the pressure outside the collapsed veins. An increase of intrathoracic pressure above atmospheric does tend to decrease venous return because then there is no collapse and the effective downstream pressure for venous flow is that of the right atrium, i.e. of the chest. (See fig. 6.1(a) and (b); fig. 6.1(a) was adapted from Holt (1969) who gives a fuller discussion.) The tendency of veins to change their cross-section markedly with variations in transmural pressure is exploited normally in the assistance of venous return by intermittent contraction of the skeletal muscle surrounding the veins. It is also exploited clinically, in that rhythmic compression of the legs of a patient who has recently undergone surgery, without changing the average flow-rate, is found to decrease significantly the incidence of deep-vein thrombosis, one of the major causes of post-operative complications. (See Scherer *et al.*, 1975a; Caro *et al.*, 1978, chapter 14.)

Mention should also be made of the vessels in the pulmonary circulation, which is a low-pressure system surrounded by approximately atmospheric pressure in the respiratory airspaces. The normal pressure in pulmonary capillaries and veins is only a fraction of a $kN\,m^{-2}$ above atmospheric, and any reduction in left atrial pressure or any increase in the air pressure in the lungs can cause collapse of these vessels. Indeed, blood vessels at the top of the lung normally are collapsed in an upright subject because the fall in hydrostatic pressure with height reduces their internal pressure below atmospheric. We should also notice that the airways themselves suffer collapse during forced expiration, with the consequence that at any particular lung volume there is a maximum expiratory flow-rate, which cannot be exceeded however much effort is made: see Clément, van de Woestijne & Pardaens (1973), Lambert & Wilson (1972, 1973), and Dawson & Elliott (1977) for theoretical approaches to this phenomenon.

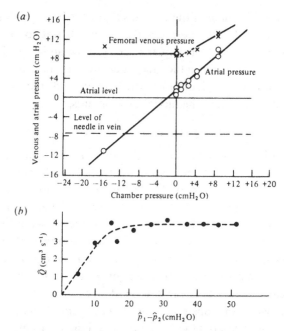

Fig. 6.1. (a) Effect of changing intrathoracic pressure on right atrial and femoral venous pressure in a dog whose trachea is connected to a breathing chamber. Intrathoracic pressure was varied by changing the pressure in the chamber from 20 cmH$_2$O above atmospheric pressure to 20 cmH$_2$O below. The hydrostatic level of the needle in the femoral vein was approximately 8 cm below the level of the right atrium. +, above atmospheric pressure; −, below atmospheric pressure. (After Holt, 1969.) (b) Measurements by Brecher (1952) representing the steady flux through the superior vena cava of a dog as a function of the pressure difference $\hat{p}_1 - \hat{p}_2$, where $\hat{p}_1$ is the pressure in the jugular vein and $\hat{p}_2$ is the pressure applied to the peripheral end. (After Rubinow & Keller, 1972.)

We saw in § 1.2 that systemic arteries do not normally suffer collapse, but those in a limb can do so when bent in a joint or when compressed by a pressure cuff. The appearance and disappearance of audible Korotkoff sounds, which are presumably a consequence of the collapse, are used as indications of systolic and diastolic arterial blood pressure.

In the next subsection we describe model experiments, performed with rubber tubes, in which the collapse process can be studied more directly than in vivo, and note particular phenomena that fluid mechanical theory should be able to explain. We follow that by discussing some of the theories that have been put forward as explanations of Korotkoff sounds, in an attempt to see what should comprise a complete explanation. The remaining sections of this chapter outline some of the theoretical work that has been done in an attempt to explain both the steady and the unsteady phenomena observed in the model experiments.

6.1.2 Model experiments

The standard experiment on collapsible tubes is that depicted in fig. 6.2 and described by Conrad (1969), Holt (1969) and Katz, Chen & Moreno (1969) among others. A segment of flexible tube is supported horizontally between two lengths of rigid tube and is contained in a chamber whose pressure, $\hat{p}_c$, can be given any chosen value. The upstream and downstream pressures in the rigid tubes, $\hat{p}_1$ and $\hat{p}_2$, can be measured, as can the flow-rate, $\hat{Q}$, through the system. Fluid is supplied from a reservoir whose height, $\hat{H}$, above the collapsible segment can be varied, as can the resistances, $\hat{R}_1$ and $\hat{R}_2$, of two variable constrictions upstream and downstream of the region of interest. Thus it is possible to vary conditions upstream, downstream and around the collapsible segment independently.

We describe first the results obtained when the chamber pressure $\hat{p}_c$ and the downstream resistance $\hat{R}_2$ are held constant, and the flow-rate through the system is varied by varying either $\hat{R}_1$ or $\hat{H}$. Fig. 6.3 presents families of graphs of $\hat{p}_1$, $\hat{p}_2$ and $\hat{p}_1 - \hat{p}_2$ against $\hat{Q}$, each curve representing a different downstream resistance $\hat{R}_2$. Fig. 6.4 shows a single curve from a family like that of fig. 6.3(c), together with sketches of the state of collapse of the flexible segment at different points on the curve. There are three distinct sections of this curve, as follows.

(I) The downstream pressure $\hat{p}_2$ increases monotonically with $\hat{Q}$, (fig. 6.3(b)) by virtue of the fixed downstream constriction (these curves are not linear, as they would be for a constant resistance ($\hat{p}_2 = \hat{R}_2\hat{Q}$), but are of the form

$$\hat{p}_2 = k_1\hat{Q}^2 + k_2 \tag{6.1}$$

since the constriction was a sharp orifice). Therefore, when the flow-rate is sufficiently large, $\hat{p}_2$ can exceed $\hat{p}_c$ by an arbitrary amount, and so the pressure everywhere in the collapsible segment

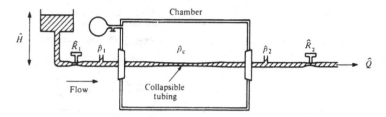

Fig. 6.2. Experimental arrangement for studying flow in a collapsible tube.

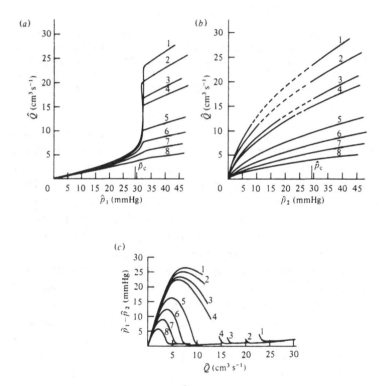

Fig. 6.3. (a) $\hat{p}_1$ as a function of $\hat{Q}$ for constant external pressure ($\hat{p}_c =$ 29.5 mmHg) and different settings of the downstream resistance. Curves correspond to different values of k_1 in (6.1) measured in mmHg $(cm^3\ s^{-1})^{-2}$: 1, 0.05; 2, 0.07; 3, 0.10; 4, 0.12; 5, 0.27; 6, 0.45; 7, 0.9; 8, 4.0. (b) $\hat{p}_2$ as a function of $\hat{Q}$. (c) Pressure drop $\hat{p}_1 - \hat{p}_2$ as a function of $\hat{Q}$; breaks in the curves represent regions of oscillation. (After Conrad, 1969.)

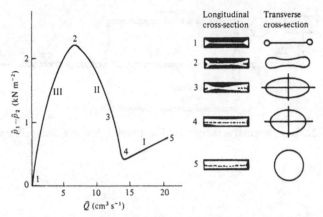

Fig. 6.4. Pressure drop across the collapsible segment as a function of flow-rate for fixed downstream resistance and external pressure. Longitudinal and transverse cross-sections at various flow-rates are also shown: 1, zero flow; 2, small flow; 3, 4, larger flows; 5, flow so large that tube has become circular. Flow is from right to left. (After Conrad, 1969.)

exceeds $\hat{p}_c$ by at least as much. Hence, according to the relation between transmural pressure and cross-sectional area (fig. 1.10(a)), the cross-section of the segment is everywhere circular and is very stiff, so that the flow within it is everywhere Poiseuille flow. Thus $\hat{p}_1 - \hat{p}_2$ is proportional to $\hat{Q}$, with an almost constant resistance.

(II) When $\hat{Q}$ is reduced below a certain critical value, the downstream pressure $\hat{p}_2$ becomes as small as, or smaller than, the chamber pressure $\hat{p}_c$, so that the cross-section of the collapsible segment begins to change shape and to collapse towards the downstream end (again according to fig. 1.10(a)). As the transmural pressure falls further, the cross-sectional area at the downstream end falls rapidly, because of the large distensibility; the collapse of the flexible segment also extends further upstream. This means that the resistance to flow rises rapidly, both because of the increase in the viscous resistance of a narrowed tube and because of the energy dissipation in the separated jet that must emerge where the tube has to widen out again as it joins the downstream rigid segment (this jet will in many cases be turbulent). Thus the pressure drop required to maintain the (gradually falling) flow-rate also rises dramatically.

(III) Finally, when the whole segment is collapsed ($\hat{p}_1 < \hat{p}_c$), its cross-section has a rather rigid dumb-bell configuration, and, as the flow-rate is reduced still further, no further change in cross-section occurs, the resistance once more becoming constant, at a value 10–100 times higher than before collapse.

If, instead of varying the upstream conditions, we vary the downstream resistance $\hat{R}_2$, so that $\hat{p}_2$ is varied independently of $\hat{p}_1$, then when $\hat{p}_2$ exceeds $\hat{p}_c$ by a sufficient amount, the flow-rate is again proportional to $\hat{p}_1 - \hat{p}_2$. However, when $\hat{p}_2$ falls below $\hat{p}_c$, collapse begins; because $\hat{p}_1$ remains high, this collapse does not extend throughout the flexible segment, and a fixed degree of collapse results. Further reduction in $\hat{p}_2$ has no effect on the flow-rate, as shown in fig. 6.5 (see fig. 6.1 for evidence of this phenomenon *in vivo*). Note the overshoot of flow-rate in fig. 6.5(*a*), reflecting the progression of the collapse process to its final state as $\hat{p}_2$ falls below $\hat{p}_c$. Physiologists have for many years used segments of collapsible tube in this way as part of the apparatus for perfusing different parts of the circulatory system at a constant flow-rate, independent of downstream pressure. The device is known as a 'Starling resistor', having been introduced by Knowlton & Starling (1912).

Varying $\hat{p}_c$ independently of $\hat{p}_1$ and $\hat{p}_2$ also has a predictable effect: if $\hat{p}_c < \hat{p}_2$, there is no collapse and $\hat{Q}$ is independent of $\hat{p}_c$; however, as $\hat{p}_c$ rises above $\hat{p}_2$ it is $\hat{p}_1 - \hat{p}_c$ not $\hat{p}_1 - \hat{p}_2$ that governs the flow-rate.

In most experiments on collapsible tubes, an important phenomenon is observed when the tube is partially collapsed (section II of fig. 6.4). That is that self-excited oscillations, in area, flow-rate and $\hat{p}_2$, are found to develop during this phase, even when the upstream reservoir, the upstream and downstream resistances, and the chamber pressure are held fixed. Both Conrad (1969) and Katz *et al.* (1969) report that, when the flow-rate is varied by adjustment of upstream conditions, oscillations occur only for sufficiently small values of the downstream resistance $\hat{R}_2$ (or the constant k_1 in (6.1)); the breaks in the curves in figs. 6.3(*b*) and (*c*) indicate that oscillations were taking place. Some of Katz *et al.*'s results indicate that when $\hat{R}_2$ is such that no oscillations take place for a given value of $\hat{p}_c$, then increasing $\hat{p}_c$ (and hence increasing the

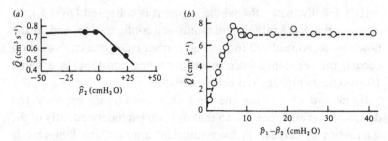

Fig. 6.5. (a) Effect on flow-rate of changing $\hat{p}_2$, keeping $\hat{p}_1$ and $\hat{p}_c$ constant. (After Holt, 1969.) (b) The same, from measurements of Brecher (1952). (After Rubinow & Keller, 1972.)

extent of collapse) does not generate oscillations. Conrad (1969) showed one set of results indicating that the frequency of the oscillations increases as the flow-rate decreases through section II of fig. 6.4. In his case the frequency range was 0.6 Hz to 1.7 Hz (his tubes had an undisturbed diameter, d, of 1.27 cm). However, J. M. Fitz-Gerald (private communication) and Conrad, Cohen & McQueen (1978) have reported a much wider range of frequencies, depending on the values of $\hat{R}_1$, $\hat{R}_2$, $\hat{p}_c$ and d and on the precise elastic properties of the tube. This is borne out by the experience of Ur & Gordon (1970) who performed experiments using both rubber tubes and excised segments of artery (diameter 0.4 cm) and who reported frequencies in the range 12 Hz to 50 Hz. Their experiments are described further in § 6.1.3.

The ultimate aim of theoretical studies is to explain the above phenomena accompanying flow in collapsible tubes; the theoretical description of the self-excited oscillations is a particularly challenging problem. In §§ 6.2–6.3 we introduce various avenues of approach that are currently proving fruitful, although in no case is the work yet complete. First, however, we look more closely at the particular phenomenon of Korotkoff sounds, and discuss the various mechanisms that have been put forward to explain them.

6.1.3 Mechanisms of Korotkoff sounds

The phenomenon of Korotkoff sounds was described in § 1.2; they are heard when a cuff is inflated round a limb until the pressure exceeds systolic arterial pressure and the arteries are occluded, and

then the cuff pressure is gradually reduced. The sounds are charac-
terised by a sharp clicking noise as cuff pressure, $\hat{p}_c$, falls below
systolic arterial pressure. This becomes louder, and may be
followed by a brief murmur as $\hat{p}_c$ is reduced further; subsequently
the sound becomes muffled, and it finally disappears when cuff
pressure is close to diastolic pressure. The high-frequency (60–
180 Hz) content of the sounds at different stages of cuff deflation
was shown in fig. 1.27.

Various apparently different mechanisms have been proposed to
explain the sounds, but one factor that they all share is the
importance of the collapsibility of the artery wall when subjected to
external compression. When the cuff pressure is significantly
greater than diastolic, but less than systolic, pressure, the artery is
occluded for most of the cycle. When the pulse arrives the artery is
forced open, subsequently to close again as the pulse pressure falls.
Gonzalez (1974) recorded the sounds accompanying this process in
a human brachial artery, both over the cuff and some way downstream, and reported a single sharp spike on the recording (the click
lasting 0.02–0.04 s), followed by a lower-amplitude noisy murmur,
lasting about 0.1 s, in most subjects. The period of the cardiac cycle
was about 1.0 s. He associated the click with a steep pressure wave
(a 'shock': § 2.1.4) formed as blood is forced through the opening
constriction, and terminated as the constriction closes again. The
time lapse between the sounds heard at the cuff and at a known
distance downstream was consistent with wavefront propagation at
the speed predicted by the theory of chapter 2. Gonzalez did not
report coherent oscillations as the artery closed again towards the
end of systole. Nor did he record frequency spectra, but the nature
of the signal, backed up by flow visualisation in models, suggested
that the murmur, which persists into diastole, is associated with
turbulence generated by the jet-like flow through the constriction in
systole. When the cuff pressure is close to systolic, the click is heard
but not the murmur, which he attributed to insufficient time for
turbulence to be generated in the jet. When cuff pressure is only a
few hundred N m^{-2} above diastolic pressure, the clicking becomes
muffled, because, according to this author, the artery does not
collapse completely. At an even lower pressure the audible sound
disappears altogether, but the precise value of cuff pressure at which

this occurs depends on the characteristics of the sound-recording equipment, and may still be above diastolic pressure. The qualitative conclusions of this author seem plausible, but it is a pity that he did not record frequency spectra for comparison with the measurements of other workers (e.g. McCutcheon & Rushm r, 1967), and with the known characteristics of intermittent turbulent flow through constrictions (Young & Tsai, 1973b).

More insight can be gained from the observations of Ur & Gordon (1970), who recorded Korotkoff sounds in the brachial artery of a dog. They do not present such a recording, but they do show direct upstream and downstream pressure measurements taken in the same animal, after death, when the artery was connected upstream to a perfusion system in order to provide a known pressure head. These measurements are compared with those made in isolated limbs, and in excised arteries and rubber tubes supported as in fig. 6.2. In each of these preparations the authors performed three different manoeuvres, and in each case all preparations gave qualitatively similar results. One was to keep upstream and cuff pressures steady, as in the experiments of Conrad (1969) and others, and in this case continuous self-excited oscillations were recorded for a wide range of upstream and cuff pressures, as in the experiments already described.

The second manoeuvre was to reduce the upstream pressure, while keeping cuff pressure constant; this manoeuvre is intended to simulate the fall in arterial pressure after peak systole, but does not model the early part of systole in which the arterial pressure rises and (if cuff pressure exceeds diastolic pressure) forces the artery open. The pressure recordings from one such experiment (in an excised artery) are shown in fig. 6.6(a), together with the sound recorded through a stethoscope. It can be seen that, as the upstream pressure falls past the value of the cuff pressure, oscillations of a fixed frequency are generated in downstream pressure and in sound recorded (oscillations in area were simultaneously observed). Aurally, the sound resembles the click of Korotkoff sounds, but these measurements differ from those of Gonzalez by virtue of the oscillations. The two sets of experiments can be reconciled, however, if account is taken of the *rate* at which upstream pressure is reduced. Ur & Gordon recorded no oscillations and no clicking sound (cf. fig. 6.6(b)) when the upstream pressure reduction in the models took place very rapidly and lasted for less than 0.25 s; occlusion of the artery also took place rapidly, as in the in-vivo experiments of Gonzalez (1974). Admittedly the diastolic pressure fall *in vivo* lasts for almost three-quarters of the cycle, or about 0.7 s in man, which exceeds the critical value of 0.25 s in Ur & Gordon's experiments. However, Ur & Gordon did not report the critical reduction time in the isolated limb or in the dead dog, although they did report an abbreviation of the time over which oscillations could be heard, in the former, as the pressure was reduced more rapidly. Since the elastic properties of an artery tethered to its surrounding tissue are different from its properties when excised, even if stretched to its in-vivo length, a change in the critical reduction time would not be surprising (tethering also, no doubt, accounts for the fact that the oscillation frequency in the dog's brachial artery is much less *in vivo* (4–8 Hz) than *in vitro* (12–50 Hz)). In summary, then, collapse of the artery in diastole is likely to be accompanied by a burst of self-excited oscillations as long as the arterial pressure does not fall too rapidly. The evidence of Gonzalez is that the oscillations are absent *in vivo*, but their prevalence in Ur & Gordon's experiments suggests that they could occur, even *in vivo*, if circumstances were suitable.

The third manoeuvre performed by Ur & Gordon was intended to simulate systole by raising cuff pressure until the artery was occluded, and then allowing it to fall while holding upstream conditions constant. However, because the fluid level in their upstream reservoir was not kept constant, the upstream pressure also fell once outflow had begun, and the simulation was not exact. Nevertheless, figs. 6.6(b) and (c) show that the results were very similar to the 'diastolic' results, in that oscillations of given frequency occurred during the pressure fall as long as the rate of fall was not too rapid.

The fact that Ur & Gordon reported no sounds during the experiments with rapid pressure reductions, while Gonzalez (1974) still reported a click *in vivo*, is presumably associated with the fact that Ur & Gordon modelled only diastole or systole, not a complete beat. Thus the brief, steep pressure pulse formed *in vivo* is likely to

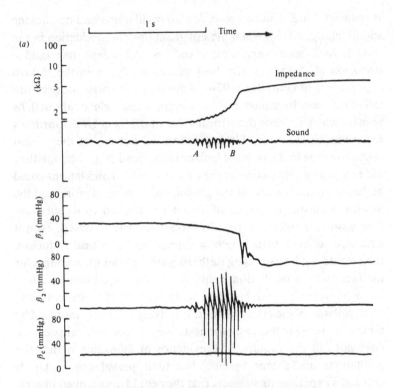

Fig. 6.6. (*a*) Oscillation and sound production in an excised segment of artery. The uppermost trace records the electrical impedance of a section of the tube including the collapsed region; this is related (but not directly proportional) to the cross-sectional area at the narrowest point. A rapid reduction in reservoir pressure, simulating diastole, produces a click consisting of a short burst of oscillation lasting about 300 ms. (*b*) Production of a click in the model by rapid reduction of chamber pressure. (*c*) More rapid reduction of the chamber pressure causes the tube to open in a very short time without significant oscillation or sound production. (After Ur & Gordon, 1970.)

be absent. Further points to be noted from Ur & Gordon's experiments are:

(*a*) High-speed cine-photography showed that waves were propagated both upstream and downstream from the oscillating constriction; these are presumably pressure waves as analysed in chapter 2.

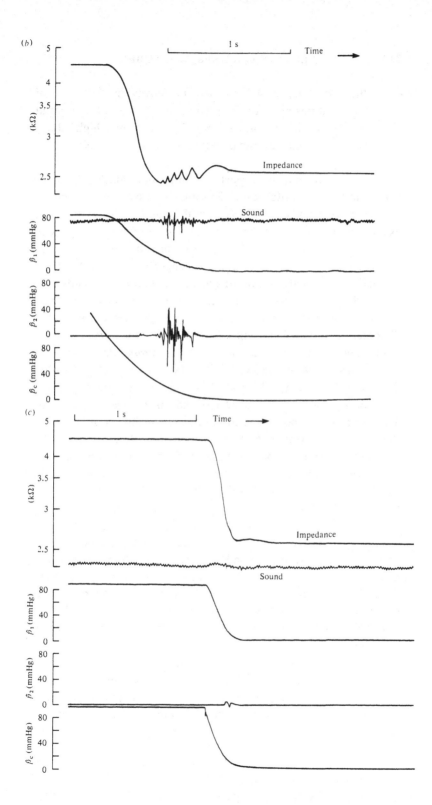

(b)

1 s Time →

(kΩ)

Impedance

p_1 (mmHg)

Sound

p_2 (mmHg)

p_c (mmHg)

(c)

1 s Time →

(kΩ)

Impedance

Sound

p_1 (mmHg)

p_2 (mmHg)

p_c (mmHg)

(b) If the upstream pressure is varied very slowly, the frequency of the oscillations also varies slowly, being 2–3 times as great at low values of the pressure difference $\hat{p}_1 - \hat{p}_c$ than at high values; this is consistent with the observations of Conrad (1969).

Ur & Gordon did not report the 'murmurs' which follow the 'clicks' in both McCutcheon & Rushmer's (1967) and Gonzalez's (1974) recordings of real Korotkoff sounds. However, close examination of one of their recordings taken with an expanded time-scale (fig. 6.7) reveals the presence of irregular, high-frequency, low-amplitude fluctuations superimposed on the basic oscillations. They also reported similar fluctuations in steady flow past a fixed constriction, when turbulence is to be anticipated. Furthermore, the canine artery used by Ur & Gordon had a diameter of about 0.4 cm, which is smaller than the human brachial artery (about 0.6 cm) so that the Reynolds number downstream of the constriction is also likely to be smaller (although Ur & Gordon did not quote velocity). That would mean that the intensity of any turbulence produced would be less than in Gonzalez's subjects. Thus there seems no reason to doubt the conclusion of Gonzalez that the murmur originates in the turbulent jet downstream of the constriction. Whether or not it occurs depends both on the Reynolds number in the artery and on the degree of constriction. In steady flow this follows from the fact that, as long as there is flow separation, turbulence occurs if the Reynolds number *at the constriction* exceeds a value that depends on the geometry of the constriction, but that can be as low as 500–600. Its persistence downstream, however, depends on the Reynolds number in the undisturbed tube. In unsteady flow, the occurrence of turbulence depends on the amount of time available for the growth of small disturbances (cf. § 5.3), and is complicated (Young & Tsai, 1973a).

The above discussion has linked Korotkoff sounds with particular events in the flow of blood through a collapsing artery, but has not given a fluid dynamical explanation of those events. One such 'explanation' has been provided by Anliker & Raman (1966), who proposed a model to predict that value of cuff pressure, above diastolic, at which the circular configuration of the artery wall first becomes unstable. The artery wall is then expected to buckle and

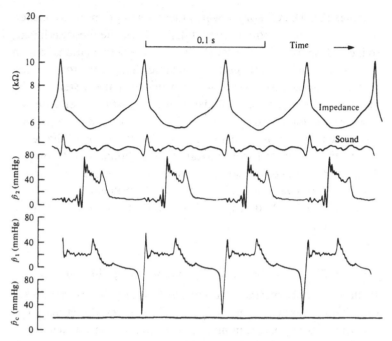

Fig. 6.7. Detailed recording of oscillation in a segment of latex tube. (After Ur & Gordon, 1970.)

generate sounds, as long as the growth-rate of the growing disturbances is sufficiently rapid for the disturbances to be observable (i.e. audible) before the arterial pressure rises again to a value at which the circular configuration is stable. The stability analysis is another example of Flügge's shell theory (cf. Shayo & Ellen (1974), discussed in § 5.3), and since Anliker & Raman ignored any effect of blood flow, the presence of blood in the artery has no effect on the buckling criterion, its inertia serving merely to determine growth-rates. The analysis predicts stability for a brachial artery, so that Korotkoff sounds must disappear, for cuff pressures less than 800 N m^{-2} (6 mmHg) above diastolic. A better prediction might be made by combining this theory with that of Shayo & Ellen (1974), in which the blood flow itself is a destabilising influence.

The trouble with such stability analyses is that they treat the blood vessel wall as homogeneous and isotropic, with no stresses in the undisturbed configuration, whereas the real artery is much more complex (§ 1.1). A theory in which these assumptions were relaxed would be extremely complicated. Furthermore, the theory is linear, so it can describe what happens only when the deformations remain quite small, and cannot describe the large changes in cross-section that take place during collapse. In order to make a start on the analysis of the dynamics of collapse and oscillation, therefore, we beg the question posed by elasticity theory, and assume *a priori* that the average transmural pressure across a cross-section of the vessel at any time is related to the local cross-sectional area through a given single-valued function such as one of those presented in fig. 1.10(*a*). This is very crude, but enables us in the next three sections to concentrate on fluid dynamical aspects of flow and oscillations in collapsible tubes.

6.2 Viscous flow in slowly varying collapsible tubes

In all the model experiments reported above, the Reynolds number of the flow was large, and it is unlikely that the direct action of viscosity was important in most of the phenomena, especially the oscillations whose period is normally much smaller than the time for viscous diffusion across the tube. Nevertheless, although none of the experiments has looked for it, there must exist, in a given experiment, a critical value of the Reynolds number below which oscillations cannot occur; this critical value should be predictable. Furthermore, there are many medium-sized or small veins and pulmonary vessels in which the Reynolds number is less than (say) 10 and that experience compressive transmural pressures. Indeed, observation of pulmonary capillaries shows that it is these smaller vessels that are normally closed at the top of the lung in a resting subject (West *et al.*, 1969). It would be of interest to see if the phenomena reported above for steady flow in large collapsible tubes are still expected to occur in smaller ones.

Work on unsteady viscous flow in collapsible tubes is still at an early stage, so this section is largely devoted to an analysis of steady (and hence also quasi-steady) flow in a finite segment of such a tube at Reynolds numbers around 1 or below. The work is taken from the paper by Wild, Pedley & Riley (1977). The analysis is based on lubrication theory and therefore requires that the cross-sectional area vary slowly with longitudinal distance. The effect of inertia is analysed as a perturbation to the basic lubrication theory solution; it is important to include it because it is not negligible in any of the (large-Reynolds-number) experiments, and without inertia the self-excited oscillations cannot develop.

The basic solution was first given, and applied to tubes of circular cross-section, by Rubinow & Keller (1972). They showed that the flow-rate, $\hat{Q}$, and the local pressure gradient, $d\hat{p}/d\hat{x}$, are related by an equation of the form

$$\hat{Q} = -\sigma \, d\hat{p}/d\hat{x}, \qquad (6.2)$$

where σ is the conductivity, which depends on cross-sectional area (as deduced from lubrication theory) and hence, from data such as that of fig. 1.10(a), on transmural pressure, $\hat{p}_{tm}$. Now $\hat{p}_{tm} = \hat{p} - \hat{p}_c$, where $\hat{p}$ is the internal pressure and $\hat{p}_c$ the external (assumed constant), and $\hat{Q}$ is independent of $\hat{x}$. Thus if $\hat{L}$ is the distance between the upstream station, where $\hat{p} = \hat{p}_1$, and the downstream station, where $\hat{p} = \hat{p}_2$, integration of (6.2) gives

$$\hat{Q} = \frac{1}{\hat{L}} \int_{\hat{p}_2}^{\hat{p}_1} \sigma(\hat{p} - \hat{p}_c) \, d\hat{p}. \qquad (6.3)$$

If $\hat{p}_2 > \hat{p}_c$, σ will be fairly large and more or less independent of $\hat{p}$ (because the distensibility is small), so $\hat{Q}$ will be approximately proportional to $\hat{p}_1 - \hat{p}_2$. On the other hand, if $\hat{p}_2 < \hat{p}_c$, σ will be very small for values of $\hat{p}$ less than about $\hat{p}_c$, and (6.3) can be approximately replaced by

$$\hat{Q} \approx \frac{1}{\hat{L}} \int_{\hat{p}_c}^{\hat{p}_1} \sigma(\hat{p} - \hat{p}_c) \, d\hat{p},$$

which is independent of $\hat{p}_2$. This is one explanation of the steady experimental results described above.

Wild et al. (1977) extended the work of Rubinow & Keller (1972) in four ways. First, they considered tubes of elliptic cross-section, in order more accurately to model their collapse. This model is incorrect when the cross-section becomes dumb-bell shaped and its area very small, but the measured pressure–area relation (fig. 1.10(a)) is rather uncertain then anyway, and a better model would not make the predictions of conductance (which is very low in these circumstances) more accurate. (Flaherty *et al.* (1972*a*) computed the shape and conductance σ of a buckled cylindrical tube, linearly elastic and uniform along its length.) Secondly, they included to first order the effects of inertia, as was done for rigid circular tubes by Manton (1971) as a perturbation to lubrication theory, and by Lee & Fung (1970) numerically. Hall (1974) also extended lubrication theory by considering unsteady flow in a slowly varying rigid tube of small eccentricity when a pulsatile pressure difference is applied across its ends. Thirdly, Wild *et al.* did not restrict attention to uniform external pressure $\hat{p}_c$, but calculated the variation in internal pressure, fluid velocity and tube cross-sectional area when the external pressure varied along the tube, in particular when it resembled that applied by a cuff of finite length (see fig. 6.12). Fourthly, previous calculations had not covered the experiment described above in which the downstream resistance is held constant. Our results do extend to this case, and, in fact, they are found to predict multiple-valued $\hat{Q}$–$\Delta\hat{p}$ curves like that of fig. 6.4. As in all the quoted work, blood was taken to be a homogeneous and Newtonian fluid, which is a good approximation in vessels of diameter greater than 100 μm, in which the shear-rate exceeds 100 s^{-1} (see § 1.1). An outline of the analysis and the results is given below.

6.2.1 *Lubrication theory and the effect of inertia*

In steady conditions the shape of the tube does not change with time, and the relation between local pressure gradient and flow in an arbitrary slowly varying elliptical tube can be calculated independently of the pressure–area relation.

We consider steady, viscous incompressible flow through a slowly varying elliptical tube of length L, defined in Cartesian coordinates $(Lx, a_0 y, a_0 z)$ by

$$y^2/a^2(x) + z^2/b^2(x) = 1 \qquad 0 \leqslant x \leqslant 1, \qquad (6.4)$$

where a_0 is a characteristic tube radius, and $a_0 a$, $a_0 b$ are the semi-major and semi-minor axes of the elliptical cross-section. The use of lubrication theory requires that

$$\varepsilon = a_0/\hat{L} \ll 1, \qquad (6.5)$$

while the Reynolds number

$$Re = U_0 a_0/\nu$$

remains $O(1)$ as $\varepsilon \to 0$; U_0 is a scale for the axial velocity component. The velocity field $\hat{\mathbf{u}}$ is scaled so that each dimensionless component is $O(1)$, and is taken to be

$$\hat{\mathbf{u}} = U_0(u, \varepsilon v, \varepsilon w)$$

with pressure

$$\hat{p} = (\rho U_0^2/\varepsilon Re)p. \tag{6.6}$$

It is convenient to work in a coordinate system in which the tube cross-section does not vary with x. Accordingly we introduce new transverse coordinates

$$\eta = y/a(x), \qquad \zeta = z/b(x),$$

so that (6.4) becomes

$$\eta^2 + \zeta^2 = 1 \qquad 0 \leqslant x \leqslant 1. \tag{6.7}$$

The full equations of motion are

Continuity

$$Du + (1/a)v_\eta + (1/b)w_\zeta = 0, \tag{6.8}$$

x-momentum

$$\varepsilon Re[u\,Du + (v/a)u_\eta + (w/b)u_\zeta]$$
$$= -Dp + (1/a^2)u_{\eta\eta} + (1/b^2)u_{\zeta\zeta} + \varepsilon^2\,D^2 u, \tag{6.9}$$

y-momentum

$$\varepsilon Re[u\,Dv + (v/a)v_\eta + (w/b)v_\zeta]$$
$$= -(1/\varepsilon^2 a)p_\eta + (1/a^2)v_{\eta\eta} + (1/b^2)v_{\zeta\zeta} + \varepsilon^2\,D^2 v, \tag{6.10}$$

z-momentum

$$\varepsilon Re[u\,Dw + (v/a)w_\eta + (w/b)w_\zeta]$$
$$= -(1/\varepsilon^2 b)p_\zeta + (1/a^2)w_{\eta\eta} + (1/b^2)w_{\zeta\zeta} + \varepsilon^2\,D^2 w, \tag{6.11}$$

where

$$D = \partial/\partial x - (a'/a)\eta\,\partial/\partial\eta - (b'/b)\zeta\,\partial/\partial\zeta,$$

and ' means d/dx. The boundary conditions are simply $u = v = w = 0$ on the wall (6.7).

We solve the problem as a power series in ε, on the assumption that $R = O(1)$. Thus we take

$$u = u_0 + \varepsilon u_1 + \varepsilon^2 u_2 + \cdots$$

with similar expressions for v, w and p. The leading terms in (6.10) and (6.11) show that p_0 is a function only of x, and the leading term of (6.9) (the equation for unidirectional motion) gives

$$u_0(x, \eta, \zeta) = [a^2 b^2 / 2(a^2 + b^2)]G_0(x)(1 - \eta^2 - \zeta^2),$$

where

$$G_0(x) = -dp_0/dx.$$

This is the elliptical-tube version of Poiseuille flow. For given functions $a(x)$ and $b(x)$, G_0 is determined by the condition that the volume flow-rate is independent of x. If U_0 is defined as the average velocity of the flow when the tube cross-section is a circle of radius a_0, and if volume flow-rate is non-dimensionalised with respect to $a_0^2 U_0$, we have

$$Q = \int\int_{\substack{\text{cross-}\\ \text{section}}} uab \, d\eta \, d\zeta = \pi.$$

Hence

$$G_0 = 4(a^2 + b^2)/a^3 b^3 \tag{6.12}$$

and

$$u_0 = (2/ab)(1 - \eta^2 - \zeta^2).$$

This is the basic lubrication theory solution, leading to a value for the conductance σ, from (6.2), of

$$\sigma = (\pi/4\mu)[\hat{a}^3 \hat{b}^3 / (\hat{a}^2 + \hat{b}^2)]. \tag{6.13}$$

In order to determine the next approximation to u and p it is necessary to calculate the leading terms of the expansions for the secondary velocities, v_0 and w_0. Equations (6.10) and (6.11) require that these quantities satisfy

$$\frac{1}{a} v_{0\eta\eta\zeta} + \frac{a}{b^2} v_{0\zeta\zeta\zeta} = \frac{b}{a^2} w_{0\eta\eta\eta} + \frac{1}{b} w_{0\eta\zeta\zeta},$$

and (6.8) and (6.13) imply

$$\frac{1}{a} v_{0\eta} + \frac{1}{b} w_{0\zeta} = \frac{2(ab)'}{a^2 b^2} (1 - \eta^2 - \zeta^2) - \frac{4}{ab} \left(\frac{a'}{a} \eta^2 + \frac{b'}{b} \zeta^2 \right).$$

The solution that satisfies the boundary conditions is

$$v_0 = (2a'/ab)\eta(1 - \eta^2 - \zeta^2), \qquad w_0 = (2b'/ab)\zeta(1 - \eta^2 - \zeta^2).$$

(6.14)

We note that the streamlines of these secondary motions are the same as those of stagnation point flow as long as a' and b' have opposite signs, since

$$\frac{v_0}{w_0} = \frac{a'\eta}{b'\zeta} = \frac{ya'/a}{zb'/b}.$$

This represents flow out along the major axis and in along the minor axis when a is increasing and b decreasing, and vice versa. When a' and b' both have the same sign, however, the secondary streamlines are those of a source or sink on the axis $y = z = 0$.

The next term in the pressure expansion, p_1, is also independent of η and ζ, from (6.10) and (6.11), so that u_1 satisfies

$$(1/a^2)u_{1\eta\eta} + (1/b^2) u_{1\zeta\zeta} = -G_1(x)$$

$$-[4(ab)'/a^3 b^3]Re(1 - \eta^2 - \zeta^2)^2,$$

where

$$G_1(x) = -dp_1/dx.$$

The solution of this that satisfies the boundary condition is

$$u_1 = Re(1 - \eta^2 - \zeta^2)(c_0 + c_1 \eta^2 + c_2 \eta^4 + c_3 \zeta^2 + c_4 \zeta^4 + c_5 \eta^2 \zeta^2),$$

(6.15)

where

$$c_0 = G_1 a^2 b^2/2Re(a^2 + b^2)$$

$$+ \alpha(33 + 608\delta^2 + 2238\delta^4 + 608\delta^6 + 33\delta^8),$$

$$c_1 = -\alpha(66 + 850\delta^2 + 1074\delta^4 + 238\delta^6 + 12\delta^8),$$

$$c_2 = -\alpha\beta(33 + 44\delta^2 + 3\delta^4),$$

$$c_3 = -\alpha(12 + 238\delta^2 + 1074\delta^4 + 850\delta^6 + 66\delta^8),$$

$$c_4 = -\alpha\beta(3 + 44\delta^2 + 33\delta^4),$$

$$c_5 = 4\alpha\beta(3 + 34\delta^2 + 3\delta^4),$$

and

$$\alpha = 2b^2 A'/45A^3 \beta (1 + 15\delta^2 + 15\delta^4 + \delta^6),$$

$$\beta = 1 + 6\delta^2 + \delta^4,$$

$$\delta = b/a, \tag{6.16}$$

$$A = ab. \tag{6.17}$$

G_1 is again determined from the volume flux condition, in the form

$$\iint u_1 \, d\eta \, d\zeta = 0$$

which unexpectedly simplifies to

$$G_1(x) = -2ReA'/A^3 \tag{6.18}$$

and therefore depends on the area change, but not on the cross-sectional shape of the tube. This part of the solution has been expressed in terms of two new variables: δ, which is a measure of the eccentricity of the ellipse and always lies between 0 and 1, and A, the dimensionless cross-sectional area. In the application to elastic tubes it is found convenient to work with these quantities and a mean diameter, represented by

$$c = a + b. \tag{6.19}$$

The effect of inertia on the lubrication solution is expressed by the perturbations εu_1 and εG_1 to the velocity and pressure gradient. Each term is in fact proportional to εRe, which is the important small parameter. To see their effect, we look at the velocities and pressure drop in a tube whose elliptical cross-section varies in a given way. The cross-sectional area is taken to vary as

$$A = 1 - (729/32)x^4(1-x)^2$$

so that the minimum area of $\frac{1}{2}$ is achieved at $x = \frac{2}{3}$. The shape of the tube is fixed by specifying the relation between a and b, i.e. between c and δ. Two different relations are used, as follows:

(i) constant perimeter (as for rubber tubes) equal to 2π, so

$$\int_0^{\frac{1}{2}\pi} [1 - (1 - \delta^2) \sin^2 \theta]^{1/2} \, d\theta = \frac{\pi}{2a} = \frac{\pi(1+\delta)}{2c}. \tag{6.20}$$

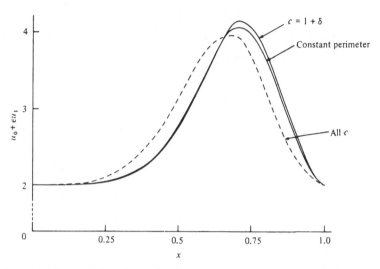

Fig. 6.8. The variation of velocity along the centre line of the rigid tube for the two relations between area and shape (represented by c). Broken curves, no inertia; continuous curves, $\varepsilon Re = 1$. (After Wild *et al.*, 1977.)

(ii) perimeter decreasing as area decreases; in view of the limited data, a simple model in which the length of the major axis remains constant as A decreases is chosen to give a qualitative indication of the behaviour of blood vessels. We thus choose

$$c = 1 + \delta = 1 + A.$$

Results are presented in three ways: the variation of u and p along the centre line of the tube (figs. 6.8 and 6.9) and the variation of u_1 across the semi-major and semi-minor axes (fig. 6.10). The constriction in each case accelerates the flow along the centre line (fig. 6.8), to a maximum velocity which is at $x = \frac{2}{3}$ when inertia is negligible ($\varepsilon Re = 0$, broken curve), but which increases and occurs further downstream as inertia becomes more important (the continuous curves are for $\varepsilon Re = 1$). The maximum pressure gradient also increases as inertia becomes more important (fig. 6.9), but occurs further upstream than in the absence of inertia, where $A' < 0$ in (6.18). It is interesting to note that the presence of inertia (mediated by the secondary motions) causes a deceleration of the flow in the centre of the tube as A decreases, but an acceleration at the edge of both the major and the minor axes (fig. 6.10). The

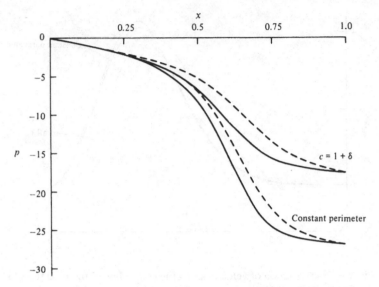

Fig. 6.9. The variation of pressure along the centre line of the rigid tube.
(After Wild *et al.*, 1977.)

reverse is true as A increases again. This is in marked contrast with
inviscid flow, where the secondary motions cause an acceleration
near the wall on the major axis (as A increases), and a deceleration
on the minor axis (Sobey, 1976a). In the case of constant major axis
$(c = 1 + \delta)$, the perimeter for a given area is smaller than in the other
case. Thus the change in velocity and the pressure drop are also
significantly smaller, because the cross-section is more nearly circular at each value of A. In this case $v_0 = 0$ (see (6.14)) and w_0 has the
same sign as $b'\zeta$: the secondary streamlines are straight, and parallel
to the minor axis.

6.2.2 Application to collapsible tubes

In this application the cross-sectional area of the tube at any value of
x is assumed to depend only on the local transmural pressure, in the
manner shown in fig. 1.10(a) and represented by the equation

$$\hat{p}_{\text{tm}} = \hat{P}_0 P(A), \qquad (6.21a)$$

where $\hat{P}_0$ is a dimensional scaling factor, and P is a dimensionless
function of A. As defined above, the area is made dimensionless by

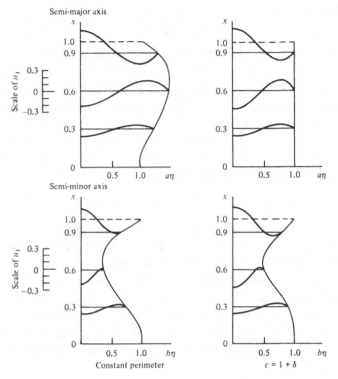

Fig. 6.10. First-order velocity profiles at three stations along the semi-major and semi-minor axes of the rigid tube. (After Wild *et al.*, 1977.)

dividing by πa_0^2, and now we choose a_0 to be the radius of the tube at the transition between circular and elliptical cross-section (i.e. the tube is circular for $A \geqslant 1$, elliptical for $A < 1$). We then fix $\hat{P}_0$ as the value of $\hat{p}_{tm}$ when the tube is just circular, i.e. by requiring that $P(1) = 1$. This elastic pressure-scale is different from the fluid dynamic pressure-scale used above (see (6.6)) so we introduce the dimensionless parameter

$$S = \varepsilon Re \hat{p}_0 / \rho U_0^2 = \pi a_0^4 \hat{p}_0 / \mu L \hat{Q}, \qquad (6.22)$$

where $\hat{Q}$ is the flow-rate through the tube. Thus, for a given tube with given elastic properties, S can also be regarded as an inverse measure of flow-rate. If we stick to the convention that all pressures are non-dimensionalised with respect to $\rho U_0^2 / \varepsilon Re$, then (6.21) becomes

$$p_{tm} = SP(A). \tag{6.21b}$$

One of the experimental situations that we are attempting to model is that of a finite length of collapsible tube, with external pressure $(\rho U_0^2/\varepsilon Re)p_c(x)$ given for all x, and with internal pressure given at the entrance, $x = 0$. Clearly then, the distribution of area with distance along the tube, and hence the relation between pressure drop and flow-rate down the tube, will depend on the transmural pressure at $x = 0$. Thus, another dimensionless parameter on which the flow will depend is

$$p_{tm}(0) = (\varepsilon R/\rho U_0^2)[\hat{p}(0) - \hat{p}_c(0)];$$

this will determine the initial cross-sectional area, $A(0)$, from (6.21b). In computing the results, it is more convenient to specify $A(0)$, and derive $p_{tm}(0)$ from that equation.

For a given flow-rate and a given distribution of area, the pressure distribution inside the tube is determined from (6.12) and (6.18). For given distributions of external and internal pressure, the cross-sectional area is determined from (6.21b). Combining the two and using (6.19), we obtain the following ordinary differential equation for $A(x)$:

$$SP'(A)\frac{dA}{dx} = -\frac{dp_c}{dx} - \frac{4(c^2 - 2A)}{A^3} + \frac{2\varepsilon Re}{A^3}\frac{dA}{dx}. \tag{6.23}$$

This is solved (by simple numerical integration) subject to the initial condition that $A(0)$ is given.

The elastic properties of the tube are represented by the function $P(A)$. This is specified to fit the experimental curve for veins given in fig. 1.10(a), in the form

$$P(A) = (1/A)[0.34(3A - 1) + 0.01(3A - 1)^5$$
$$+ 0.1(3A - 1)\,e^{-5A}/A^2]. \tag{6.24}$$

Both curves are plotted in fig. 6.11. The first two terms within the square brackets are of the form specified by Rubinow & Keller (1972), and the last term is added to ensure that dA/dx remains finite as $A \to 0$ in (6.23), for otherwise negative areas are predicted and the model breaks down. Note that the area has reduced to

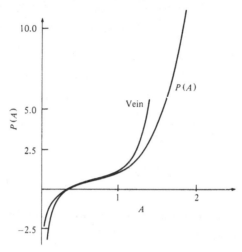

Fig. 6.11. Comparison of the function $P(A)$ with the transmural pressure–area relation for the canine vena cava (as shown in fig. 1.10(a)). (After Wild et al., 1977.)

one-third of its reference value ($A = \frac{1}{3}$) when the transmural pressure is zero, which is appropriate for veins but not for rubber tubes. Note too that this pressure–area relation is more complicated, and probably less accurate, than that used in § 6.3 below, but it is retained so that Wild et al.'s results can be quoted directly. The qualitative behaviour of the results, at the present fairly low Reynolds numbers, will not be affected

The external pressure p_c could be taken to be constant. Instead we choose a form representative of a cuff inflated over part of the length, in the form

$$p_c(x) = ST\{1 - \exp[1 - 1/4(x - \tfrac{1}{2})^2]\}. \tag{6.25}$$

This is plotted in fig. 6.12; p_c is zero at the two ends of the collapsible segment ($x = 0$ and 1), and takes the maximum value ST at the mid-point, $x = \frac{1}{2}$. This corresponds to a dimensional cuff pressure of $\hat{T} = \rho U_0^2 ST / \varepsilon Re = \hat{P}_0 T$ (see (6.22)). Thus, if the flow-rate is varied in a given tube with a given cuff pressure, T remains constant as S is varied.

The relation between the area and the shape of the cross-section must be specified, as in the last section. We shall present results only

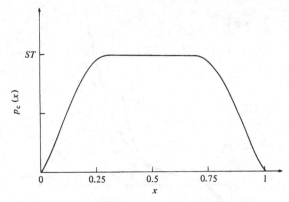

Fig. 6.12. The external pressure, $p_c(x)$, in the form of an inflated cuff. (After Wild *et al.*, 1977.)

for the case in which the major axis of the ellipse remains constant and the perimeter decreases during collapse (case (ii) above), i.e. in which

$$c = 2A^{1/2} \quad \text{for } A \geqslant 1 \text{ (circular)},$$
$$c = 1 + A \quad \text{for } A < 1 \text{ (elliptical).}$$

(6.26)

Wild *et al.* (1977) also plotted results for the case $c = 2$, and they were qualitatively very similar.

Integration of (6.23) can proceed when the following parameters are specified: εRe, S, T, $A(0)$; the last of these determines $p_{tm}(0)$ from (6.21b), and this is the same as $p(0)$ because $p_c(0) = 0$. Table 6.1 gives the radii, lengths, mean velocities, Reynolds numbers and

Table 6.1. *Values of parameters in veins*

Vein	$\hat{A}$ (cm^2)	$\hat{L}$ (cm)	U_0 (cm s^{-1})	Re	εRe
Inferior vena cava	0.5	30	15–40	400–1000	7–17
Medium-sized vein	0.2	2	1–10	10–100	1–10
Venule	0.002	0.15	0.2–0.5	0.02–0.05	0.0003–0.0007

values of εRe (where ε is taken to be the radius–length ratio of the whole vein) for three typical canine veins of different sizes. We can see that the present theory is likely to be inapplicable to large veins, but applicable to medium-sized and small veins where εRe remains below 2. The pressure–area relation given by (6.24) (and fig. 6.11) was derived for large veins, and is therefore unlikely to be completely accurate when applied to small veins, even when scaled with a different value of A_0, but no further information is available. According to fig. 6.11, $\hat{P}_0$ takes a value close to $1.0 \, \text{kN m}^{-2}$, and we therefore choose this to be its value. The numbers given in table 6.1 then show that the parameter S (see (6.22)) can vary over a very wide range, from about 30 to about 10^4. The results presented below do not cover such a wide range, because they prove to be fairly insensitive to S when S is large, since the hydrodynamic pressure drop is then small compared with that required to make a significant change in tube area. T is taken to vary between 0 and 5, at which value the maximum cuff pressure is $5\hat{P}_0$. The initial area $A(0)$ is taken to be greater than, equal to, and less than the area ($A = 1$) at which the cross-section begins to become elliptical.

In fig. 6.13 we plot the cross-sectional area A as a function of distance x along the tube for various values of the flow-rate parameter S. The continuous curves are with inertia ($\varepsilon Re = 1$) and the broken curves without ($\varepsilon Re = 0$). The external pressure is constant ($T = 0.5$) and the initial area is constant ($A(0) = 1.1$). The main feature of the results is that there is a critical flow-rate (proportional to S^{-1}) above which the area becomes very small just downstream of the peak external pressure, and does not recover further downstream where the external pressure returns to zero. The critical value of S, say S_0, is about 30 from fig. 6.13. The results are clearly very sensitive to the presence of inertia near this critical flow-rate, and predictions that ignore it are likely to be in error. The present predictions are also inaccurate for εRe as large as 1, when $S \approx S_0$, as can be seen from the large difference between the zero-inertia and the non-zero-inertia curves, but they indicate the qualitative effect of inertia. Error is also introduced by the rapid rate of change of A with x, which indicates that the effective value of ε may not be very small.

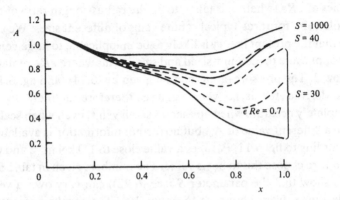

Fig. 6.13. Cross-sectional area, A, of the elastic tube as a function of x for various values of the flow-rate parameter, S. $c = 2\sqrt{A}$ for $A > 1$, $c = 1 + A$ for $A \leq 1$. Broken curves, zeroth-order solution ($\varepsilon Re = 0$); continuous curves, with first-order inertial correction ($\varepsilon Re = 1$). (After Wild *et al.*, 1977.)

It is clear that the pressure drop required for a given flow-rate (i.e. the resistance) will be significantly greater for $S < S_0$ than for $S > S_0$. This is confirmed by fig. 6.14, in which the pressure is plotted against x for the same case as shown in fig. 6.13.

In fig. 6.15, the area at the downstream end of the tube, $A(1)$, is plotted against the external pressure, T, again for various values of S and for $\varepsilon Re = 0, 1$. This indicates how the critical value S_0 increases with T, so that smaller and smaller flow-rates are required to generate collapse as T increases. Once more we see that the effect of inertia is small except when S is close to S_0.

In order to make some comparison with the experiments described in § 6.1, we wish to plot the dimensional pressure drop, $\hat{p}_1 - \hat{p}_2$, as a function of flow-rate, $\hat{Q}$, for given values of external pressure, $\hat{p}_c$, and for given conditions upstream and down. In dimensionless terms, this requires that we plot $(p_1 - p_2)/S = (\hat{p}_1 - \hat{p}_2)/\hat{P}_0$ against S^{-1} for fixed T. In fig. 6.16 we present such plots for various values of T, in conditions where the upstream area $A(0)$, and hence the upstream pressure $\hat{p}(0)$, are held constant; for these curves $A(0) = 1.1$ and we have considered only the case of constant major axis. These curves are clearly the same shape as

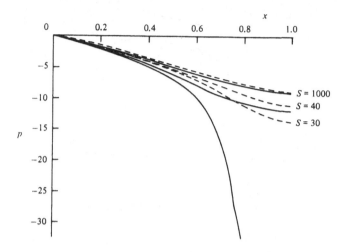

Fig. 6.14. The variation of pressure along the centre line of the elastic tube for various values of S. $c = 2\sqrt{A}$ for $A > 1$, $c = 1 + A$ for $A \leqslant 1$. Broken curves $\varepsilon Re = 0$; continuous curves, $\varepsilon Re = 1$. (After Wild *et al.*, 1977.)

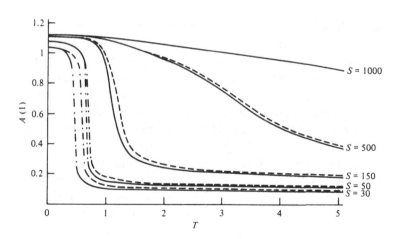

Fig. 6.15. Cross-sectional area at the downstream end of the tube, $A(1)$, as a function of the external pressure, T, for various values of S. $c = 2\sqrt{A}$ for $A > 1$, $c = 1 + A$ for $A \leqslant 1$. Broken curves, $\varepsilon Re = 0$; continuous curves, $\varepsilon R = 1$; dot–dash parts of curves, points that have not been computed. (After Wild *et al.*, 1977.)

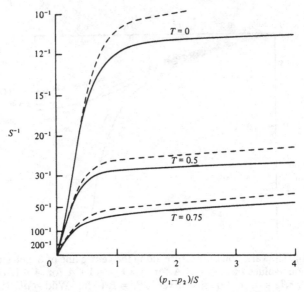

Fig. 6.16. Pressure drop along the tube as a function of the flow-rate, S^{-1}, for constant upstream area and pressure. The curves are plotted for three different values of external pressure, and for $\varepsilon Re = 0$ (broken curves) and $\varepsilon Re = 0.5$ (continuous curves). (After Wild *et al.*, 1977.)

those shown in figs. 6.1(*b*) and 6.5(*b*) and clearly demonstrate that an increase in driving pressure above a critical value does not increase the flow-rate. In order to make quantitative comparison with experiment, we must estimate the values of the dimensionless constants for Brecher's (1952) experiments on both the canine vena cava and rubber tubes. All the necessary data are not given by him, but we can take the radius of the vena cava to be about 0.5 cm (tables 1.1 and 6.1), and suppose a reasonable value of ε to be 0.1 (the axial length-scale is thus taken to be 5 cm, shorter than the vessel length, but possibly an overestimate for a typical length-scale during collapse). A flow-rate, $\hat{Q}$, of 4 cm^3 s^{-1} (fig. 6.1(*b*)) means an average velocity of 5 cm s^{-1}, so when $\nu = 4 \times 10^{-6}$ m^2 s^{-1} as for blood, the Reynolds number $Re \approx 62$ and $\varepsilon Re \approx 6.2$. Furthermore, if we take $\hat{P}_0 = 1$ kN m^{-2}, as for the vena cava in the experiments of Moreno *et al.* (1970) (fig. 1.10(*a*)), the parameter S is seen to be about 2500. The pressure drops in fig. 6.1(*b*) of 0–50 cmH$_2$O, i.e. of 0–5 kN m^{-2}, are equivalent to values of $(p_1 - p_2)/S$ of 0–5. Thus the

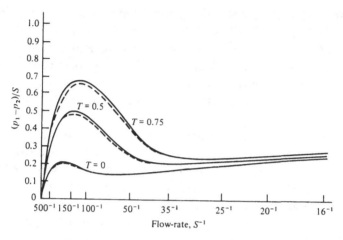

Fig. 6.17. Pressure drop along the tube as a function of the flow-rate, S^{-1}, for constant downstream resistance. The curves are plotted for three different values of external pressure and for $\varepsilon Re = 0$ (broken curves) and $\varepsilon Re = 0.5$ (continuous curves). (After Wild *et al.*, 1977.)

curve of fig. 6.1(*b*) has similar abscissa to that of fig. 6.16, but is squashed down to very small values of the ordinate; in the experiments the external pressure was presumably associated with respiration, and T would have been in the neighbourhood of 0.75. A similar conclusion is reached in the case of Brecher's rubber-tube experiments (fig. 6.5(*b*)), since he used tubes comparable in size with the vena cava. The value of $\hat{P}_0$ was presumably smaller in this case (cf. $\hat{P}_0 \approx 0.2$ kN m^{-2} from fig. 1.10(*a*) or $\hat{P}_0 \approx 0.5$ kN m^{-2} from fig. 1 of Katz *et al.* (1969)) but the flow-rates were larger (6 cm^3 s^{-1}) and S will therefore have been nearly as large as in the vena cava.

The discrepancy in scale between the theoretical and experimental curves must be principally a consequence of the fact that the theory has been developed for $\varepsilon Re \ll 1$, while in the experiments $\varepsilon Re \approx 6$. Thus the viscous pressure drop in the experiments must have been much less than the dynamic pressure changes experienced by the flow as it passed through the constriction, and the theory cannot be relevant. The fact that the curves have qualitatively the same shape is a consequence of the fact that the constriction causes enhanced pressure drop at high Reynolds number as well as at low Reynolds number. The mechanism is of course different, probably being associated with separation of the flow from the constriction, rather than enhanced viscous stresses (see § 6.3). If the dynamic pressure drop were used instead of the viscous in the non-dimensionalisation, the corresponding values of S would be reduced by a factor εRe, which reduces the 2500 estimated above to about 400, and therefore puts the experiments and theory on a similar scale.

In fig. 6.17 we plot the same quantities as in fig. 6.16, but for a different experiment. This time we keep the downstream resistance $\hat{R}_2$ constant, as in the experiments of Conrad (1969), which led to figs. 6.3 and 6.4. Constant $\hat{R}_2$ means a constant ratio between $\hat{p}_2$ and $\hat{Q}$, or (in dimensionless terms) a constant value of $p_2 = \varepsilon \pi a_0^3 \hat{p}_2 / \hat{Q}$. This means that the one-point boundary condition to be applied to (6.23) is that $A(1)$ is given by

$$P[A(1)] = p_2/S.$$

The higher the flow-rate, the higher is S^{-1} and the higher is $A(1)$, so that collapse is less; integration of (6.23) will determine $A(0)$ and hence p_1. In fig. 6.17, the constant value of p_2 is taken to be 50, which when $\varepsilon = 0.07$ and $a_0 = 0.63$ cm corresponds to a downstream resistance of 91 N m^{-2} per cm^3 s^{-1}, i.e. of 6.9 mmHg per 10 cm^3 s^{-1}, which is comparable with the smallest of the downstream resistances used by Conrad (see fig. 6.3(b)). The 'inertial' curves of fig. 6.17 are for $\varepsilon Re = 0.5$. The main features of this figure are the facts that $\Delta \hat{p}$ is a multiple-valued function of $\hat{Q}$, as in the experiments, and that this phenomenon is independent of the presence of inertia. As in the case of fig. 6.16, however, the quantitative comparison is poor, because of the large value of S appropriate to the experiments: in Conrad's system ($a_0 = 0.63$ cm, $\hat{P}_0 \approx 0.5$ kN m^{-2}, $\varepsilon = 0.07$), a flow-rate of 2.5 cm^3 s^{-1} (see fig. 6.3(c)) corresponds to $S \approx 10^4$. Once more the discrepancy can be attributed to the dominance of the dynamic pressure drop over the viscous pressure drop in Conrad's high-Reynolds-number experiments ($\varepsilon Re \approx 8.1$).

When self-excited oscillations occur in collapsible tubes, they do so on the descending section of the curve of pressure drop against flow-rate, as shown in fig. 6.3(c). It is on the corresponding section of the curves in fig. 6.17 that attention is currently being focussed in an analysis of unsteady flow whose aim is to predict a critical value of εRe above which oscillations can occur. The basic lubrication theory in the unsteady case is the same as in the steady case, but calculation of the effects of inertia (which must be included for oscillations to be possible) presents a major problem. The problem is that in order to calculate the first-order secondary velocities (such as those in (6.14) for the steady case) it is necessary to know the velocity components on the elliptical boundary, which varies slowly in both time and space. Calculation of the normal velocity component is straightforward, but that of the tangential component requires a more detailed knowledge of the elastic properties of the tube wall than we have previously assumed. One possibility is to assume that the length of every element of the cross-section's perimeter varies in direct proportion to the total perimeter (and would therefore remain virtually constant in the case of a rubber tube); thus the paths of each element could be calculated and their velocities deduced. Another possibility would be to assume linear elasticity, and treat the wall as a membrane in which the local tension is everywhere proportional to the radius of curvature (since the internal pressure is uniform across the cross-section in both the zeroth- and the first-order approximations). However, both approaches lead to a considerable increase in complexity and we have restricted ourselves initially to tubes whose cross-section remains circular (as in the steady theory of Manton (1971)) but that still have pressure–area relations of the form of fig. 6.11. The fluid mechanics is now simple, but leads to partial differential equations for $p_0(x, t)$, $p_1(x, t)$ and $A(x, t)$, instead of the ordinary differential equation (6.23), and these have not yet been solved.

6.3 A lumped-parameter model for self-excited oscillations

6.3.1 *Physical mechanisms*

As will become clear in the rest of this chapter, there are several different mechanisms that could each drive the self-excited oscillations observed in the model experiments. In proposing a theory for the phenomenon, therefore, it is not enough to demonstrate that one particular mechanism *can* cause oscillations in some experiments; it is necessary, instead, to consider one particular experiment and to construct a theory for it, firmly based on sound physical principles, which will show to what extent the several mechanisms contribute to the oscillations in that case. The models to be described in this section and the next have not yet reached that stage, but it will be seen that there are at least four possible mechanisms by which steady flow can become unstable, and the quantitative details suggest that any one of them may be relevant to the observed oscillations of Conrad (1969). Any particular experiment, therefore, should be analysed in terms of all of these mechanisms, not just one. The exposition of the models reveals gaps in the 'sound physical principles' on which all such models must be based, demonstrating that a complete theory is still some way off, because the gaps must be filled, either by theory or by experiment.

The models concern the experiment depicted in fig. 6.2, and described on pp. 304–7, in which a length of collapsible tube is supported between two rigid tubes, each having constant resistance. Suppose that initially the flow is steady and the chamber pressure is zero or negative, so that the collapsible segment is distended and has an approximately constant, circular cross-section. We assume that this configuration is stable to small disturbances. Now let us consider what happens if at time $t = 0$ the chamber pressure is suddenly raised to a new constant value, $\hat{p}_c$.

If $\hat{p}_c$ is sufficiently small, then the transmural pressure $\hat{p}_{tm}$ will everywhere remain greater than the critical value below which collapse begins. As can be seen from fig. 1.10(a), that value is close to zero for rubber tubes, though not for veins; detailed calculations of the critical value were made by Anliker & Raman (1966) in their analysis of the initiation of buckling in a fluid-filled cylindrical shell of linearly elastic material.

If $\hat{p}_c$ is large enough, however, $\hat{p}_{tm}$ will exceed the critical value and the tube will begin to collapse at its downstream end. Since the Reynolds number is large, the time-scale for the collapse will be much shorter than the viscous diffusion time, and the perturbation to the initial flow will be effectively inviscid. Therefore the flow will be accelerated through the narrowed segment, and the fluid pressure will fall in the manner described by Bernoulli's equation (even when the unsteady, reactive pressure is taken into account). Note that the reactive pressures generated in the fluid by the collapse process will be propagated both upstream and downstream by a pressure wave in the tube wall, whose speed can be calculated from the theory of chapter 2. When the collapsible segment is short, and the upstream tube effectively rigid, the propagation will be virtually instantaneous. The pressure fall at the constriction will accentuate the collapse, which will continue until *either* a new, stable equilibrium is reached (because the tube is very stiff when almost completely collapsed) *or* some other effect comes into play. In fact, various 'other effects' must come into play whether or not there is a theoretical stable equilibrium.

The first such effect is the development of significant energy losses as the collapse proceeds. The inviscid disturbance to flow in the core will cause viscous boundary layers to develop on the walls, which can initially, no doubt, be analysed by the theory of Smith (1976*d*) and Duck (1979). When the cross-sectional area of the constriction becomes very small, and hence the velocity of the flow through it becomes large, the boundary layers will fill the cross-section and the viscous energy dissipation in this narrow region of high shear will become important. In a quasi-steady, quasi-parallel flow model this could be represented by the high resistance of a narrow tube, as described by (6.13). Probably more important, however, is the fact that unless the wall slope is kept extremely small (e.g. by a large longitudinal wall tension) the flow will separate at approximately the narrowest point, and an asymmetric jet will emerge into the downstream rigid tube. There may also be upstream separation (Smith, 1977*b*), but since the experiments generally indicate that the narrowing of the tube upstream of the narrowest point is much more gradual than the subsequent widening (fig. 6.18), this is not inevitable, and in any case the head loss associated with it will not be as dramatic as with the downstream separation. The downstream jet will also normally become turbulent unless the Reynolds number at the constriction is below about 500 (cf. p. 314). Now the processes of flow separation and jet instability each take a short but finite time, related to the convection time-scale $\hat{a}_0/\hat{U}$, where $\hat{a}_0$ is the undisturbed tube radius and $\hat{U}$ the velocity through the constriction. On such a time-scale, therefore, we expect significant energy losses to develop, in addition to the direct viscous losses already discussed. These losses will cause a

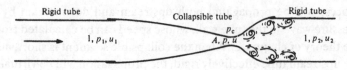

Fig. 6.18. Sketch of collapsible segment specifying the areas, pressures and velocities at three stations, as required by the lumped-parameter model.

sharp pressure drop downstream of the constriction and a dramatic reduction in the flow-rate. This will in turn generate a rise in pressure at, and upstream of, the constriction, and this pressure rise will first slow down the collapse, then reverse it. There will normally be an equilibrium state in which there is significant energy loss and some degree of collapse (see § 6.3.3) below. If this equilibrium is stable, the system will tend to it (unless there is sufficient inertia in the motion of the wall and the fluid for the area to return to its open position, and possibly beyond, in which case the separation bubble would be swept away and the cycle would be free to start again). If the equilibrium is unstable, on the other hand, the system will either tend to another equilibrium that exists at the same values of the parameters, or oscillations will occur because the tube is so stiff at both very large and very small areas that the area cannot either increase or decrease indefinitely.

Circumstantial support for the importance of flow separation in the self-excited oscillations observed by Ur & Gordon (1970) comes from their detailed discussion. They report the following sequence of events in an oscillation whose period is approximately 70 ms, based on the recordings of fig. 6.7 corrected for the time taken by pressure waves to propagate to the measuring sites (the collapsible segment was 25 diameters long). About 3 ms before maximal closure of the constriction, there began a large (200 mmHg) and rapid drop in downstream pressure, lasting about 5 ms; this may be interpreted as the development of separation. Ur & Gordon associate it with deceleration of the column of fluid downstream of the constriction, which is, of course, an inevitable correlate of the pressure fall. As the tube closes there is also a sharp, but less large (55 mmHg), rise in upstream pressure, lasting about 2 ms, which is presumably the reactive pressure rise also mentioned above. Following maximal closure of the tube, the downstream pressure returns, within 8 ms, to a level of 20–25 mmHg, and the upstream pressure falls to about 35 mmHg in 22 ms. These are elevated above the undisturbed state because of the distension of the upstream segment. Such relatively gradual pressure changes are associated with the re-opening of the tube, which is maximally open (having overshot its undisturbed configuration) after 32 ms. In the next 36 ms there is a gradual reduction in tube area, before sudden collapse again takes place and the cycle restarts.

Implicit in the above discussion is the assumption that the elastic behaviour of the tube at each cross-section can be described by the same pressure–area curve, such as that in fig. 1.10(a). It has been pointed out by J. M. Fitz-Gerald (personal communication) that this is unlikely to be true if the tube is either thin-walled or quite long, when the segment of flexible tube upstream of the constriction may play an important role. As the flow at the constriction is slowed down by the energy losses caused by separation, there will tend to be more flow into the upstream segment than out of it, so that it will become distended. This distension will then tend to pull open the constricted segment through longitudinal tensile forces. If the longitudinal tension in the undisturbed wall is sufficiently great, and if the wall is sufficiently thin, then the restoring force associated with this tension could dominate that associated with the resistance to transverse bending as the cross-section changes shape (cf. § 1.1.4); a numerical model incorporating this possibility has recently been developed by Collins & Tedgui (1979). The model to be presented later in this section will ignore longitudinal tension, and is therefore likely to be relevant only to relatively short, relatively thick-walled tubes. Future developments require the incorporation of longitudinal tension and longitudinal curvature into the model on the lines proposed by Collins & Tedgui (1979), by Griffiths (1975c) and by J. M. Fitz-Gerald. However, it is unlikely to affect the basic fluid mechanics that underlie the oscillations predicted by the lumped-parameter model outlined below.

What lumped-parameter models inevitably do ignore are phenomena directly associated with the continuous variation of cross-sectional area and velocity in the real system. Thus if the flow in the longitudinally slowly varying segment upstream of a steady constriction is intrinsically unstable to a local mechanism, then the steady state will be unstable even if the lumped-parameter model predicts stability. This is shown to be a real possibility by the analysis of § 6.4, where an instability mechanism analogous to that of roll-waves in open channel flow (Dressler, 1949) is shown to operate in elastic tubes.

Another potentially important effect, associated with the continuously varying system, has also been ignored hitherto. It comes from the fact that an elastic tube can support pressure waves, as described in chapter 2. 'Choking' is therefore predicted to occur if the local fluid speed anywhere becomes equal to the local speed of propagation of such waves. This effect is also investigated briefly in § 6.4, and may well be relevant in many experiments in which oscillations have been observed.

6.3.2 *Mathematical formulation*

The detailed, non-linear, three-dimensional, unsteady fluid mechanics of the collapse process is dauntingly complex, and in order to formulate a tractable model some drastic simplifications have to be made. Here we reduce the whole system to a third-order set of ordinary differential equations with the consequence, as already indicated, that many real phenomena cannot be described. Several interesting phenomena are described, however, and it is hoped that, if the model is carefully derived, they may correspond to some of those observed experimentally. The model is similar in many respects to that proposed by Katz *et al.* (1969), but there are important differences, so that although the present system is of lower order, it is somewhat more soundly based. Moreover, the very fact of its being of lower order means that greater analytical (and hence physical) understanding can be obtained; in view of the variety of behaviour that is predicted for different values of the parameters, this is essential. Another similar model has been proposed by Schoendorfer & Shapiro (1977), who analysed a collapsible tube device that might act as a vocal source for laryngectomised patients. The oscillations of the collapsible tube walls would replace those of the vocal chords as the basis for speech.

The most drastic assumption is to assert that the geometrical configuration of the collapsible segment can be completely specified by a single variable, the cross-sectional area, $\hat{A}$, at the narrowest part of the constriction, close to the downstream end (fig. 6.18). We also assert that the tube-wall elasticity can be fully taken into account by specifying a single-valued functional relation between the transmural pressure, $\hat{p}_{tm}$, and $\hat{A}$. So we beg all the elastic questions not only by choosing a given function relating $\hat{p}_{tm}$ and $\hat{A}$ at each value of $\hat{x}$, the axial distance, but also by saying that the narrowest portion of the tube is representative' of the whole segment. This is equivalent to the statement that the configuration of the tube, the longitudinal shape as well as the transverse shape, is completely specified when $\hat{A}$ is specified; no independent distension of the upstream segment relative to the constriction is possible. Thus the presence or absence of longitudinal tension and/or curvature in the wall is not important, except in its effect on the pressure–area relation. The main advantage of the assumption, of course, is that the final differential equation for $\hat{A}$ is ordinary (in $\hat{t}$) not partial (in $\hat{t}$ and $\hat{x}$), and can be solved relatively simply.

In the equations to be presented, the area $\hat{A}$ is made dimensionless with respect to the upstream and downstream area $\hat{A}_0$, and all velocities are non-dimensionalised with respect to an arbitrary scale, $\hat{U}_0$. Pressures and time are scaled with respect to $\rho \hat{U}_0^2$ and to $\hat{a}_0/\hat{U}_0$ respectively, where $\hat{a}_0$ is the radius of the rigid tubes and ρ is the fluid density. The collapsible segment is depicted in fig. 6.18 with the dimensionless area, pressures and velocities marked at the three stations to be considered: upstream (suffix 1), downstream (suffix 2) and at the constriction (no suffix). The dimensionless external pressure is p_c. The straight lines represent the rigid tubes, upstream and downstream. The only independent variable is the dimensionless time, t; the equations governing this model can be written down as follows.

Elasticity. As already discussed, the elastic properties of the collapsible segment are represented by a single equation relating the transmural pressure at the narrowest point to the cross-sectional area there:

$$p - p_c = P(A). \tag{6.27}$$

If quantitative agreement with experiment is hoped for, the function $P(A)$ should be measured independently for every piece of collapsible tube used; graphs like those of fig. 1.10(a) would be obtained. Here, however, we shall use a simple functional form to represent rubber tubes (in the same way that (6.24) was used to represent veins). We shall incorporate into it the similarity solution derived by Flaherty *et al.* (1972a) for the relation at small values of A when there is no longitudinal variation. This is based on the fact that when the opposite walls of the tube are in line contact (fig. 6.19(a)) the original radius of the tube cannot be relevant to the pressure–area relation. The only length-scale available is a typical radius of curvature, say R. Now the analysis of bending of the tube wall (§ 1.1.4) shows that

$$1/R = M/EI, \qquad (6.28)$$

where M is the bending moment (per unit length of the tube), I is the moment of inertia of the wall cross-section (per unit length) and E is the Young's modulus of the material, assumed constant. Of these quantities only M depends on R, and is proportional to the bending force (per unit length) times a distance, i.e. to $\hat{p}_{tm}R$ times R. Thus (6.28) shows that the transmural pressure, $\hat{p}_{tm}$, is proportional to R^{-3}, which is proportional to $\hat{A}^{-3/2}$. We follow Shapiro (1977) and use the simplest form for $P(A)$ that is consistent with this result as $A \to 0$ and that represents a change to circular cross-section (and greater stiffness) at zero transmural pressure (cf. fig. 1.10(a)); this is

$$
\begin{aligned}
P(A) &= P_0[1 - (A_c/A)^{3/2}], & A < A_c, \\
&= 10P_0(A/A_c - 1), & A > A_c,
\end{aligned}
\right\} \qquad (6.29)
$$

and is plotted in fig. 6.19(b). Here A_c is the (dimensionless) area at which the transition takes place; in an experiment one might expect A_c to be slightly less than 1, on the assumption that a little distension is required to fit the flexible tube over the rigid ones, but in all subsequent calculations we shall take $A_c = 1$. This cannot have great qualitative significance. P_0 is equal to a dimensional pressure-scale, $\hat{P}_0$, divided by $\rho \hat{U}_0^2$; its value will be chosen later. The factor 10 in the second of equations (6.29) is a rather arbitrary representation of the greater stiffness of the circular cross-section.

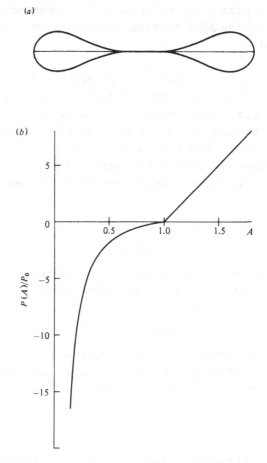

Fig. 6.19. (a) Shape of tube cross-section when there is line contact between opposite sides and a similarity solution for $P(A)$ exists. (After Flaherty et al., 1972a.) (b) Graph of the pressure–area relation (6.29).

In some experiments it may be important to include the inertial and viscoelastic properties of the tube wall. In that case, the extra terms

$$\eta\dot{A} + I_w\ddot{A}$$

should be added to the right-hand side of (6.27), where η and I_w are constants representing wall viscosity and inertia respectively, and an overdot means d/dt. These terms do not increase the order of the system, but merely add to the number of imprecisely known constants in the model; they are therefore set to zero in all present calculations.

Conservation of mass. Equations are required to relate the difference between the inflow and outflow at the ends of the collapsible segment to the rate of change of its volume. Recognising that in most experiments the collapse occurs close to the downstream end of the flexible segment, we suppose that all volume change occurs upstream of the narrowest point. Thus the flow-rate through the constriction is equal to the flow-rate downstream,

$$uA = u_2, \tag{6.30}$$

while, for the upstream segment

$$u_1 - uA = l\dot{A}_v, \tag{6.31}$$

where l is the dimensionless length of the upstream segment and A_v is its average area. In the absence of detailed experimental data, we take

$$A_v = \tfrac{1}{2}(1 + A), \tag{6.32}$$

but it must be emphasised that this is somewhat arbitrary. Note too that there would be no difficulty in including some downstream compliance in (6.30), with a much smaller value of l, for example, but in the interests of simplicity we omit it.

Momentum equations in the rigid tubes. The upstream and downstream rigid tubes may be long, straight tubes, along which the pressure drop is proportional to the flow-rate, or they may incorporate variable constrictions (cf. Conrad, 1969) which would mean that at all but the smallest flow-rates the pressure drop would be proportional to the square of the flow-rate. They will probably be a combination of the two, and we choose to represent them by resistances, $\hat{R}_1$ and $\hat{R}_2$, that are non-decreasing functions of the magnitudes of the velocities, $\hat{u}_1$ and $\hat{u}_2$, through them. In most of the numerical calculations, however, $\hat{R}_1$ and $\hat{R}_2$ are taken to be constants. The fluid in the rigid tubes will also have inertia, and we choose to represent this linearly from the start. If we suppose that the flow enters the upstream rigid tube from a reservoir of constant dimensionless head, P_1, and that it leaves the downstream rigid tube into atmosphere at zero pressure, the momentum equations for these two tubes are

$$P_1 - p_1 = R_1 u_1 + I_1 \dot{u}_1, \qquad (6.33)$$

$$p_2 = R_2 u_2 + I_2 \dot{u}_2, \qquad (6.34)$$

where $R_{1,2} = \hat{R}_{1,2}/\rho \hat{U}_0$ and $I_{1,2}$ are corresponding dimensionless inertances.

Momentum/energy equation downstream of the constriction. We have already emphasised the importance of the unsteady head loss associated with the development of a separated, probably turbulent, jet downstream of the constriction as its area decreases, and its subsequent disappearance as the tube opens up again. Since some of the observed oscillations had periods as low as 0.02 s (Ur & Gordon, 1970), comparable with a typical convection time $\hat{a}_0/\hat{U}_0$, the phase relation between the unsteady head loss and (say) the velocity $\hat{u}$ at the constriction is likely to be important. The use of a quasi-steady relation is unlikely to be truly representative. However, this is one of the areas of fluid mechanics about which almost nothing is yet known[†] (except for some limited results on sinusoidally oscillating flow through a fixed constriction by Young & Tsai (1973b)), and we have no alternative but to use the quasi-steady relation here.

For steady separated flow through a sharp constriction, the integral form of the momentum equation can be applied to the fluid between the constriction and a downstream station at which the mean velocity profile is approximately flat (where the jet has spread to the side-walls: fig. 6.20). When the jet is turbulent, this is not far downstream. Use of the equation requires a knowledge of the pressure on the downstream facing walls of the constriction; this must be approximately uniform and equal to the pressure at the constriction, for otherwise the lateral pressure gradient would not permit an approximately parallel-sided jet to form. The momentum

---

[†] Dr T. W. Secomb and Dr C. D. Bertram at the University of Cambridge are currently working to fill this gap, both theoretically and experimentally.

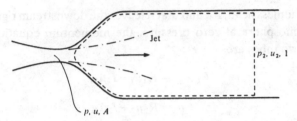

Fig. 6.20. Control surface for use with the integral momentum equation, leading to (6.35).

equation then gives

$$p + Au^2 = p_2 + u_2^2,$$

on the assumption that the longitudinal viscous shear force on the tube walls makes a negligible contribution. Together with (6.30) this gives

$$p - p_2 = -f(A)u_2^2, \tag{6.35}$$

where

$$f(A) = 1/A - 1, \quad \text{if } A < 1, u_2 > 0. \tag{6.36a}$$

Young & Tsai (1973a) made some measurements on steady flow through symmetric and asymmetric smooth constrictions of two different areas, and fitted their results to the equation

$$p_1 - p_2 = K_1 u_2 + \tfrac{1}{2} K_2 u_2^2 (1/A - 1)^2,$$

which is equivalent to the above if there is no head loss upstream of the constriction, if $K_1 = 0$ and if $K_2 = 1$. Young & Tsai's results indicated different values of K_2, which, however, always lay in the range 0.9–1.2. The value of K_1 depended strongly on both the area and the shape of the constriction, but this linear contribution to $p_1 - p_2$, although not negligible, was relatively small (<20%) for a severe constriction ($A = 0.11$) at Reynolds numbers of the order of 1000. In some subsequent experiments on constrictions made by boring small cylindrical holes through cylindrical plugs, Seeley & Young (1976) found rather higher values of K_2 (1.3–1.8), but the earlier, smoothly varying constrictions are expected to be more applicable here. We therefore retain (6.35) and (6.36a), but we shall subsequently permit the incorporation of a linear term into the upstream head loss.

At times when the tube is distended, with $A \geq 1$, $(6.36a)$ is inappropriate, because the flow would not be separated. Then we may assume negligible head loss, so that (6.35) can still be used, but with

$$f(A) = \tfrac{1}{2}(1/A^2 - 1) \leq 0, \quad \text{if } A \geq 1, u_2 > 0. \qquad (6.36b)$$

Finally, it is possible that during large-amplitude oscillations, the flow through the constriction might briefly reverse its direction (this has been observed by Dr C. D. Bertram, personal communication). In this case, too, the flow would converge to the constriction, and there would be negligible quasi-steady head loss:

$$f(A) = \tfrac{1}{2}(1/A^2 - 1) > 0, \quad \text{if } A < 1, u_2 < 0. \qquad (6.36c)$$

Momentum/energy equation upstream of the constriction. If the possibility of upstream separation is neglected, as suggested above, then the only quasi-steady head loss between station 1 and the constriction will come about through direct viscous action. The simplest assumption, appropriate for a short length of flexible tube, is to neglect it altogether, and also to neglect the inertia of the fluid in the upstream segment. In that case we have

$$p_1 + \tfrac{1}{2}u_1^2 - (p + \tfrac{1}{2}u^2) = 0. \qquad (6.37a)$$

However, this is likely to be inappropriate for long flexible segments, for very slight constrictions (where $f(A)$ is small), or for very severe constrictions, when there is a considerable length of collapsed tube with a very small area and a very high resistance. In the latter case especially there will be a significant viscous head loss. To the right-hand side of $(6.37a)$ we therefore add a term $\tfrac{1}{4}lR(A)u$, where $\tfrac{1}{4}l$ is an arbitrary representation of the length of the collapsed segment, and $R(A)$ is a dimensionless measure of the resistance per unit length of the collapsed segment with area A.

If the cross-section of the flexible tube upstream of the narrowest point (immediately after which there is a rapid expansion) is assumed to be a slowly varying ellipse, then an appropriate description of the quasi-steady resistance is provided by lubrication theory, from (6.13). When non-dimensionalised as in the present section, this gives

$$R(A) = (4/Re)(\delta + 1/\delta)/A, \qquad (6.38)$$

where $Re = \hat{a}_0 \hat{U}_0/\nu$ is a constant Reynolds number, and δ is the ratio of the minor to the major axis of the ellipse. For a distended tube, with $A > 1$, δ is equal to 1 and (6.38) reduces to Poiseuille's formula, $R(A) = 8/ReA$. When $A < 1$, however, the relation between δ and A must be specified, as in § 6.2. For rubber tubes, the assumption of constant perimeter is appropriate, as given by (6.20), whose right-hand side can be rewritten as $\frac{1}{2}\pi(\delta/A)^{1/2}$. The left-hand side is the complete elliptic integral of the second kind $E(1-\delta^2)$ (Abramowitz & Stegun, 1965, p. 590). Now if $0 \leqslant \delta \leqslant 1$, then $1 \leqslant E(1-\delta^2) \leqslant 1.57$. Thus a rough, but not unreasonable, approximation in circumstances where δ and A vary widely is to treat E as a constant; here we set $E^2 = 2$, and replace (6.20) by

$$\delta = 8A/\pi^2 \quad \text{when } A < 1. \tag{6.39}$$

When the upstream part of the flexible segment is quite long, the inertia of the fluid in it may not be negligible. We account for it by adding to the right-hand side of (6.37a) another term $l\dot{u}_v$, where u_v is an average fluid speed, which we take, again arbitrarily, to be the speed at the location where the area is A_v (see (6.32)), assumed to be about half-way along the segment. Thus

$$u_v A_v - uA = \tfrac{1}{2} l\dot{A}_v. \tag{6.40}$$

The final form of (6.37a) is thus

$$p_1 + \tfrac{1}{2}u_1^2 - (p + \tfrac{1}{2}u^2) = \tfrac{1}{4}lR(A)u + l\dot{u}_v. \tag{6.37b}$$

This completes the formulation of the lumped-parameter model. It will be used in three ways: to examine what equilibrium states are possible, to investigate their stability, and to compute the form of the non-linear oscillations that may arise when they are unstable. Whenever numerical values of the dimensionless parameters are required, they will be chosen as far as possible to be comparable with those pertaining to Conrad's (1969) experiments, and will be derived as and when they are needed. Because all parameters cannot be inferred precisely, the results will be only of qualitative value. However, when the model is applied to future experiments, the parameters must be worked out carefully so that quantitative validation may be possible.

6.3.3 *Equilibrium states*

Computation of the equilibrium states should lead to graphs, such as those of fig. 6.3(c) or 6.4(a), of the dimensionless pressure drop across the collapsible segment, $p_1 - p_2$, against the dimensionless flow-rate, Q. For each point on such a curve, therefore, we fix Q and compute the steady value of A that is associated with it. In the steady state u_1, u_2 and uA are all equal to Q (see (6.30) and (6.31)), and A is computed from the equation obtained by eliminating p and p_2 from (6.27), (6.34) and (6.35), i.e. from

$$R_2 Q - f(A)Q^2 = p_c + P(A), \qquad (6.41)$$

where $P(A)$ and $f(A)$ are given by (6.29) and (6.36a or b) respectively. Once A has been computed, p_1 is calculated from (6.37a or b) and P_1 from (6.33).

The roots of (6.41) can be assessed qualitatively from fig. 6.21, where the right-hand side is plotted against A as the continuous curve and there are several plots of the left-hand side, corresponding to various values of Q and of the other parameters. It can be seen that the equation has either one or three roots. It always has at least one root, because (a) $P(A)$ tends to infinity as A tends to infinity, while $f(A)$ tends to $-\frac{1}{2}$, and (b) $P(A)$ (i.e. $-A^{-3/2}$) tends to $-\infty$ more rapidly as A tends to 0 than $-f(A)$ (i.e. $-A^{-1}$). Further, if $R_2 Q > p_c$ there is always one root greater than 1; this state always turns out to be stable, and we can identify it with the state at large flow-rates when the tube is distended. This will correspond to section I of the graph in fig. 6.4: $p_1 - p_2$ will be zero if (6.37a) is used, but not quite zero if (6.37b) is used.

As $R_2 Q$ is slowly reduced below p_c, the equilibrium value of A will fall below 1, i.e. the tube will begin to collapse. The collapse itself will be smooth and gradual if (6.41) has only one root for all Q, as can be seen from the sequence of three broken curves in fig. 6.21. However, if there are three roots when $R_2 Q > p_c$, the collapse will be catastrophic, since the first pair of roots will suddenly cease to exist, and the only possible equilibrium will have a very small value of A. The gradual collapse can conceivably occur quasi-steadily, but the catastrophic collapse cannot possibly do so. The criterion that distinguishes between gradual and catastrophic collapse is whether the slope at $A = 1_-$ of the left-hand side of (6.41) is less than or

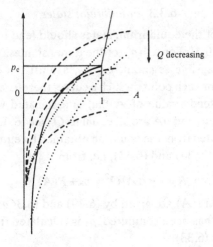

Fig. 6.21. Sketch of the curves representing the two sides of (6.41), plotted against cross-sectional area, A. Continuous curve, right-hand side; broken curves, left-hand side when (6.42) is satisfied, different curves representing different values of Q; dotted curves, left-hand side when (6.42) is not satisfied.

greater than that of the right-hand side, when $R_2Q = p_c$. In the case of constant R_2, this means that the collapse will be gradual if

$$R_2 \geqslant p_c/(1.5P_0)^{1/2} \qquad (6.42)$$

and catastrophic otherwise. There do not seem to have been any previous theoretical or experimental studies that distinguish between these two modes of collapse, although in their models of forced expiration, Clément et al. (1973) considered only catastrophic collapse, while Lambert & Wilson (1972) implicitly considered only gradual collapse. As we shall see, the behaviour of the collapsing tube, as represented for example by the graph of A against t, depends critically on whether (6.42) is satisfied or not.

In order to compute the details of these equilibrium states, we need to assign numerical values to the parameters P_0, p_c, R_2 and l. A complete survey of all possible values has not yet been undertaken; only values roughly appropriate to Conrad's (1969) experiments have been used. The internal radius of Conrad's uncollapsed tube, $\hat{a}_0$, was 0.0063 m, and its length was 0.089 m; we shall take $l = 10$. Conrad employed flow-rates of up to 3×10^{-5} m^3 s^{-1} (see fig. 6.3);

we choose the velocity-scale, $\hat{U}_0$, to be 0.15 m s^{-1}, corresponding to a flow-rate of 1.9×10^{-5} m^3 s^{-1}, so that the pressure-scale $\rho \hat{U}_0^2$ is 22.5 N m^{-2} ($\rho = 10^3$ kg m^{-3} for water). The Reynolds number, Re, is 860 ($\nu = 1.1 \times 10^{-6}$ N m^{-2} s for water). The external pressure used for all the experiments of fig. 6.3(c) was 29.5 mmHg, which corresponds to a value of $p_c = 180$; in most computations we have used $p_c = 200$, which tends to increase the maximum values of $p_1 - p_2$ somewhat above those shown in fig. 6.3(c).

We can obtain a rough estimate of P_0 by using the theory of Flaherty *et al.* (1972a), which was translated into the form of a graph of transmural pressure against area by Shapiro (1977). Shapiro's fig. 1, for example, shows that $A = 0.21$ when $\hat{p}_{tm} = -10E(h/\hat{a}_0^3)/12(1 - \sigma^2)$, where E and σ are the Young's modulus and Poisson's ratio of the tube wall and h is its thickness. Conrad (1969) reported that his tubes had $E \approx 1.6 \times 10^5$ N m^{-2} and $h = 0.93$ mm, and rubber has a Poisson's ratio close to 0.5, so that in dimensionless terms $P = -25.4$ when $A = 0.21$. Using (6.29), therefore, we obtain $P_0 = 2.7$. In all computations we have used the slightly stiffer value of 4; this has the effect of diminishing the maximum of the $p_1 - p_2$ versus Q graph.

Finally we must choose $R_2 = \hat{R}_2/\rho \hat{U}_0$. In most calculations this is taken to be a constant for convenience, although most of Conrad's downstream resistance came from an adjustable constriction across which the pressure drop was equal to $k_2 \hat{Q}^2$ plus a constant. This change turns out to have no important effect, and we estimate a suitable value for R_2 from curve 4 of fig. 6.3(b). Taking $\hat{p}_2 = 15$ mmHg at a flow-rate of 10 cm^3 s^{-1} we obtain $R_2 \approx 170$. In fact the values we use below are $R_2 = 150, 75, 50$. Note that the smallest of these values is less than $p_c/(1.5P_0)^{1/2}$ (see (6.42)), while the other two are above it. Also, the smaller R_2 is, the higher we expect the maximum of the $p_1 - p_2$ versus Q curve to be (fig. 6.3(c)).

Results of the equilibrium computations are given in figs. 6.22(a) and (b). The continuous curves are those obtained using the full equation (6.37b), while the broken curves are derived by neglecting viscous resistance in the collapsible segment itself (see (6.37a)). Inclusion of the viscous resistance increases the pressure drop, especially at low flow-rates, where the concave nature of the broken curves is almost converted to the approximately linear form

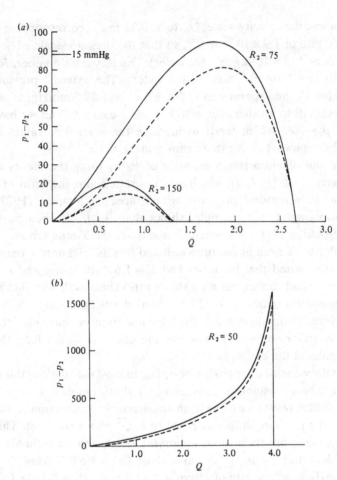

Fig. 6.22. Graphs of $p_1 - p_2$ against Q, as calculated from the lumped-parameter model; $P_0 = 4$, $p_c = 200$, $l = 10$. Continuous curves, $R(A)$ given by (6.38); broken curves, $R(A) = 0$. (a) $R_2 = 75$ and $R_2 = 150$, for which (6.42) is satisfied. (b) $R_2 = 50$, for which (6.42) is not satisfied.

observed experimentally (fig. 6.3(c)). For values of Q greater than p_c/R_2, the pressure drop across the collapsible segment is, of course, zero when viscous resistance is not incorporated, and turns out to be negligibly small even when it is incorporated. In fact, the continuous curves in fig. 6.22(a), representing the cases of gradual collapse ($R_2 = 150, 75$), bear a close qualitative resemblance to the experimental curves, including the effect of reducing the downstream resistance, which is to increase the value of the maximum pressure drop and to extend the range of flow-rates for which collapse occurs. An exactly similar effect is obtained by increasing the chamber pressure, p_c, as is observed experimentally and as is easily predictable from (6.41). We might note here that there is no qualitative difference if the downstream resistance is made non-linear by writing $k_2 Q$ for R_2, except that the broken curves become even more concave at small flow-rates.

For quantitative comparison with experiment, the results in fig. 6.22(a) must be converted into dimensional terms. The case $R_2 = 150$, chosen to correspond roughly with curve 4 of fig. 6.3, has a maximum value of $p_1 - p_2$ of about 20, which represents 450 N m^{-2} or 3.4 mmHg; this is much less than the value of about 22 mmHg recorded in fig. 6.3(c). Furthermore, the value of Q at which the tube becomes circular, about 1.33, corresponds to a flow-rate of about 25 cm^3 s^{-1}, significantly bigger than the observed value of 15 cm^3 s^{-1}. The latter discrepancy is reduced by taking the more appropriate values of p_c and R_2 derived above (180 and 170 respectively) when the critical flow-rate is reduced to 20 cm^3 s^{-1}, but that reduces the maximum value of $p_1 - p_2$ still further. Even when the smaller value of 2.7 is used for P_0, the maximum value of $p_1 - p_2$ is restored to only about 23. It is only when P_0 is reduced to the low value of 1.0 that the maximum value of $p_1 - p_2$ becomes comparable with the observations: $(p_1 - p_2)_{max} = 127 \approx 21$ mmHg (fig. 6.23).[†] If R_2 is reduced to 75, then the maximum pressure drop becomes comparable with those observed with the original values of P_0 and p_c (4 and 200 respectively), but then the range of flow-rates for which collapse occurs extends far beyond those observed (fig. 6.22(a)). Thus *close* quantitative agreement with Conrad's experiments can be achieved only by adjusting the parameters away from the values chosen *a priori*; this may be because the parameters were not correctly chosen originally, but is equally likely to stem from the crudeness of the present model. The good qualitative

---

[†] After this book had gone to press, Mr Conrad informed me that there was a misprint in his 1969 paper and that the actual thickness of the tube wall was much less than the value quoted. Hence P_0 should be considerably smaller than 1, and agreement between these calculations and Conrad's experiments is unlikely to be quantitatively very good.

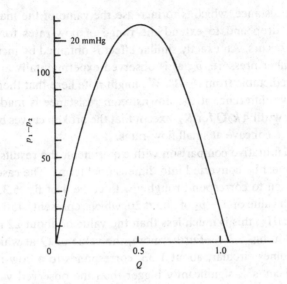

Fig. 6.23. Graph of $p_1 - p_2$ against Q, as calculated from the lumped-parameter model; $P_0 = 1$, $p_c = 180$, $R_2 = 170$, $l = 10$.

agreement, on the other hand, makes one reasonably confident that the chief physical mechanisms underlying the equilibrium states have been incorporated in the model, at least when it predicts gradual collapse.

However, a completely different picture is presented when R_2 is sufficiently low for catastrophic collapse to be predicted (fig. 6.22(b)). Here the shape of the $p_1 - p_2$ versus Q graph is quite different, and the maximum value of $p_1 - p_2$ is far greater, representing a dimensional pressure drop of 270 mmHg. Furthermore, although the equilibrium value of A is an increasing function of Q for small Q, as in the case of gradual collapse, in this case A starts to decrease again when Q exceeds 2.6. Nothing like this has ever been reported experimentally, which suggests strongly that the predicted equilibrium is unstable and that oscillations should be seen.

6.3.4 Stability and oscillations

The stability of an equilibrium and whether or not oscillations develop when it is unstable depend on what quantity is held constant as the equilibrium is perturbed. In an experiment the most natural such quantity is P_1, the overall driving pressure head, while another possibility is u_1, the flow-rate through the upstream rigid segment; the two would be equivalent if the upstream resistance R_1 were effectively infinite. For most of the present calculations we take P_1 to be constant.

To examine the stability of an equilibrium we denote the equilibrium values of all variables by the suffix zero, and suppose that they are all subjected to a small perturbation with exponential time variation. Thus we set

$$A = A_0 + a' \, e^{\sigma t}, \qquad u = u_0 + u' \, e^{\sigma t}, \qquad u_2 = Q_0 + u_2' \, e^{\sigma t}$$

etc. and substitute into the governing equations (6.27) and (6.29)–(6.40), neglecting non-linear terms in the perturbation amplitudes a', u' etc. These perturbation amplitudes are then eliminated from the equations, and the result is a cubic equation for σ. If any root of this equation has positive real part, then the equilibrium is unstable. The cubic obtained when the viscous pressure drop in the collapsible segment is not ignored (i.e. when (6.37b) is used, not (6.37a)), and R_1 and R_2 are taken to be constant, is

$$C_3 \sigma^3 + C_2 \sigma^2 + C_1 \sigma + C_0 = 0, \tag{6.43}$$

where the coefficients are given by

$$C_3 = \tfrac{1}{2} l I_2 [I_1 + l/(1 + A_0)],$$

$$C_2 = \tfrac{1}{2} l I_2 [R_1 - Q_0 - 4 Q_0/(1 + A_0)^2]$$
$$\quad + \tfrac{1}{2} l [I_1 + l/(1 + A_0)](R_2 - 2 Q_0 f_0),$$

$$C_1 = \tfrac{1}{2} l (R_2 - 2 Q_0 f_0)[R_1 - Q_0 - 4 Q_0/(1 + A_0)^2]$$
$$\quad + I_2 [P_0' - u_0^2/A_0 + \tfrac{1}{4} l u_0 (R_0' - R_0/A_0)]$$
$$\quad + [I_1 + 2l/(1 + A_0)](Q_0^2 f_0' + P_0'),$$

$$C_0 = (R_2 - 2 Q_0 f_0)[P_0' - u_0^2/A_0 + \tfrac{1}{4} l u_0 (R_0' - R_0/A_0)]$$
$$\quad + (R_1 - Q_0 + u_0/A_0 + \tfrac{1}{4} l R_0/A_0)(Q_0^2 f_0' + P_0').$$

In these equations $f_0, f_0', P_0', R_0, R_0'$ represent the relevant functions evaluated at $A = A_0$: $f_0 = f(A_0)$, $P_0' = P'(A_0)$, etc.

Whether or not the roots of a cubic have positive real part can be determined from *Routh's criterion*, which is very clearly expounded by Porter (1967). In the case when C_3 is positive, as here, this can be stated as follows:

(a) If $C_0 < 0$, the equation has at least one positive real root; that is, the instability can be classed as an exponential instability, in which the disturbance grows monotonically, without oscillation.

(b) If $C_0 > 0$ and either $C_2 < 0$ or $C_2 C_1 - C_3 C_0 < 0$ (or both) the equation has two roots with positive real part; these roots may be either real or complex. If they are complex, the instability will be in the form of a growing oscillation. If all the Cs are positive, and $C_2 C_1 - C_3 C_0 < 0$, then the two roots with positive real part *must* be complex.

It can be shown after a little algebra that

$$C_0 = (P_0' + Q^2 f_0') \, dP_1/dQ,$$

where dP_1/dQ is the rate at which the overall driving pressure increases with the flow-rate, i.e. the *overall* resistance of the system. Furthermore, the factor $P_0' + Q^2 f_0'$ is positive at all the equilibrium areas A_0 for which the slope of the continuous curve in fig. 6.21 exceeds that of the broken or dotted curve, i.e. for all equilibria except the middle one when there are three. This middle equilibrium has $C_0 < 0$, and is therefore unstable wherever $dP_1/dQ > 0$, which is the normal situation. The other equilibria, however, have $C_0 < 0$ if and only if $dP_1/dQ < 0$, i.e. exponential instability will result whenever the resistance of the system is negative. This is a familiar result from electrical circuit theory; attention has been drawn to it by both Conrad (1969) and Griffiths (1975). However, this instability does not normally lead to oscillation. At values of P_1 for which $dP_1/dQ < 0$ there are three possible equilibrium values of Q. Although the middle one is unstable, the system will tend to one of the others if that is stable, and the one with $A > 1$ always is stable.

The use of Routh's criterion, however, demonstrates that instability may arise when the overall resistance is not negative, a result that does not previously seem to have been noticed in the present context. This instability must lead to oscillation because there is no other equilibrium at the given value of P_1. The coefficients of (6.43) are too complicated for criteria based on the system's parameters to become immediately obvious, but numerical calculation of the roots of (6.43) confirms the possibility of such an instability for realistic parameter values.

Before the roots of (6.43) can be computed, it is necessary to choose values for the additional parameters, I_2, I_1 and R_2. If the pressure difference required to give an average acceleration $d\hat{u}_2/d\hat{t}$ to the fluid in the downstream rigid tube is $\hat{I}_2\, d\hat{u}_2/d\hat{t}$, then $I_2 = \hat{I}_2/\rho\hat{a}_0$. We estimate $\hat{I}_2$ as the mass of fluid in the downstream segment divided by its cross-sectional area, so that $I_2 = \hat{l}_2/\hat{a}_0$, where $\hat{l}_2$ is the length of the segment. Conrad (1969) did not report the value of $\hat{l}_2$, but his diagrams suggest that it was about $2\frac{1}{2}$ times the length of the collapsible segment. We therefore choose $I_2 = 25$. Conrad also did not report any details of the upstream rigid segment, because his rather simplified theory did not reveal its possible importance. In accordance with what would be the case if both rigid segments were long and straight, we have chosen to keep the ratios I_1/I_2 and R_1/R_2 equal to each other, at a value β, say. We have then chosen to examine three different values of β, equal to 1.5, 1.0 and 0.5.

The results of the stability calculation, for the equilibrium states represented by the continuous curves in figs. 6.22 and 6.23, are as follows.

(a) $P_0 = 4$, $p_c = 200$, $R_2 = 150$ (the lower curve in fig. 6.22(a)). This case is stable for all Q and for all three values of β.

(b) $P_0 = 4$, $p_c = 200$, $R_2 = 75$ (the upper curve in fig. 6.22(a)), and $\beta = 1.5$ so that $R_1 = 1.5R_2$. There is a 'negative-resistance' (non-oscillatory) instability for a small range of flow-rates just below the value above which the tube remains circular: $2.51 < Q < 2.67$. There are no other instabilities. This result agrees with the experimental evidence that instability arises only for sufficiently small values of R_2.

(c) The same values of P_0, p_c, R_2 but smaller values of β ($= 1.0$, 0.5). In each case the negative-resistance instability arises for a comparable range of flow-rates: $2.43 < Q < 2.67$ for $\beta = 1.0$ and $2.32 < Q < 2.67$ for $\beta = 0.5$. However another, oscillatory instability becomes manifest for a range of smaller flow-rates, owing to the fact that $C_2C_1 - C_3C_0 < 0$ ($C_2 > 0$). The range of flow-rates is $1.39 < Q < 1.90$ for $\beta = 1.0$ and $0.77 < Q < 2.32$ for $\beta = 0.5$, which in each case overlaps the rising part of the $p_1 - p_2$ versus Q curve in fig. 6.22(a). Neither Conrad (1969) nor Katz et al. (1969) observed oscillations in such circumstances, no doubt because the values of R_1 applicable to their experiments were too large. This prediction of an instability that does not depend on negative resistance suggests an interesting avenue of future experimental research.

(d) $P_0 = 4$, $p_c = 200$, $R_2 = 50$ (fig. 6.22(b)). This is the case of catastrophic collapse according to the equilibrium calculation. Here the overall resistance increases with Q for all Q up to 4, at which value it falls dramatically but above which there is a stable equilibrium with $A > 1$. Nevertheless, even with $\beta = 1.5$, the equilibria in most of the range $0 < Q < 4$ are unstable, with $C_2 C_1 - C_3 C_0 < 0$, $C_2 > 0$. Only the smallest flow-rates, $0 < Q < 0.62$, lead to stable equilibrium. This was foreseen above, and would explain why graphs like fig. 6.22(b) are not obtained in practice.

(e) $P_0 = 1$, $p_c = 180$, $R_2 = 170$ (fig. 6.23). This is the case for which the equilibrium predictions have the closest quantitative agreement with Conrad's experiments. No instability is predicted for $\beta = 1.5$, but a negative-resistance, and hence non-oscillatory, instability is predicted for $\beta = 1.0$, when Q lies in the range $0.88 < Q < 1.05$. (This range would coincide closely with the gap in curve 4 of fig. 6.3(c) if the scale for Q were adjusted to make the critical flow-rates for collapse ($Q = p_c/R_2$) coincide. As it is, the predicted p_c/R_2 corresponds to a flow-rate of 20 cm^3 s^{-1}, not 15 cm^3 s^{-1} as measured.)

(f) Finally we consider what happens if either or both of R_1 and I_1 are very large: this would be one way, in practice, of ensuring that the upstream velocity, u_1, remained constant. In this case the coefficients in (6.43) are given approximately by

$$C_3 = \tfrac{1}{2} l I_1 I_2, \qquad C_2 = \tfrac{1}{2} l (I_2 R_1 + I_1 R_2 - 2 I_1 Q_0 f_0),$$
$$C_1 = \tfrac{1}{2} l R_1 (R_2 - 2 Q_0 f_0) + I_1 (Q_0^2 f_0' + P_0'),$$
$$C_0 = R_1 (Q_0^2 f_0' + P_0').$$

We have already seen that $Q_0^2 f_0' + P_0' > 0$ for all equilibria of

interest, so Routh's criterion predicts instability only if either
$C_2 < 0$ or $C_2 C_1 - C_3 C_0 < 0$, i.e. only if

$$R_2 - 2 Q_0 f_0 < 0. \tag{6.44}$$

This is the same condition that we would have obtained by
setting u_1 equal to a constant at the outset of the stability
analysis. It can also be regarded as a negative-resistance
instability in that

$$R_2 - 2 Q_0 f_0 = dp_0/dQ_0,$$

so that instability is predicted if the net resistance of those parts
of the system downstream of the constriction is negative.
Examination of the equilibrium equation (6.41) shows that
(6.44) is also equivalent to the condition $dA_0/dQ_0 < 0$, and the
only one of the examples computed above that satisfies this is
the case of catastrophic collapse (fig. 6.22(b)) for $2.6 < Q < 4.0$.
The examples plotted in figs. 6.22(a) and 6.23 would be stable if
u_1 were held constant. However, there is no *a priori* reason why
this instability should be associated only with catastrophic
collapse.

The above results show that the values of the resistances and
inertances of the rigid tubes have a strong effect on whether the
flows are unstable and on what type of instability is predicted. It is
also likely that they will influence the non-linear motion that
develops after instability: as reported by both Conrad (1969) and
Ur & Gordon (1970), a range of oscillation frequencies can be
obtained by adjusting the parameters. On the other hand, the *shape*
of the oscillations, as indicated by the graphs of A and Q against t,
for example, will be dominated by the non-linearities and are
unlikely to be greatly affected by the rigid-tube parameters.

A few preliminary examples, computed by Dr C. D. Bertram, are
shown in figs. 6.24–6.26. The standard parameters, $P_0 = 4$, $p_c = 200$, $l = 10$, are the same in each case, while R_2 and β are different.
The plots in parts (a) are of A against t, while those of parts (b) are
of u_2 against t. Fig. 6.24 shows the oscillations that develop in a case
of gradual collapse when the equilibrium suffers an oscillatory
instability ($R_2 = 75$, $\beta = 0.5$, $P_1 = 259$ corresponding to $Q = 1.5$ in
fig. 6.22(a)), while fig. 6.25 shows the corresponding graphs for a

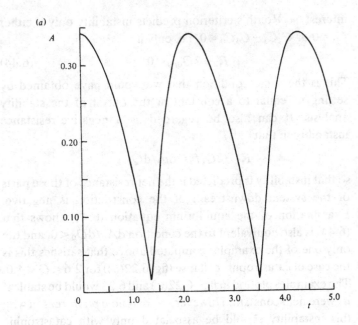

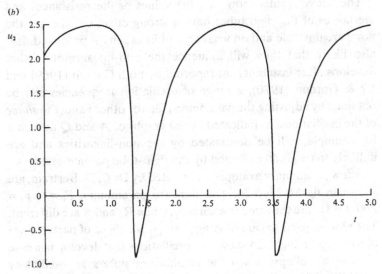

Fig. 6.24. Steady state oscillations, represented by graphs of (a) A against t and (b) u_2 against t. $R_2 = 75$, $\beta = 0.5$, $P_1 = 259$.

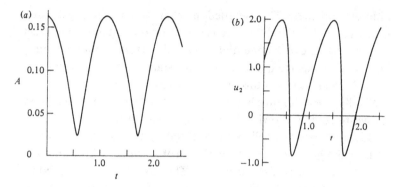

Fig. 6.25. As fig. 6.24, with $R_2 = 50$, $\beta = 1.5$, $P_1 = 209$.

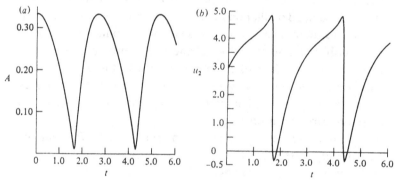

Fig. 6.26. As fig. 6.24, with $R_2 = 50$, $u_1 = 3$ (fixed).

case of catastrophic collapse ($R_2 = 50$, $\beta = 1.5$, $P_1 = 209$ corresponding to $Q = 1.0$ in fig. 6.22(b)). The oscillations have different amplitudes and frequencies, and somewhat different shapes, although in each case the tube tends to remain open, with a relatively high velocity through it, for a longer fraction of the cycle than it remains almost closed (this is especially true for the case depicted in fig. 6.24). Note too that in each case there is a brief period of downstream flow reversal shortly after the time at which the area of the constriction is narrowest.

Fig. 6.26 shows the corresponding curves for an unstable case in which u_1 is held constant ($R_2 = 50$, $u_1 = 3$). Here the area curves do not look very different, but the downstream velocity varies in a quite different manner. The detailed mechanics have not yet been investigated, but the difference is no doubt associated with the fact that this is a case of exponential, not oscillatory, instability. We may note that, as expected, the negative-resistance instabilities for cases with P_1 fixed do not in this model lead to oscillations.

Finally, the difference between gradual and catastrophic collapse is investigated by allowing P_1 to fall gradually from a large value at which the tube is uncollapsed, and stable, to one for which it is collapsed. Fig. 6.27 shows the variation of A with t in three different cases; in each case $dP_1/dt = -0.1$. The first is a case of gradual collapse, where no instability is predicted at any stage, and no oscillations are observed (fig. 6.27(a)). The second is also a case of gradual collapse, but one in which negative resistance, and hence instability, is predicted for a small range of flow-rates; this manifests itself as a damped oscillation as the collapse proceeds (fig. 6.27(b)). Finally, we show in fig. 6.27(c) a case of catastrophic collapse, with $R_2 = 50$. In this case our rather elementary computer programme broke down for values of R_1 comparable with those used in the other cases, because A was predicted to fall rapidly to a value as small as the accumulated numerical errors. The computations were therefore carried out for a very large value of R_1 ($\beta = 10^6$), which constrains u_1 to be approximately proportional to P_1, and the quasi-steady-state oscillations that develop are comparable with those for fixed u_1 shown in fig. 6.26.

Although these results are only preliminary, they do indicate that the present crude, lumped-parameter model is capable of describing a wealth of physical phenomena, not all of which have yet been fully understood or realised experimentally. There is clearly scope for a great deal of further work, both theoretical and experimental, even without taking into account those aspects of unsteady flow in collapsible tubes that cannot be described by a lumped model.

6.4 Other mechanisms of instability

The above analysis has shown that certain mechanisms by which oscillations in collapsible tubes are excited can be described by a lumped-parameter model. In this section we examine other mechanisms that cannot be so described. The discussion is based on

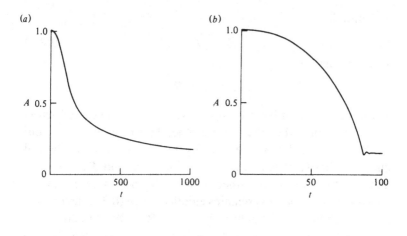

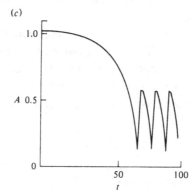

Fig. 6.27. Graphs of A against t for gradually falling values of P_i. (a) $R_2 = 150$, $\beta = 1.5$: gradual collapse, stable. (b) $R_2 = 75$, $\beta = 1.5$: gradual collapse, unstable. (c) $R_2 = 50$, $\beta = 10^6$: catastrophic collapse, unstable.

the equations governing approximately one-dimensional flow in a slowly varying collapsible tube, as set out in chapter 2. We let $A(x, t)$, $p(x, t)$ and $u(x, t)$ be the dimensionless cross-sectional area, pressure and velocity in the segment of tube upstream of the narrowest point. (Downstream of that point the tube area opens out very rapidly again as it joins to the downstream rigid tube and the one-dimensional model proposed here will break down.) The relevant equations are then those of elasticity

$$p - p_c = P(A), \tag{6.45}$$

of continuity

$$A_t + (uA)_x = 0, \qquad (6.46)$$

and of momentum

$$u_t + uu_x = -p_x - R(A)u. \qquad (6.47)$$

The last term in (6.47) represents the viscous force, averaged across a cross-section of the tube; a more general form could be used, but we have assumed that the velocity profile is quasi-steady and quasi-parallel so that this term is linear in u, and $R(A)$ can be deduced from lubrication theory. If we further assume that the cross-section of the tube remains elliptical, then (6.38) can be used for $R(A)$; however, the only properties of $R(A)$ that are important here are that it is positive and that it increases more rapidly than $1/A$ as A decreases.

There are a number of other factors that may be important in other experiments. These include variation of external pressure, p_c, with x, a longitudinal component of body force such as gravity (which is actually equivalent to a variation of p_c with x), and variation with x of the elastic properties of the tube, so that (6.45) is replaced by

$$p - p_c = P(A, x).$$

The equilibrium states of such more general systems have been exhaustively studied by Shapiro (1977), and a variety of steady and unsteady phenomena associated with them have been described by Griffiths (1971, 1975). However, the main points can be made with reference to the simple set of equations (6.45)–(6.47).

In looking for equilibrium conditions we set the time derivatives equal to zero and eliminate p to obtain

$$A_x/A = -R(A)u/(c^2 - u^2), \qquad (6.48)$$

where

$$c = [AP'(A)]^{1/2} \qquad (6.49)$$

is the dimensionless speed of infinitesimal pressure waves along the tube when its cross-sectional area is A (cf. (2.4)). We shall describe flows for which $u < c$ or $u > c$ as subcritical or supercritical respectively, by analogy with free surface liquid flow in a uniform channel

(Shapiro (1977) thoroughly examined the analogies between flow in collapsible tubes and both channel flow and compressible gas flow in rigid tubes). If we suppose that the flow is subcritical at the point where it enters the collapsible segment, then (6.48) shows that A is a decreasing function of x. Now the flow-rate Q is constant, so $u = Q/A$ must be an increasing function of x. For the model chosen here, in which $P(A)$ is given by (6.29) with $A_c = 1$, c is given by

$$c^2 = 10P_0 A, \quad A > 1,$$
$$= 1.5P_0/A^{3/2}, \quad A < 1, \qquad (6.50)$$

so for $A < 1$, c is also an increasing function of x, but it does not increase as rapidly as u. Thus if the tube is long enough, a point will eventually be reached at which $u = c$. At this point $-A_x$ is predicted to be infinite, so the model must have broken down before then; we conclude that a steady flow with the prescribed value of Q is impossible. This is the phenomenon of 'choking' familiar from gas dynamics.

Suppose that the pressure far upstream, P_1, is gradually raised from a value at which steady, subcritical flow is possible everywhere upstream of the narrowest point. As P_1 increases, Q will also increase until *either* instability occurs according to the lumped model, and we suppose that this does not happen, *or* the flow becomes critical at the narrowest point. For higher values of P_1 the steady flow cannot be maintained and oscillations must ensue. There is no possibility, as there sometimes is in channel flow or gas dynamics, of a continuous passage through the critical point with a hydraulic jump or shock (or 'elastic jump': Dawson & Elliott, 1977) downstream, because the numerator of (6.48) cannot be made to go to zero when the denominator does. As P_1 is raised above the critical value, the flow will remain almost critical at the narrowest point, and a pressure wave will propagate upstream to reduce the incoming flow-rate. However, since P_1 is held fixed, this will be reflected and the process will constantly repeat itself in an oscillatory manner. No complete model of such oscillations has yet been constructed, although Reyn (1974) has done some preliminary calculations on a model that neglects all energy dissipation except that occurring in 'shocks' as and when they develop.

An estimate of whether choking will occur in a particular case can be made if we assume that the lumped-parameter model is valid downstream of the narrowest point; the area at that point will then be given in terms of Q by (6.41). Of the examples computed above, choking occurs in some cases that were predicted to be stable as well as some unstable cases (e.g. $R_2 = 75$, $Q = 1.5$ from fig. 6.22(a), which is stable if $\beta = 1.5$ but unstable if $\beta = 0.5$ or 1.0), and does not occur in others, including cases that were predicted to be unstable (e.g. $R_2 = 75$, $Q = 1.0$, $\beta = 0.5$; or $R_2 = 50$, $Q = 1.3$, $\beta = 1.5$). Choking is not found at any flow-rate when $R_2 = 150$. We conclude that oscillations in a given experiment may involve either choking or a lumped-model instability or both; Conrad's (1969) experiments are particularly difficult to interpret in this context.

Finally, let us assume that choking does not occur, and that upstream of the constriction the equations (6.45)–(6.47) have equilibrium solutions $A = A_0(x)$, $p = p_0(x)$, $u = u_0(x)$. Can these be unstable by a mechanism that is not included in the lumped model? In order to investigate this, we suppose that the equilibrium solutions are sufficiently slowly varying with x that, at each x, they may be regarded as uniform for the purpose of a stability analysis. We then set

$$A = A_0 + a'(x, t)$$

with similar expressions for p and u, substitute into (6.45)–(6.47), linearise, and eliminate p' and u'. The resulting equation for a' is then

$$[\partial/\partial t + (u_0 + c_0)\, \partial/\partial x][\partial/\partial t + (u_0 - c_0)\, \partial/\partial x]a'$$
$$+ R(A_0)(\partial/\partial t + c_1\, \partial/\partial x)a' = 0, \qquad (6.51)$$

where c_0 is the wave speed, from (6.49), evaluated at area A_0, and

$$c_1 = u_0[1 - A_0 R'(A_0)/R(A_0)]. \qquad (6.52)$$

Equation (6.51) is the equation that describes roll-waves in open channels (see Dressler, 1949; and Whitham, 1974, p. 85). It follows immediately that the equilibrium flow is stable if and only if

$$u_0 - c_0 < c_1 < u_0 + c_0.$$

In our case $u_0 < c_0$, because choking is assumed not to occur, so the first inequality is identically satisfied. The second inequality then predicts that instability will occur if

$$u_0/c_0 > -R(A_0)/A_0 R'(A_0); \qquad (6.53)$$

the right-hand side of this is positive because R is a decreasing function of A.

If the tube cross-section remained circular at all times, then $R \propto 1/A$ and the right-hand side of (6.53) would be equal to 1. In that case instability would be predicted to set in at the same fluid velocity as choking; any velocity lower than c_0 would be stable. However, in our case R increases more rapidly than $1/A$ as A decreases, so the right-hand side of (6.53) is less than 1, and instability by this roll-wave mechanism will occur at a velocity less than the wave speed. Thus choking will never be the sole cause of oscillation; roll-wave instability must occur first.

If the cross-sectional area is very small, then (6.38) and (6.39) show that $R \propto 1/A^2$, with the consequence that the critical velocity is one-half of the wave speed:

$$u_{\text{crit}} = \tfrac{1}{2} c_0 = 0.61 P_0^{1/2} A_0^{-3/2}.$$

Some of the maximum equilibrium velocities calculated from the above lumped model exceed this value (e.g. the unchoked case $R_2 = 75$, $Q = 1.0$ from fig. 6.22(a), which is stable according to the lumped model if $\beta = 1.5$ but not if $\beta = 0.5$) while others, including cases that are unstable by the lumped mechanism, do not (e.g. $R_2 = 50$, $Q = 0.65$, $\beta = 1.5$ from fig. 6.22(b)). Thus there certainly will be circumstances in which the roll-wave instability is the dominant cause of oscillation, and others in which the lumped-parameter mechanism dominates.

As we have already remarked, roll-wave instability will always occur at a lower velocity than choking, and thus, in the particular experiment modelled here, should be regarded as a more important source of unsteady behaviour. The only other well-known example of this kind of instability, roll-waves themselves in channel flow, is not widely observed and not regarded as very important. This is doubtless because in that case they occur only at supercritical speeds, and supercritical channel flows more often break down through single hydraulic jumps or bores.

The next stage of this theoretical work must be to combine the lumped-parameter model for flow downstream of the constriction with the quasi-one-dimensional analysis of this section for flow upstream of it. This will result in a non-linear set of hyperbolic partial differential equations that will have to be solved numerically, although one hopes that the above discussion will provide a useful framework. Other fundamental physical effects that should be incorporated into the model as soon as a clear mathematical description of them can be formulated are the intermittent flow separation and energy loss at an oscillating constriction, and the effects on the tube pressure–area law of longitudinal curvature and longitudinal tension in the wall. Careful experiments must also be done to test all the predictions of the present models and their successors; in particular, the model parameters must be recorded sufficiently accurately that it is at last possible to distinguish between the different mechanisms of instability and of oscillation. Finally, of course, a fuller understanding of the mechanics of collapsible tubes will enable us to make more confident interpretations of physiological phenomena such as flow limitation in forced expiration, venous return, and Korotkoff sounds.

ANALYSIS OF A HOT-FILM ANEMOMETER

A.1 Introduction

All the velocity profiles measured in arteries (and reported in chapter 1), almost all the profiles measured in models or casts of arterial junctions (chapter 5) and all direct measurements of wall shear-rates in models have been obtained by the use of a hot-film anemometer (or its close relation, an electrochemical shear probe). Therefore it is important to understand how such a device operates, particularly since the main justification for the detailed theoretical analysis of flow in bends and bifurcations (chapters 3 to 5) rests on the claim that hot-film anemometry is not at present capable of the accurate measurement of unsteady wall shear in arteries.

A constant-temperature hot-film anemometer consists of a thin metallic (usually gold) film mounted flush with the surface of an insulated solid probe, which is inserted into the fluid whose velocity is to be measured. The temperature of the film is maintained by an electronic feedback circuit at a fixed value, T_1, slightly higher than the temperature of the fluid, T_0, which is also assumed to be constant. The power required to maintain it is proportional to the rate at which heat is lost to the fluid, which is in turn related to the velocity of the fluid flowing past the probe. In steady flow, this latter relation is obtained by calibration in known flows, after which the probe can, in principle, be used in any steady flow of the same fluid. In order to use such a probe unambiguously in unsteady flow, it is necessary that the same relation between heat loss and fluid velocity should obtain at all times, i.e. that the behaviour of the probe should be quasi-steady. This appendix consists of a review of the theoretical progress that has been made in understanding the response of a hot-film anemometer in steady and in unsteady flow. We shall concentrate exclusively on fluid mechanical aspects, and shall therefore assume from the start that the electronics faithfully record the rate of heat loss from the probe as a function of time, that the

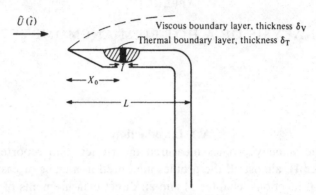

Fig. A.1. Sketch of a hot-film anemometer probe. The dark rectangle represents the film; the shaded region around it represents the insulating substrate. Lengths $\hat{l}$, L, X_0 are defined. The temperature of the film is T_1, while that of the oncoming fluid is T_0. (After Pedley, 1976b.)

film itself is of uniform thickness and uniform, high conductivity (so that the temperature at its surface is uniform and constant) and that the thermal conductivity of the insulated substrate is much lower than that of the ambient fluid so that heat loss through the substrate is negligible. The last assumption was shown by Bellhouse & Schultz (1967) to be satisfied in a conducting liquid such as water or blood, but not in air.

The type of probe most commonly used to measure velocity profiles in arteries is depicted in fig. A.1. The insulating substrate in which the rectangular film is embedded is mounted on the surface of a hypodermic needle that has been bent into an L-shape so that the point can be aligned with the flow after insertion through an artery wall; such a probe has been used by Schultz *et al.* (1969), Seed & Wood (1970a, b, 1971), Nerem & Seed (1972), Clark & Schultz (1973), Clark (1974), Nerem *et al.* (1974a) and others. We denote the streamwise length of the *probe* by L, the streamwise length of the *hot-film* by $\hat{l}$, the transverse breadth of the film by b, and the distance of the leading edge of the film from the leading edge of the probe by X_0; the values of L, $\hat{l}$ and X_0 in two of these studies are given in table A.1. These studies, by Seed & Wood (1970b) and by Clark (1974), are the two in which careful calibration measurements were made in known unsteady flows of different amplitudes

Table A.1

Author	$\hat{l}$ (cm)	X_0 (cm)	L (cm)	Temperature of ambient water (°C)	Prandtl no. ν/κ	γ (see text)
Seed & Wood (1970b)	0.01	0.15	0.3	37	4.6	0.17
Clark (1974)	0.02	0.25	0.5†	20	6.9	0.22

† Clark did not report the value of L, but since he did not use the probe in reversing flow, it is irrelevant. We choose $L = 0.5$ cm so that X_0/L can be taken equal to 0.5 in each case.

and frequencies: see § A.4 below. In each of these studies, b was about $2.5\hat{l}$. (Note that hot-film probes were first developed to measure the skin friction (or shear-rate) on given surfaces such as the outside of solid bodies or the inside of tubes (Liepmann & Skinner, 1954); in this case the film is mounted directly onto the surface in question. The electrochemical technique differs only in that solute is transported from an electrode rather than heat from a film.)

The object of theoretical analysis is to predict the rate of heat transfer from the film to the ambient fluid. This requires a knowledge of the temperature field in the neighbourhood of the hot-film, which is expected to be determined by a balance between advection and diffusion, assuming that free convection is negligible.† Given this, the advecting flow over the film is independent of the temperature field; in the example depicted in fig. A.1 it can be determined by an analysis of the viscous boundary layer over the probe. In the case of wall shear probes in models of arteries, the advecting flow would be that discussed in chapters 3 to 5.

In the studies to be reviewed, a number of simplifying assumptions are made, as follows.

(i) The only property of the flow field that influences the heat transfer is the wall shear-rate on the film, $\hat{S}$. This implies that the thickness, δ_T, of the region where the temperature differs significantly from T_0 (the thermal boundary layer) is much less than the transverse length-scale for variation of the longitudinal velocity (i.e. the thickness, δ_V, of the viscous boundary layer on the probe), with the result that (a) the local velocity profile is linear and (b) the normal component of velocity can be neglected. In other words, if $\hat{x}$ is the Cartesian coordinate parallel to the film in the direction of the wall shear, $\hat{y}$ is perpendicular to the film, and $\hat{z}$ is perpendicular to the other two (fig. A.2), then the velocity field ($\hat{u}$, $\hat{v}$, $\hat{w}$) is given by

$$\hat{u} = \hat{S}\hat{y}, \qquad \hat{v} = \hat{w} = 0. \tag{A.1}$$

---

† Free convection will be negligible if the dimensionless parameter $F = U_0^2/\beta_0 g \hat{l} \Delta T$ is very large, where U_0 is a typical advection velocity, β_0 is the volumetric coefficient of expansion of the fluid, g is the gravitational acceleration, and ΔT is the temperature difference between the film and the fluid (Ostrach, 1964). In the experiments referred to above, ΔT was always less than 5 °C, to avoid damage to the blood, β_0 for water is about 3.7×10^{-4} °C^{-1}, so with $\hat{l} = 0.02$ cm (table A.1), F is about 27 even for a velocity as small as 1 cm s^{-1}.

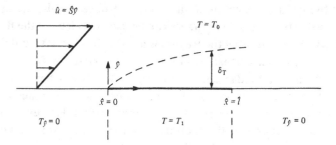

Fig. A.2. Two-dimensional model of the hot-film, embedded in the region $0 \leqslant \hat{x} \leqslant \hat{l}$ of the infinite plane $\hat{y} = 0$, with a uniform shear flow over it.

When both the velocity field over the probe and the temperature field over the film are given by boundary layer theory, this assumption requires that $\hat{l}/X_0 \ll \min(1, \sigma)$, where σ is the Prandtl number of the fluid (Pedley, 1972a); $\sigma > 1$ for liquids like water and blood.

(ii) The wall shear rate, $\hat{S}$, is independent of $\hat{x}$ over the film $(0 \leqslant \hat{x} \leqslant \hat{l})$; this also requires that $\hat{l}/X_0 \ll 1$. Both this and the previous assumption seem to be reasonably well satisfied for the probe dimensions listed in table A.1 (Pedley 1972a, 1976b).

(iii) The flow over the probe is effectively two-dimensional, with the consequence that, however the wall shear, $\hat{S}$, might vary with time, it never develops a component in the $\hat{z}$-direction, and does not vary with $\hat{z}$. This too is reasonably well satisfied by the velocity probes of fig. A.1 as long as they are carefully aligned with the flow, since they are approximately cylindrical in cross-section. It is also exactly satisfied for a wall shear probe in a long straight pipe where the flow is unidirectional, but may not be so near bends or bifurcations on account of the secondary motions.

(iv) In applying the theory of the hot-film to a particular velocity probe, we assume that the flow over the probe is the same as that over a finite flat plate, i.e. that (a) there is no longitudinal pressure gradient and (b) the effect of transverse probe curvature is negligible. Condition (a) will be satisfied if the probe is not yawed and is approximately cylindrical, while (b) requires that the radius of the cylindrical mounting be large compared with the viscous boundary layer thickness. This is not well satisfied by the probes under investigation, but the error is easy to assess, as shown in § A.4.3.

(v) In making application to the velocity probes, we also assume that viscous boundary layer theory can be used to calculate the flow field, which requires that the Reynolds number based on X_0 and the free-stream velocity be large. This is also discussed in § A.4.

(vi) A further fundamental assumption, which has been made throughout this book, is that the non-homogeneous character of blood does not affect the behaviour of the hot-film. However, the *maximum* thickness of the thermal boundary layer over the hot-film is only about 40 μm when the blood velocity is steady at 1 m s^{-1} (taking $\hat{l} = 0.01$ cm; see Pedley, 1972a), and this is only five times the diameter of a red cell. Thus the assumption is almost certainly false, and the analysis below is not directly applicable when blood, not water, is flowing past the probe. Possible ways in which the red cells could influence heat transfer in steady flow are discussed by Seed & Wood (1970b) and by Clark (1974); the former found that the slope of the (linearised) graph of anemometer output against velocity is the same for whole blood as for water, whereas a crude theory indicates that the slope for blood should be less by about 30% (the slope for blood plasma *was* as predicted). However, no way of analysing the effect is at present known to the author. Experimentalists usually overcome the difficulty by calibrating their instruments in blood from the animal being studied, or at least from another animal of the same species.

That is a complete list of the assumptions made concerning the flow field over the hot film. In order to analyse the temperature field, two additional assumptions are made:

(vii) The temperature field over the film is two-dimensional; since $b/\hat{l}$ is only about 2.5 for the probes under discussion, this is likely to be a source of considerable error, which has so far not been examined theoretically (see § A.6).

(viii) the film length l is sufficiently large for thermal boundary layer theory to be applicable in calculating heat transfer. As we shall see (§ A.4) this is also unlikely to be accurate, and improved theories for steady flow are presented in §§ A.5 and A.6.

Making all the above assumptions except (viii), we can now give a mathematical formulation of the problem to be solved (with reference to fig. A.2). A fluid of constant thermal diffusivity κ

occupies the region $\hat{y} > 0$, and flows in the $\hat{x}$-direction with velocity given by (A.1), where the wall shear, $\hat{S}$, is a known function of time, $\hat{t}$. The temperature of the fluid far from the wall is T_0; the regions $\hat{x} < 0$ and $\hat{x} > \hat{l}$ of the wall consist of insulating material, while the region $0 \leqslant \hat{x} \leqslant \hat{l}$ is maintained at temperature T_1. If $T(\hat{x}, \hat{y}, \hat{t})$ is the fluid temperature, then the equations and boundary conditions from which it is to be determined are

$$T_{\hat{t}} + \hat{S}(\hat{t})\hat{y}T_{\hat{x}} = \kappa(T_{\hat{x}\hat{x}} + T_{\hat{y}\hat{y}}) \qquad (A.2)$$

and

$$\left.\begin{array}{l} T \to T_0 \quad \text{as } |\hat{x}^2 + \hat{y}^2| \to \infty, \\ T = T_1 \quad \text{on } \hat{y} = 0, 0 \leqslant \hat{x} \leqslant \hat{l}, \\ T_{\hat{y}} = 0 \quad \text{on } \hat{y} = 0, \hat{x} < 0 \text{ and } \hat{x} > \hat{l}. \end{array}\right\} \qquad (A.3)$$

The object of the theory is to calculate the rate of heat transfer from the film, per unit length in the $\hat{z}$-direction, equal to

$$\hat{Q}(\hat{t}) = -\rho C_p \kappa \int_0^{\hat{l}} T_{\hat{y}}|_{\hat{y}=0} \, d\hat{x}, \qquad (A.4)$$

where ρ and C_p respectively are the density and the specific heat at constant pressure of the fluid.

A.2 The steady boundary layer solution

We begin by considering the simplest case, that in which the flow is steady, so that $\hat{S}$ is constant and the first term in (A.2) is absent, and in which the film is long enough for the boundary layer approximation to be made, so that $T_{\hat{x}\hat{x}}$ is negligible compared with $T_{\hat{y}\hat{y}}$. This means that the equation becomes parabolic in $\hat{x}$, and the solution at a given value of $\hat{x}$ depends only on conditions at smaller values; in other words, the heated film can be treated as if it were semi-infinite. Balancing the advective and the diffusive terms shows that a lateral length-scale for the temperature variation (the thermal boundary layer thickness, δ_T) is proportional to $(\kappa\hat{x}/\hat{S})^{1/3}$. Thus the boundary layer approximation will be valid over most of the film if this is much less than $\hat{x}$ for most $\hat{x}$, i.e. if the Péclet number, l^2, is large, where

$$l^2 = \hat{S}\hat{l}^2/\kappa; \qquad (A.5)$$

in fact, it is shown in §§ A.5 and A.6 that l^2 must exceed about 400 for accurate prediction of the heat transfer (to within 2%).

If we introduce dimensionless variables

$$\theta = \frac{T - T_0}{T_1 - T_0}, \qquad x = \hat{x}\left(\frac{\hat{S}}{\kappa}\right)^{1/2}, \qquad y = \hat{y}\left(\frac{\hat{S}}{\kappa}\right)^{1/2}, \qquad \text{(A.6)}$$

the boundary layer equation becomes

$$y\theta_x = \theta_{yy} \qquad \text{(A.7)}$$

with boundary conditions

$$\theta = 1 \quad \text{on } y = 0, \qquad \theta \to 0 \quad \text{as } y \to \infty. \qquad \text{(A.8)}$$

The solution of this problem, first obtained by Lévêque (1928), is

$$\theta = \theta_0(\eta) = 1 - c_0 \int_0^\eta e^{-s^3} \, ds, \qquad \text{(A.9)}$$

where

$$c_0 = 1 \Big/ \int_0^\infty e^{-s^3} \, ds = 1/\Gamma(\tfrac{4}{3}) = 1.120$$

and

$$\eta = y(9x)^{-1/3}. \qquad \text{(A.10)}$$

If

$$q(x) = -\theta_y|_{y=0} = c_0(9x)^{-1/3}, \qquad \text{(A.11)}$$

then the dimensionless heat transfer from the film is

$$Q = \frac{\hat{Q}}{\rho C_p (T_1 - T_0)\kappa} = \int_0^l q(x) \, dx \qquad \text{(A.12)}$$

$$= \tfrac{1}{2} 3^{1/3} c_0 l^{2/3}. \qquad \text{(A.12a)}$$

As well as being the square root of the Peclet number, l can be seen to be the dimensionless length of the hot-film. For a given hot-film, however, $l^{2/3}$ is proportional to $\hat{S}^{1/3}$, i.e. the heat transfer from the hot-film, and hence the output from the anemometer, is proportional to the one-third power of the local wall shear. This is found experimentally (Liepmann & Skinner, 1954; see also fig. A.17). Since in a steady, flat-plate boundary layer the wall shear is proportional to the three-halves power of the free-stream velocity, the heat transfer from a film mounted on a probe like that in fig. A.1

should be proportional to the one-half power of the velocity; this too is found experimentally (see Seed & Wood (1970b) for example).

In addition to the boundary layer, it is useful for future reference (§ A.5) to consider the thermal wake of the hot-film, for $\hat{x} \gg \hat{l}$. Since the downstream wall is an insulator, all the heat put into the fluid by the film will be convected downstream in the wake, so that the rate of heat flux in the $\hat{x}$-direction in the wake must be equal to $\hat{Q}$. In dimensionless terms this implies that

$$\int_0^\infty \theta(x, y)y \, dy = Q. \qquad (A.13)$$

The temperature distribution in the wake, to which boundary layer theory will always be applicable sufficiently far downstream, can be obtained by noticing (a) that the balance of advection and diffusion still means that η is the appropriate similarity variable and (b) that the integral constraint (A.13) therefore requires θ to take the form $x^{-2/3}$ times a function of η. By transforming from (x, y) to (x, η), (A.7) becomes

$$\theta_{\eta\eta} + 3\eta^2\theta_\eta - 9\eta x\theta_x = 0. \qquad (A.14)$$

The boundary condition at the wall in the wake is $\theta_\eta = 0$, and the solution of the required form turns out to be

$$\theta = Ax^{-2/3} e^{-\eta^3}. \qquad (A.15)$$

The integral constraint then shows that

$$A = Q/9^{2/3}c_1, \qquad (A.16)$$

where

$$c_1 = \int_0^\infty s \, e^{-s^3} \, ds = \tfrac{1}{3}\Gamma(\tfrac{2}{3}).$$

When Q is given by (A.12a), this means that $A = c_0 l^{2/3}/6c_1$, but the main interest in the result (A.16) is that its validity is independent of whether or not boundary layer theory is applicable in the region $0 < x < l$, i.e. independent of the value of l. In fact, for fairly short films it proves to be simpler to calculate the temperature in the far wake than the temperature gradient over the whole film, so that Q is derived from (A.16) instead of (A.12) (see § A.5). For unsteady flows, however, there is no integral constraint like (A.13), so (A.12) has to be used directly.

A.3 The unsteady boundary layer with non-reversing shear

Here we retain the boundary layer approximation but allow the wall shear to vary with time, restricting it only by the requirement that it remains positive, so that $\hat{x} = 0$ is the 'leading edge' of the hot-film at all times. Thus the film is still effectively semi-infinite, and the solution will be independent of $\hat{l}$. We suppose that the wall shear varies with time according to

$$\hat{S}(\hat{t}) = \hat{S}_0 S(\Omega \hat{t}),$$

where $\hat{S}_0$ is a dimensional scale factor and Ω is a typical frequency of the time variation. The theory for this problem was given by Pedley (1972a), and follows closely the method presented in § 3.2.2 for the unsteady viscous boundary layer on a flat plate. We again use the non-dimensionalisation (A.6) with $\hat{S}_0$ for $\hat{S}$ and introduce the dimensionless time $t = \Omega \hat{t}$. The boundary layer equation now becomes

$$\omega \theta_t + S(t) y \theta_x = \theta_{yy}, \qquad (A.17)$$

where $\omega = \Omega/\hat{S}_0$; this equation has to be solved subject to the boundary conditions (A.8).

It is clear, by analogy with the viscous case, that if either x or the dimensionless frequency, ω, is sufficiently small, the other being fixed, the temperature field will be quasi-steady, while if either is large the oscillatory and the mean components will become uncoupled. The relevant combination of x and ω turns out to be the quantity

$$x_1 = \omega (9x)^{2/3}. \qquad (A.18)$$

A.3.1 Small x_1

We seek an expansion in powers of x_1, whose leading term represents the quasi-steady solution, and therefore make the transformation $(x, y, t) \rightarrow (x_1, \eta_1, t)$ where

$$\eta_1 = y[S(t)/9x]^{1/3}. \qquad (A.19)$$

The transformed equation is

$$\theta_{\eta_1\eta_1} + 3\eta_1^2\theta_{\eta_1} - 6\eta_1 x_1\theta_{x_1} = \frac{x_1}{S^{2/3}(t)}\left[\theta_t + \frac{\dot{S}(t)}{3S(t)}\eta_1\theta_{\eta_1}\right],$$

and we seek a solution in powers of x_1:

$$\theta(x_1, \eta_1, t) = \sum_{n=0}^{\infty} x_1^n\theta_n(\eta_1, t). \tag{A.20}$$

The first term is of course the quasi-steady solution (A.9):

$$\theta_0(\eta_1, t) \equiv \theta_0(\eta_1);$$

the next two terms are (Pedley, 1972a)

$$\theta_1 = \beta(t)\theta_{11}(\eta_1), \qquad \theta_2 = \beta^2(t)\theta_{21}(\eta_1) + \dot{\beta}(t)S^{-2/3}(t)\theta_{22}(\eta_1),$$

where

$$\beta(t) = \tfrac{1}{3}\dot{S}(t)S^{-5/3}(t)$$

and the functions θ_{11}, θ_{21}, θ_{22} satisfy the following ordinary differential equations:

$$\theta_{nm}'' + 3\eta_1^2\theta_{nm}' - 6n\eta_1\theta_{nm} = F_{nm}(\eta_1),$$

$$F_{11} \equiv \eta_1\theta_0', \qquad F_{21} \equiv \eta_1\theta_{11}', \qquad F_{22} \equiv \theta_{11}.$$

These can be solved either analytically, in terms of confluent hypergeometric functions, or numerically, which is more convenient. The quantities needed for calculation of the heat transfer are the gradients of these functions at $\eta_1 = 0$, and these are

$$\theta_{11}'(0) = 0.143, \qquad \theta_{21}'(0) = -0.00243, \qquad \theta_{22}'(0) = -0.0118.$$

The dimensionless heat transfer per unit length of the film,

$$q(x, t) = -\theta_y|_{y=0}, \tag{A.21}$$

is thus given by

$$(9x)^{1/3}q = S^{1/3}(t)\{c_0 - x_1\beta(t)\theta_{11}'(0)$$
$$- x_1^2[\beta^2(t)\theta_{21}'(0) + \dot{\beta}(t)S^{-2/3}(t)\theta_{22}'(0)] + \cdots\}, \tag{A.22}$$

of which the first term is the quasi-steady result. The total heat transfer, $Q(t)$, from a hot-film of dimensionless length l is obtained by integrating (A.22) over the range $0 \leqslant x \leqslant l$.

The above expansion shows that the quasi-steady solution is accurate at a given value of x as long as

$$\lambda = |x_1\beta(t)| = \left|\frac{3^{1/3}\hat{x}^{2/3}\,\mathrm{d}\hat{S}/\mathrm{d}\hat{t}}{\kappa^{1/3}\hat{S}^{5/3}(\hat{t})}\right| \qquad (\text{A.23})$$

is always less than 1 (cf. (3.3)). The series itself will be a useful asymptotic expansion if the $O(x_1^2)$ term is always much less than the $O(x_1)$ term, and in that case the first two terms of (A.22) will be a good approximation to the almost quasi-steady solution. Pedley (1972a) made a number of computations for sinusoidally oscillating shear,

$$S(t) = 1 + \alpha_1 \sin t, \qquad (\text{A.24})$$

and these suggest that the first two terms of (A.22) are sufficiently accurate if λ is less than 0.5, and the quasi-steady result is accurate if λ is less than 0.1. An example in which the maximum value of λ is 0.49 is given in fig. A.3, in which the quantity $(9x)^{1/3}q$ is plotted against t over a complete cycle (here $x_1 = 2$, $\alpha_1 = 0.5$). The three curves were computed from the one- two- and three-term expansions of (A.22) respectively; it can be seen that the two- and the three-term expansions differ by very little, whereas they both have a marked phase lag behind the quasi-steady term. The case $x_1 = 0.1$, $\alpha_1 = 0.8$ (not shown here) is an example in which the maximum value of λ is 0.097, and in which the one- and the two-term expansions differ by less than 2% for all t.

A.3.2 Large x_1

Here we examine the case of high frequency, or large distance from the leading edge, in which the temperature field is far from quasi-steady. We restrict attention from the start to sinusoidal oscillations in shear, about a non-zero mean, with $S(t)$ given by (A.24). This case may not be relevant to the practical use of a hot-film ane-mometer, since that requires approximately quasi-steady behaviour, but could be useful in interpreting particular experimental results (especially when the electrochemical technique is being used, since κ is much smaller for solutes than for heat, and hence x_1 and λ are larger: see (A.23)). We shall give only a brief

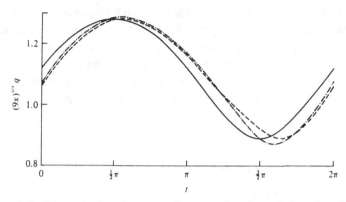

Fig. A.3. Dimensionless heat transfer as a function of time (small-x_1 expansion). Here $(9x)^{1/3}q$ is plotted against t over a complete cycle for the case $x_1 = 2$, $\alpha_1 = 0.5$. The three curves represent one, two and three terms of (A.22) (continuous, dot–dash and broken curve, respectively).

outline of the solution, since it is similar to, but rather simpler than, that in § 3.2.2.

When $\alpha_1 = 0$, the shear, and hence the temperature field, is steady, and the relevant y-coordinate is η (see (A.10)). If there were no mean shear, on the other hand, heat would be confined to a thermal Stokes layer, of thickness $(\kappa/\Omega)^{1/2}$, and the appropriate y-coordinate would be

$$\zeta = \hat{y}(\Omega/\kappa)^{1/2} = x_1^{1/2}\eta.$$

We therefore seek matched asymptotic expansions for θ in powers of $x_1^{-1/2}$, with η as the outer y-variable, and ζ the inner one.

In terms of the outer variables, (A.17) is

$$\theta_t = x_1^{-1}[\theta_{\eta\eta} + (1 + \alpha_1 \sin t)(3\eta^2\theta_\eta - 6\eta x_1\theta_{x_1})], \qquad \text{(A.25)}$$

and the boundary conditions are that $\theta \to 0$ as $\eta \to \infty$ and that the expansion should match to an inner expansion as $\eta \to 0$. We propose the outer expansion

$$\theta = \sum_{n=0}^{\infty} x_1^{-n/2}\tilde{\theta}_n(\eta, t);$$

this is substituted into (A.25) to derive the equation for $\tilde{\theta}_n$, which is of the form

$$\tilde{\theta}_{nt} = F(\eta, t),$$

where F depends only on $\tilde{\theta}_{n-2}$. The solution for $\tilde{\theta}_n$ thus consists of a known function of η and t, plus an as yet arbitrary function of η, say $f_n(\eta)$; this function is determined (apart from a multiplicative constant a_n) by the requirement that secular terms are absent from $\tilde{\theta}_{n+2}$. The first few terms of the outer expansion, taking into account the outer but not the inner boundary condition, are found to be

$$\tilde{\theta}_0 = \theta_0(\eta)(1 + a_0),$$

where θ_0 is the steady solution (A.9),

$$
\left.
\begin{aligned}
\tilde{\theta}_1 &= a_1\, e^{-\eta^3}[1 - 3G\eta + \tfrac{1}{2}\eta^3 + O(\eta^4)], \\
\tilde{\theta}_2 &= 3c_0\alpha_1 \cos t\, \eta^2\, e^{-\eta^3}(1 + a_0) + a_2\, e^{-\eta^3}, \\
\tilde{\theta}_3 &= a_3\eta\, e^{-\eta^3} + a_1[\cdots], \\
\tilde{\theta}_4 &= 6c_0\alpha_1 \sin t\, e^{-\eta^3}(1 - 3\eta^3)(1 + a_0) \\
&\quad - \tfrac{3}{4}c_0\alpha_1^2\, e^{-\eta^3}[3 \cos 2t(4\eta^3 - 3\eta^6)(1 + a_0) \\
&\quad + 2 + 6\eta^3 - 9\eta^6] + a_2[\cdots] \\
&\quad + a_4\, e^{-\eta^3}[1 - 6G\eta - \eta^3 + O(\eta^4)],
\end{aligned}
\right\}
\qquad \text{(A.26)}
$$

where

$$G = \Gamma^2(\tfrac{2}{3})/\Gamma^2(\tfrac{1}{3})$$

and certain functions multiplying a_1 and a_2 have been omitted since these constants are subsequently shown to be zero. Further details are given by Pedley (1972a), who took the expansion up to $n = 7$.

In terms of inner variables, (A.17) is

$$\theta_{\zeta\zeta} - \theta_t = x_1^{-1/2}(1 + \alpha_1 \sin t)6\zeta\theta_{x_1}, \qquad \text{(A.27)}$$

and the inner boundary condition is $\theta = 1$ at $\zeta = 0$. We seek an inner expansion of the form

$$\theta = \sum_{n=0}^{\infty} x_1^{-n/2}\Theta_n(\zeta, t),$$

and the outer boundary condition on the Θ_n is obtained by rewriting the outer expansion in terms of ζ, and expanding again in powers of $x_1^{-1/2}$. This matching serves to determine most of the a_n as well as the functions Θ_n. When $n = 3m$ (m an integer), however, the a_n are not determined, because $f_n(\eta)$ is an eigenfunction (for example, it can be seen from the expression for $\tilde{\theta}_3$ in (A.26) that $f_3(0) = 0$). As in the viscous case, the presence of such eigenfunctions is to be expected, since it is only through them that upstream conditions can influence the expansion in inverse powers of x_1. Pedley (1972a) found that $a_0 = a_1 = a_2 = a_5 = a_7 = 0$, $a_4 = \frac{3}{2}c_0\alpha_1^2$, and

$$\Theta_0 \equiv 1, \qquad \Theta_1 \equiv -c_0\zeta, \qquad \Theta_2 \equiv \Theta_3 \equiv \Theta_6 \equiv 0,$$

$$\Theta_4 = \tfrac{1}{4}c_0\zeta^4 + a_3\zeta + 3c_0\alpha_1\cos t\zeta^2$$

$$+ 6c_0\alpha_1[\sin t - e^{-\zeta/\sqrt{2}}\sin(t - \zeta/\sqrt{2})],$$

$$\Theta_5 \equiv -\tfrac{3}{2}a_4 G\zeta;$$

he also calculated Θ_7, which is fairly complicated. The dimensionless heat transfer per unit length, $q(x, t)$ of (A.21), is given by

$$(9x)^{1/3}q = -x_1^{1/2}\theta_\zeta|_{\zeta=0}$$

$$= c_0 - x_1^{-3/2}[3\sqrt{2}c_0\alpha_1(\cos t + \sin t) + a_3] + x_1^{-2}\cdot\tfrac{9}{4}c_0\alpha_1^2 G$$

$$- x_1^{-3}\{a_6 + 18\sqrt{2}c_0\alpha_1[\cos t(1 + 2\sqrt{2} - 2a_3/3c_0)$$

$$- \sin t(1 + 2a_3/3c_0)] - 9c_0\alpha_1^2 \sin 2t(8\sqrt{2} - 7)\}$$

$$+ O(x_1^{-7/2}). \tag{A.28}$$

It can be seen that the unsteadiness of the wall shear affects the heat transfer only at $O(x_1^{-3/2})$; this is in contrast to the viscous case where the *leading* term in the skin friction expansion (see (3.30)) is oscillatory. The eigenfunctions, represented by the constants a_3 and a_6, also first have an effect on the mean heat transfer at $O(x_1^{-3/2})$, but do not affect the oscillatory heat transfer until $O(x_1^{-3})$. The numerical results of Pedley (1972a) suggest that the $O(x_1^{-3})$ term is a small correction (and therefore the $O(x_1^{-3/2})$ expansion can be used accurately) for $x_1 > 20$ and for values of α_1 up to at least 0.8. Fig. A.4 illustrates this for $x_1 = 20$ and $\alpha_1 = 0.5$, with a_3 and a_6 set equal to zero.

The values of a_3 and a_6 etc. should be determined by some sort of matching with the small-x_1 expansion. However, overlap between the two expansions is much less good than in the viscous case (fig. 3.4), as can be seen from fig. A.5, where $(9x)^{1/3}q$ is plotted against t for $x_1 = 6.0$, $\alpha_1 = 0.5$. The two continuous curves represent the first two terms of (A.22) and (A.28) respectively, while the broken curve represents all of (A.28) with $a_3 = a_6 = 0$. The value of $x_1 = 6.0$ is that at which the amplitudes of the heat-transfer oscillations, according to the two two-term expansions, are approximately equal, but it can

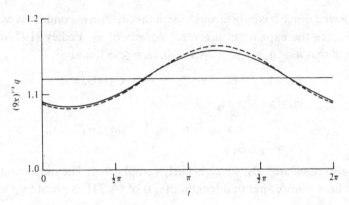

Fig. A.4. Dimensionless heat transfer as a function of time (large-x_1 expansion). Here $(9x)^{1/3}q$ is plotted against t for the case $x_1 = 20$, $\alpha_1 = 0.5$. The straight line represents the leading term of (A.28), while the two curves represent the $O(x_1^{-3/2})$ and $O(x_1^{-3})$ approximations (continuous and broken curve, respectively) with $a_3 = a_6 = 0$.

be seen that the phases are not in close agreement (use of the full equation (A.28) improves the phase agreement, but the large difference between the broken and the continuous curves shows how inaccurate the large-x_1 expansion is at this value of x_1). Pedley (1972a) suggested that an indication of the value of a_3 could be obtained by choosing it so that the means of the two expansions should be the same as the chosen 'overlap' value of x_1. From two terms of the expansions (A.28) and (A.22) this choice gives

$$c_0 - a_3 x_1^{-3/2} = \frac{c_0}{2\pi} \int_0^{2\pi} (1 + \alpha_1 \sin t)^{1/3} \, dt.$$

The value of a_3 thus obtained depends on α_1; for $x_1 = 0.6$ we have

$$a_3 = \begin{cases} 0.037 & \text{for } \alpha_1 = 0.2, \\ 0.247 & \text{for } \alpha_1 = 0.5, \\ 0.743 & \text{for } \alpha_1 = 0.8, \end{cases}$$

and it can be seen from (A.28) that such values of a_3 will have only a small effect on the heat transfer. It should be emphasised, however, that any such determination of a_3 (or a_6) will be very inaccurate; improved accuracy can be achieved only by a full numerical solution of the boundary layer equation.

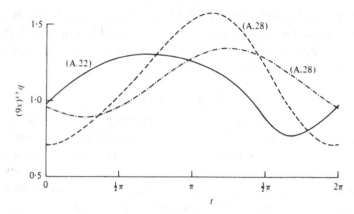

Fig. A.5. Comparison of the small- and large-x_1 expansions: $(9x)^{1/3}q$ plotted against t for the case $x_1 = 6$, $\alpha_1 = 0.5$. The continuous and dot–dash curves represent the two-term expansions of (A.22) and (A.28), respectively, while the broken curve represents all of (A.28).

A.4 A hot-film in reversing flow

We have seen that for a hot-film anemometer to be useful its response must be quasi-steady (the heat transfer being given by (A.12)), and that this requires λ, in (A.23), to remain less than about 0.1 for the majority of the cycle (in periodic flow). If the wall shear over the film, $\hat{S}$, approaches zero, then λ cannot remain small and the heat transfer cannot be quasi-steady. Now blood flow in large arteries reverses its direction at least twice each beat (fig. 1.17 or fig. 3.3), and therefore so does $\hat{S}$ (indeed, $\hat{S}$ will reverse before the centre-line velocity, as does the shear on the artery wall: fig. 3.6). How then can a hot-film anemometer be used to measure blood flow? The answer, of course, is that it cannot be employed near the time of shear reversal, but that it will be useful if λ exceeds 0.1 most of the time, whatever the direction of the shear (note that the heat transfer will remain positive during reversal, and the signal will therefore be rectified; a single hot-film cannot determine the flow direction).

In order to assess the extent to which the heat transfer is quasi-steady, and to predict how the departures from quasi-steady behaviour will manifest themselves, we need to know both what $\hat{S}$ is as a function of time and what the heat transfer is while λ exceeds

0.1. We begin by showing how the approximate method of Pedley (1976a), outlined above in § 3.2.3, can be used to calculate $\hat{S}(\hat{t})$, and then describe a similar approximate method for estimating the heat transfer while the shear is reversing.

A.4.1 The shear on the probe

The basis of the approximate theory of Pedley (1976a) was the realisation that, when the stream velocity passes through or close to zero, the flow at a given position, $\hat{x}$, on a flat plate will represent a diffusive balance between local inertia forces and viscosity. However when, after a reversal, fluid that has passed the leading edge arrives at $\hat{x}$, the diffusive flow will give way to an approximately quasi-steady flow in which viscous forces are balanced by convective inertia. In chapter 3 this idea was applied to flow in the entrance of the aorta, represented as a semi-infinite flat plate; only when the stream velocity is in the positive-$\hat{x}$ direction can there be an approximately quasi-steady flow. Here, however, we represent the anemometer probe as a finite flat plate of length L, and the film over which we need to know the shear is taken to be at its mid-point ($x_0 = \frac{1}{2}L$ in fig. A.1). For the sake of definiteness, and for comparison with the experiments of Clark (1974) and Seed & Wood (1970b), we take the free-stream velocity to vary sinusoidally with time, according to

$$\hat{U}(\hat{t}) = U_0(1 + \alpha \cos \Omega \hat{t}), \qquad (A.29)$$

where the amplitude parameter, α, exceeds 1 for a reversing flow (fig. A.6).

We suppose that, at some time close to that of peak forward velocity ($\hat{t} = 0$ in fig. A.6), there is an approximately quasi-steady boundary layer with leading edge at $\hat{x} = 0$ (here $\hat{x} = 0$ is measured from one end of the *probe*, not of the *film* as in the rest of this appendix). If the free stream reverses at time $\hat{t} = \hat{t}_A$, then the sequence of events will be as follows. Some time before reversal, at $\hat{t} = \hat{t}_{1A}(\hat{x})$ say, an approximately diffusive flow will take over; this will persist through reversal until at some later time, $\hat{t} = \hat{t}_{2A}(\hat{x})$, fluid particles that have passed the other end of the probe ($\hat{x} = L$) arrive at $\hat{x}$. Then we expect a new, approximately quasi-steady boundary layer to take over, with its leading edge at $\hat{x} = L$. This will persist

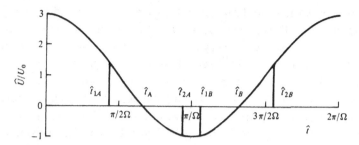

Fig. A.6. Sinusoidal velocity $\hat{U}(\hat{t})$ given by (A.29), for the case $\alpha = 2.0$. Times of flow reversal are $\hat{t}_A$, $\hat{t}_B$. Values of $\hat{t}_{1A,B}$ are also plotted for the case $x = 0.5$, $\omega' = 1.0$ in (A.31).

until the reversed flow has itself decelerated and is approaching zero, when another diffusive flow takes over, and so on.

Various modifications of this sequence of events may arise. For example, if the period of flow reversal is short-lived, it may be that fluid particles that have passed $\hat{x} = L$ never arrive at $\hat{x}$; in that case, the diffusive flow will persist until after the second reversal ($\hat{t} = \hat{t}_{2B}$ in fig. A.6). It may also be that there are values of $\hat{x}$ at which the flow is never approximately quasi-steady, as for values of (dimensionless) x greater than 0.25 in § 3.2.3. In that case the flow will consist of a steady boundary layer superimposed on an oscillatory Stokes layer with little interaction, as analysed in § 3.2.3; this situation does not occur in the examples worked out below. Finally, there will be non-reversing flows ($\alpha < 1$) in which the quasi-steady solution breaks down for a time, and a diffusive solution must be interposed.

The changeover times $\hat{t}_{1A}$, $\hat{t}_{2A}$ etc. are calculated in the same way as in (3.36). That is, $\hat{t}_{1A}$, $\hat{t}_{2A}$ are given by

$$x = \int_{t_{1A}}^{t_A} U(t)\,dt, \qquad 1 - x = \int_{t_A}^{t_{2A}} U(t)\,dt, \qquad (A.30)$$

where $\hat{U}$, $\hat{x}$ and $\hat{t}$ have been non-dimensionalised with respect to U_0, L and L/U_0 respectively, so that

$$U(t) = 1 + \alpha \cos \omega' t$$

and

$$\omega' = \Omega L / U_0. \qquad (A.31)$$

Similar equations hold for t_{1B} and t_{2B}, except that x and $1-x$ are interchanged. The changeover times defined in this way were shown by Pedley (1976a) to be very close to those for which ε (equal to $|x\dot{U}/U^2|$ or $|(1-x)\dot{U}/U^2|$: see (3.16)) is equal to 0.5, the value below which the approximately quasi-steady boundary layer solution is known to be accurate. During the periods of approximately quasi-steady flow with leading edge at $x=0$ $(0 \leqslant t \leqslant t_{1A}, t_{2B} \leqslant t \leqslant 2\pi$, etc.), the velocity in the boundary layer at any x will be given by (3.34). During the periods of reversed quasi-steady flow $(t_{2A} \leqslant t \leqslant t_{1B})$, it will be given by a similar equation, but with $1-x$ for x and the sign of the second term changed:

$$u = U(t)\{f_0'(\eta_2) - [(1-x)\dot{U}/U^2]f_{11}'(\eta_2)\},$$

where $\eta_2 = y[-U(t)/2(1-x)]^{1/2}$ and $y = \hat{y}(U_0/\nu L)^{1/2}$. During the diffusive periods $(t_{1A} < t < t_{2A}$, etc.) u will be given by (3.33). In each diffusive phase, the virtual time origin of the diffusion, t_0' in (3.33), will be determined by the requirement that the displacement thickness of the diffusive boundary layer is equal to that of the approximately quasi-steady layer from which it takes over, at the takeover time t_{1A} or t_{1B} (cf. (3.35)).

During the first flow reversal, therefore, the dimensionless wall shear, $S = \hat{S}(\nu L)^{1/2}/U_0^{3/2}$, will, at a particular value of x, be given by the following equations

$$S = \begin{cases} \dfrac{U^{3/2}(t)}{(2x)^{1/2}}\left[f_0''(0) + \dfrac{x\dot{U}}{U^2}f_{11}''(0)\right] & \text{for } 0 \leqslant t \leqslant t_{1A}, \quad (A.32a) \\[2ex] [\pi(t-t_0')]^{-1/2}\left\{U(t_0') + 2(t-t_0')\displaystyle\int_0^1 \dot{U}[t-\lambda^2(t-t_0')]\,d\lambda\right\} \\[1ex] \hspace{6cm} \text{for } t_{1A} < t < t_{2A}, \quad (A.32b) \\[2ex] U(t)\left[\dfrac{-U(t)}{2(1-x)}\right]^{1/2}\left[f_0''(0) - \dfrac{(1-x)\dot{U}}{U^2}f_{11}''(0)\right] \\[1ex] \hspace{6cm} \text{for } t_{2A} \leqslant t \leqslant t_{1B}. \quad (A.32c) \end{cases}$$

Similar equations will hold for the second reversal. We present numerical results for various values of ω' and two values of α but for only one value of $x(=0.5)$ because the hot-film is assumed to be at the middle of the finite probe surface ($X_0 = \frac{1}{2}L$ in fig. A.1). Nothing unexpected happens at other values of x (Pedley, 1976a).

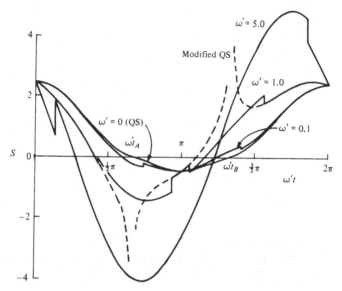

Fig. A.7. Dimensionless wall shear-rate, S, calculated from (A.32), and plotted against $\omega't$ for different values of ω' (0, 0.1, 1.0, 5.0) with $\alpha = 2.0$ and $x = 0.5$. Also plotted (broken curve) is the modified quasi-steady solution, (A.32a) and (A.32c), for $\omega' = 1.0$. (After Pedley, 1976a.)

In fig. A.7, S is plotted against $\omega't$ for the case $\alpha = 2.0$ and $x = 0.5$, with ω' taking the values 0 (quasi-steady), 0.1, 1.0, and 5.0. Also plotted for the case $\omega' = 1.0$ is the modified quasi-steady solution, i.e. the expressions (A.32a and c), without the intervening diffusive solution (A.32b); this is, of course, singular at the times when U reverses. However, if it is accepted that this modified quasi-steady solution is accurate in the regions for which it is used (and for which $\varepsilon \leqslant 0.5$), then it is clear that the approximate solution (A.32) is a much better indication of wall shear than the quasi-steady one. This is because it incorporates the known facts that both the relative amplitude of the wall shear and its phase lead over the stream velocity increase with frequency (Pedley (1972b) and fig. 3.4). The jumps in the approximate curves at the changeover times indicate that the predictions are not very accurate at the highest frequency ($\omega' = 5.0$), when the relatively inaccurate diffusive solution is used for a large fraction of the cycle, but in the subsequent application to experiment ω' never exceeds 1.0.

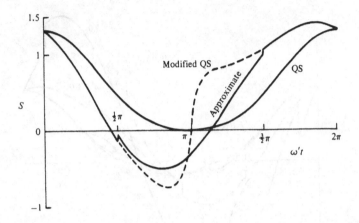

Fig. A.8. Graphs of S against $\omega' t$ for the case $\alpha = 0.98$, $x = 0.5$, $\omega' = 1.0$. Continuous curves are the quasi-steady solution (QS) and the approximate solution derived from (A.32). The broken curve is the modified quasi-steady solution. (After Pedley, 1976a.)

Fig. A.8 shows the results for a case in which the free stream does not reverse ($\alpha = 0.98$), but for which the flow has a diffusive character for about half the cycle ($\omega' = 0.1$, $x = 0.5$). Once more it can be seen that the present approximate solution departs markedly from the quasi-steady result, predicting wall shear reversal significantly ahead of the minimum stream velocity, and indicating a maximum negative value of shear equal to about 35% of the maximum positive value.

A.4.2 *The heat transfer from the film*

Now that the wall shear over the hot-film, $\hat{S}(\hat{t})$, is known, we can, according to our model of the anemometer, use the thermal boundary layer equation (A.17) to calculate the temperature distribution over the film and hence the heat transfer from it. When the shear over the film is not close to zero and the parameter λ in (A.23) is small (or the corresponding parameter with $\hat{l} - \hat{x}$ for $\hat{x}$ when $\hat{S}$ is negative), then the temperature field at the value of $\hat{x}$ in question will be approximately quasi-steady, and the heat transfer per unit length of the film will be given by the first two terms of (A.22). When the shear comes close to zero or reverses, however, so that λ exceeds 0.5, then (A.22) will not be accurate. Instead, we might expect the temperature field to be represented approximately by a diffusive balance between the terms $\omega \theta_t$ and θ_{yy} in (A.17) (recall that $\omega = \Omega/\hat{S}_0$). On the basis of this expectation, Pedley (1976b) worked out an approximate theory for the thermal boundary layer over the film that closely parallels the above theory for the viscous layer over the probe; the salient features are outlined below. Note that in this subsection, the non-dimensionalisation reverts to that of (A.6), with $t = \Omega \hat{t}$, as in § A.3.

Suppose that $S(t)$ reverses at $t = t_R$ and is positive for $t < t_R$. Then we assert that there are times $t_{1R}(x)$ and $t_{2R}(x)$ such that the dimensionless heat transfer per unit length, $q(x)$, is given by the first two terms of (A.22) when $t < t_{1R}$, and by the corresponding expression

$$q = [-S(t)/9(l-x)]^{1/3}[c_0 - x_2\beta_2(t)\theta'_{11}(0)], \qquad \text{(A.33)}$$

where

$$x_2 = \omega[9(l-x)]^{2/3} \quad \text{and} \quad \beta_2(t) = \tfrac{1}{3}[-\dot{S}(t)][-S(t)]^{-5/3},$$

when $t > t_{2R}$ and $x = l$ is the new leading edge of the film. In between, we propose a purely diffusive solution, in which

$$\theta = \text{erfc } \eta_0, \qquad \eta_0 = \tfrac{1}{2}y\{\omega/[t - t_{0R}(x)]\}^{1/2}, \qquad \text{(A.34)}$$

where t_{0R} is a virtual origin of the diffusive solution, analogous to the quantity t'_0 in (3.33) and (A.32). Thus

$$q = [\omega/\pi(t - t_{0R})]^{1/2} \quad \text{for } t_{1R} < t < t_{2R}. \qquad \text{(A.35)}$$

The choice of the t_{0R}, t_{1R} and t_{2R} is less clear-cut than the corresponding choice in the viscous case. In that case it was argued that the diffusive solution would take over from the initial quasi-steady solution, at a given value of x, when the influence of the leading edge ceased to be felt there, i.e. when fluid particles that had passed the leading edge first failed to arrive at x before being swept away by the reversing flow. This defined t_{1A} (see (A.30)). Similarly, the new quasi-steady solution would take over from the diffusive solution when fluid particles that had passed the new leading edge, travelling with the free-stream velocity, $U(t)$, first arrived at x (defining t_{2A}). In the present case, however, there is no unique free-stream velocity because the flow consists of a uniform shear, and the velocity is proportional to y. In order to apply a similar condition it is necessary to fix the convection velocity by picking a particular value of y. Pedley (1976b) proposed that the most sensible value of y would be one that was representative of the boundary layer thickness at the changeover time in question. It would thus correspond to a given value of η_1 (or η_2), say $\bar{\eta}_1$ (or $\bar{\eta}_2$), which Pedley chose to be the value at the 'heat thickness' of the thermal boundary layer, i.e. at the centre of mass of the temperature field:

$$\bar{\eta}_1 = \int_0^\infty \eta_1 \theta \, d\eta_1 \Big/ \int_0^\infty \theta \, d\eta_1, \qquad (A.36)$$

where η_1 is given by (A.19) and θ by the first two terms of (A.20). Then t_{1R} would be the solution of

$$x = \frac{y}{\omega} \int_{t_{1R}}^{t_R} S(t) \, dt \quad \text{when } y = \bar{\eta}_1 \left[\frac{9x}{S(t_{1R})} \right]^{1/3},$$

i.e. of

$$[\tfrac{1}{9} x^2 S(t_{1R})]^{1/3} = \bar{\eta}_1 \frac{1}{\omega} \int_{t_{1R}}^{t_R} S(t) \, dt.$$

Similarly

$$[-\tfrac{1}{9}(l-x)^2 S(t_{2R})]^{1/3} = \bar{\eta}_2 \frac{1}{\omega} \int_{t_R}^{t_{2R}} [-S(t)] \, dt. \Bigg\} \qquad (A.37)$$

From this choice of t_{1R} and t_{2R}, the selection of t_{0R} arises naturally: it should be chosen to make the 'heat thickness' of the diffusive layer continuous at $t = t_{1R}$. Using (A.22), (A.34) and (A.36), we obtain

$$(\pi/4\omega^2)(t_{1R} - t_{0R}) = [9x/S(t_{1R})]^{2/3} \bar{\eta}_1^2. \qquad (A.38)$$

This is equivalent to the choice in the viscous case, in which the displacement thickness, and hence the mass-flux deficit, is continuous.

Other choices of $\bar{\eta}_1$, $\bar{\eta}_2$, and t_{0R} are, of course, possible, but the present choice is reasonably self-consistent, and has the added merit that the value of λ, from (A.23), or its equivalent for reversed shear, is close to 0.5 at the changeover times, so that at least the approximately quasi-steady solution is accurate for $t < t_{1R}$ and $t > t_{2R}$. In future applications, in fact, the author would probably use the less cumbersome criterion $\lambda = 0.5$ for choosing the changeover

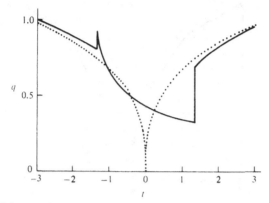

Fig. A.9. Dimensionless heat transfer per unit length, q, as a function of time, t, for uniformly decelerating shear, at $x = 0.5$. Dotted curve is quasi-steady; continuous curve is the approximate solution proposed here. (After Pedley, 1976b.)

times (as is necessary anyway for shear variations $S(t)$ that do not reverse, but that do approach zero); the results presented below, however, from Pedley (1976b), were obtained with the choices in (A.37).

One indication of the usefulness of this approximate theory can be obtained from the presence or absence of wide discontinuities in heat transfer at a particular x as a function of t. In the viscous case (fig. A.7) the discontinuities are not very great, at least for $\omega \leqslant 1.0$. In the present case, however, the discontinuity at the second changeover time, t_{2R}, is considerable. Fig. A.9 shows a plot of the dimensionless heat transfer, q (see (A.21)), against t at $x = 0.5$ for the very simple case of a uniformly decelerating shear, $S(t) \equiv -t$ (in this case we are free to choose the time-scale $1/\Omega$ in such a way that $\omega = 1$). It can be seen that although the choice of t_{0R} in (A.38) leads to a very small jump at $t = t_{1R}$, there is a very large discontinuity at $t = t_{2R}$. This is because there is no freedom to take account of the increasing convection from the trailing edge until $t = t_{2R}$. It must be accepted that the present method leads to an underestimate of heat transfer for a period just after shear reversal.

The discontinuities are less apparent in the curves of total film heat transfer, Q (see (A.12)), against time, as can be seen from fig. A.10. Once again the dotted curve represents the quasi-steady

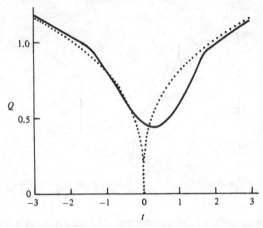

Fig. A.10. Dimensionless heat transfer Q from the whole film as a function of t for uniformly decelerating shear. Dotted curve is quasi-steady; continuous curve is the present approximate solution. (After Pedley, 1976b.)

solution and the continuous curve represents the present approximate solution. From the previous discussion we can expect an underestimate of heat transfer after the shear reversal, at $t = 0$; nevertheless, we can clearly see that a considerable departure from quasi-steady behaviour is to be expected from just before reversal ($t = -0.1$) to some considerable time after ($t = 1.4$, say).

Before applying this theory to the hot-film anemometer, with wall shear given by curves such as those in figs. A.7 and A.8, we give in fig. A.11 the results of applying it to a sinusoidal shear variation, with $S(t)$ given by (A.24). This constitutes an extension to larger amplitudes of the small-x_1 theory of § A.3.1 and fig. A.3. The different curves in each of figs. A.11(a), (b) and (c) represent different values of the relevant dimensionless parameter, called ω_1 by Pedley (1976b), and equal to $\omega l^{2/3}$; this is just $9^{-2/3}$ times the value of x_1 (see (A.18)) when $x = l$.

The results for $\alpha_1 = 10.0$ (fig. A.11(a)) show that in this case the two shear reversals are independent, with a period of approximately quasi-steady (but reversed) heat transfer in between. Each reversal looks like the single reversal in fig. A.10, with both t and Q appropriately rescaled. For $\alpha_1 = 2.0$ (fig. A.11(b)), however, the reversed quasi-steady heat transfer is not attained between the two reversals, at least when $\omega_1 = 1.0$. This indicates that some part of the film (near $x = 0$) experiences purely diffusive heat transfer for the

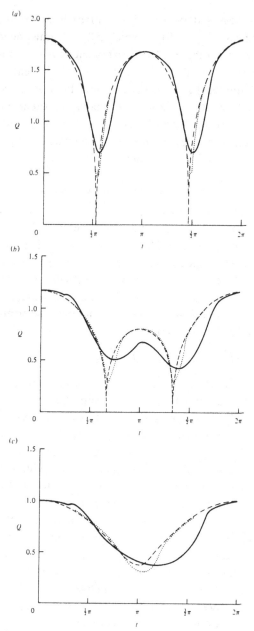

Fig. A.11. Graphs of dimensionless heat transfer, Q, against t for oscillatory shear from (A.24). Broken curves, quasi-steady ($\omega_1 = 0$); dotted curves, $\omega_1 = 0.1$; continuous curves, $\omega_1 = 1.0$. (a) $\alpha_1 = 10.0$, (b) $\alpha_1 = 2.0$, (c) $\alpha_1 = 0.9$. (After Pedley, 1976b.)

whole of the reversed phase. Fig. A.11(c) is an example of a non-reversing case ($\alpha_1 = 0.9$), in which diffusion must nevertheless take over on much of the film for a part of each cycle (of course, the region very near $x = 0$ will always have quasi-steady heat transfer in this case). All the results of fig. A.11 show a slight phase lag behind the quasi-steady solution at periods of maximum shear (associated entirely with the second term in (A.22) and (A.33)), and a rather larger phase lag near times of minimum heat transfer. However, the latter may merely reflect the inaccuracy inherent in the method.

A.4.3 Comparison with experiment

We are now in a position to apply the theory in an attempt to simulate the unsteady calibration experiments of Seed & Wood (1970b) and of Clark (1974). Each author used a sinusoidally varying free-stream velocity, as given by (A.29), so the dimensionless shear must be calculated from (A.32) and will vary as in figs. A.7 and A.8. The dimensional scale for the wall shear-rate, $\hat{S}_0$, is equal to $U_0^{3/2}/(\nu L)^{1/2}$; the dimensionless parameter important in calculating the heat transfer is thus

$$\omega l^{2/3} = (\Omega L/U_0)(\hat{l}/L)^{2/3}(\nu/\kappa)^{1/3} = \omega'\gamma. \qquad (A.39)$$

Here $\omega' = \Omega L/U_0$ is the dimensionless frequency parameter that determines the relation between the wall shear and the stream velocity (cf. (A.31) and figs. A.7 and A.8), and since $X_0 = \frac{1}{2}L$, this is equal to twice the Strouhal number ($St = \Omega X_0/U_0$) defined by Clark (1974). The quantity γ depends only on probe design (through $\hat{l}/L$) and on the Prandtl number, ν/κ, of the ambient fluid, which varies significantly with temperature; the values of γ in the two sets of experiments being modelled are given in table A.1.

The results of the theory will be presented in terms of the velocity which would be inferred from the heat transfer measurements if the quasi-steady relation between velocity and heat transfer were assumed (Clark (1974), fig. 9, presented his measurements in this way). The complete cycle is examined in five cases, as listed in table A.2. Seed & Wood (1970b) reported some measurements in reversing flow and some in non-reversing flow; however, their data were presented in terms of the ratio between actual probe output and the output that would have been measured at the known

Table A.2

Author	α	ω'	Flow reverse?	Shear reverse?	Fig. no.
Seed & Wood	0.98	0.28	No	Yes	A.12(a)
Seed & Wood	2.8	0.75	Yes	Yes	A.12(b)
Clark	0.31	0.44	No	No	A.13(a)
Clark	0.56	0.80	No	Yes	A.13(b)
Clark	0.68	0.80	No	Yes	A.13(c)

instantaneous velocity in steady flow. When the latter becomes very small, inferring velocities from their data becomes very inaccurate and the position of the points becomes uncertain. Therefore only two of their cases are chosen. Clark did not examine reversing flow, but in two of the three cases he presented, the shear on the probe did reverse, and the experiments provide a reasonable test of the theory. The two Seed & Wood cases are presented in figs. A.12(a) and (b); the three Clark cases are presented in figs. A.13(a), (b) and (c). In each case the actual velocity waveform is also shown (so too, in fig. A.12(b), is the rectified form of it, which a perfectly quasi-steady anemometer would measure).

Figs. A.12(a) and (b) show reasonable qualitative agreement between the theory and Seed & Wood's experiments, especially near the points of flow reversal, although in each case the apparent velocity inferred from their data when the actual velocity is very low is enormous, and must be regarded as uncertain. Not enough experimental points were given in each cycle to constitute a good test of the theory. In the approximately quasi-steady regimes the predictions show a slight phase lead over the experiments, which rather follow the exactly quasi-steady curve. This phase lead comes from the phase lead of wall shear over free-stream velocity. Apparently the heat transfer in practice lags behind the wall shear more than is predicted by this theory.

Fig. A.13(a) shows an example in which the shear stress does not approach close enough to zero for a diffusive regime to appear at all. It is included in order to show that the phase lead of theory over experiment is quite pronounced here too (about $\frac{1}{8}\pi$). Fig. A.13(b)

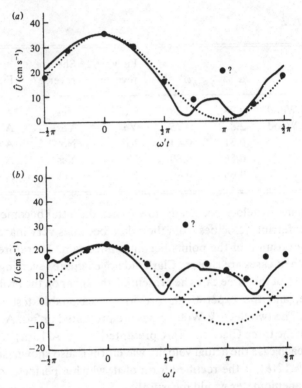

Fig. A.12. Dimensional velocity, $\hat{U}$, plotted against $\omega't$ for two cases from Seed & Wood's (1970b) experiments. Dotted curves represent the actual velocity, which would be measured by a perfectly quasi-steady instrument (including the rectified signal during flow reversal in (b)). Continuous curves represent the present predictions of the velocity that would be recorded by the instrument. Filled circles are measured data. (a) $\alpha = 0.98$, $\omega' = 0.28$; (b) $\alpha = 2.8$, $\omega' = 0.75$. (After Pedley, 1976b.)

shows a case in which the flow does not reverse but the shear does. A comparison between the theoretical curve and the points represented by open triangles shows excellent agreement, apart from a slight underestimate of the maximum heat transfer in the approximately quasi-steady regime. Unfortunately, however, the open triangles are not the experimental points, which are in fact represented by closed circles; the open triangles are the same points given a phase shift of $\frac{1}{4}\pi$. In other words, the phase lead, remarked on above, has become considerable, but the shape of the heat-transfer response, especially near minimum velocity, is very well predicted. Fig.

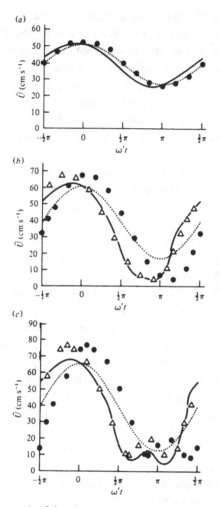

Fig. A.13. As figure A.12 for three cases from Clark's (1974) experiments. (a) $\alpha = 0.31$, $\omega' = 0.44$; (b) $\alpha = 0.56$, $\omega' = 0.80$; (c) $\alpha = 0.68$, $\omega' = 0.80$. The open triangles in (b) and (c) are the measured points given a phase lead of $\frac{1}{4}\pi$. (After Pedley, 1976b.)

A.13(c) shows another case of even larger-amplitude non-reversing flow. Again the agreement between theory and the open triangles is quite good (apart from underestimating the maxima of heat transfer), and again these represent a phase lead of $\frac{1}{4}\pi$ over the experimental points. Note that the $\frac{1}{8}\pi$ in fig. A.13(a) and the $\frac{1}{4}\pi$ in figs.

A.13(b) and (c) represent an approximately constant *time* lead, independent of frequency. Possible reasons for the phase lead are discussed below, but we should note that it was also remarked on by Pedley (1972a), who considered only the first departures from quasi-steady behaviour, and is therefore not an aberration introduced by the inaccuracy of the diffusive solution near times of flow reversal. Indeed, the discussion of fig. A.10 suggests that that inaccuracy would tend to cause a phase *lag* in heat transfer, not a phase *lead*.

In most of the cases he studied, Clark did not calculate the apparent velocity throughout the cycle, merely at the times of maximum and minimum probe output. He then plotted the ratio of the apparent velocity amplitude to the actual velocity amplitude against the Strouhal number, $St(= 0.5\omega')$; a value significantly different from 1 indicated that the quasi-steady calibration was inapplicable. Fig. A.14 shows his results (closed circles) together with the predictions of the present theory for six of his cases (open circles) and four of Seed & Wood's (open triangles); the two closed triangles are Seed & Wood's experimental results corresponding to the open triangles at the same values of St. In cases where the heat transfer shows a second maximum (as in fig. A.12(b) and fig. A.13(c)), this is interpreted as measuring a negative velocity, even when the free stream does not reverse. The results show good agreement between experiment and theory up to a value of St of about 1.0; above that value the theory is not directly applicable. There is considerable scatter in both sets of results at any given value of St, especially around 0.3, which is associated with the fact that the result depends on amplitude as well as on frequency. Nevertheless, the theory confirms the experimental finding that the quasi-steady calibration cannot be used for $St > 0.2$. Even for smaller values of St, the theoretical results (e.g. point A on fig. A.14) underline the fact that the quasi-steady calibration will break down if the amplitude of the oscillation is sufficiently large that the shear on the probe reverses.

In an attempt to discover the cause of the discrepancy between theory and experiment it is important to examine those assumptions, made in § A.1, that have not been adequately dealt with. These are assumption (iv), that the flow over the probe is similar to

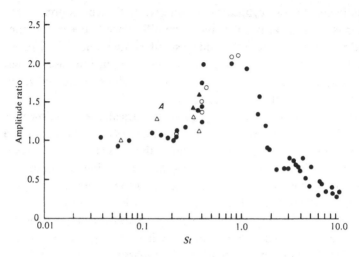

Fig. A.14. Ratio of amplitude of probe output to amplitude of actual oscillating velocity plotted against Strouhal number, $St(=0.5\omega')$. Filled circles, Clark's experiments; filled triangles, Seed & Wood's experiments; open circles and open triangles, present theory. (After Pedley, 1976*b*.)

that over a finite flat plate; assumption (v), that this flow can be calculated using viscous boundary layer theory; assumption (vii), that the temperature field over the hot-film is effectively two-dimensional; and assumption (viii), that this temperature field can be calculated using thermal boundary layer theory. Our discussion of these assumptions should centre on whether (*a*) the phase lead of wall shear would in practice be less than predicted or (*b*) the heat transfer lag would be increased.

Assumption (iv). The probes are approximately cylindrical in cross-section, and at the station occupied by the film the steady boundary layer thickness is about 0.01–0.02 cm for velocities of 20–80 cm s^{-1}, while the cylinder diameter is about 0.045 cm (in Clark's experiments). This means that the quasi-steady wall shear-rate is greater by about 30% than on a flat plate (Rosenhead, 1963, p. 450). The argument by which the unsteady shear is predicted to have a phase lead over the outer velocity, because the flow near the wall responds more readily to the unsteady pressure gradient than that far away, is unaffected by the cylindrical geometry. If the correction to the quasi-steady shear is unchanged, the difference between the one- and two-term expansions could be significantly reduced. The phase lead could be reduced from $\frac{1}{4}\pi$ to about $\frac{1}{6}\pi$ in the case of figs. A.13(b) and (c); this does not explain the whole discrepancy, but clearly deserves further investigation.

If the probe is yawed, or not quite cylindrical, then the flow may be subject to a pressure gradient. Pedley (1972b) showed that a favourable pressure gradient cuts down the phase lead of shear over outer velocity. For two-dimensional flow impinging symmetrically on a 90° wedge, the relative magnitude of the term producing the phase lead (the second term of the expansion whose first term is quasi-steady) is reduced by about 75%. However, it is inconceivable that the present probes induce such a strong pressure gradient and this factor cannot explain the discrepancy.

Assumption (v). The Reynolds number based on the mean velocity and the distance of the film from the leading edge of the probe, $\overline{Re} = U_0 X_0 / \nu$, is only about 1000 in Clark's experiments (fig. A.13). Thus viscous boundary layer theory may not be adequate to predict the shear over the film, but it is not easy to see how this inadequacy would reduce the phase lead of the shear over the outer velocity. Indeed, since the viscous region would be thicker than in boundary layer theory, one might expect the phase difference to be, if anything, greater. Another argument against this as the explanation of the discrepancy is the fact that one would expect the effect to increase as Re falls, whereas the discrepancy is less in Seed & Wood's experiments (fig. A.12), where $\overline{Re} \approx 450$.

Assumption (vii). The hot-films in both sets of experiments are only about 2.5 times longer in the cross-stream direction than in the streamwise direction, so lateral end-effects are likely to be important. The heat transfer in steady flow is likely to be increased by the lateral diffusion in a manner that is almost flow-independent, but the effect on the phase is difficult to assess. It probably does increase the phase lag, in the same way as diffusion through the substrate increases the phase lag through an increase in 'thermal inertia' (Bellhouse & Schultz, 1967). This is the second of the possible explanations that cannot be ruled out as the explanation of the phase lag, and clearly warrants further theoretical and experimental study.

Assumption (viii). Thermal boundary layer theory is inadequate for predicting steady heat transfer from a hot-film if the Péclet number, $l^2 = \hat{S}_0 l^2 / \kappa$, is too small (less than about 400, as we shall see in § A.5). For Clark's experiments the mean Péclet number is about 1000, but for Seed & Wood's it is only about 200, so the effect of departures from boundary layer theory should be considered. The problem is not susceptible to immediate intuitive solution. On the one hand, the presence of axial diffusion increases the effective length of the thermal boundary layer, which would indicate a greater time lag between heat transfer and wall shear than that predicted by boundary layer theory. On the other hand, we shall see that in steady flow it increases the net heat transfer-rate in a manner only slightly dependent on the flow, which would suggest that the effect of unsteadiness would be less. Once again, however, if this effect were responsible for the discrepancy, one would expect a greater, not a smaller, discrepancy in Seed & Wood's experiments than in Clark's.

Of the potential fluid mechanical reasons for the unwanted phase lead, the ones that require further study are three-dimensional effects in the velocity and temperature fields, and the fact that the hot-film may be too short for thermal boundary layer theory to hold. We cannot be certain of the importance of any of these without further research. The only one that has received any detailed analysis is the last, and that only in steady shear flow; this analysis is presented in the next two sections. Apart from these reasons, the explanation can only be in probe construction or electronics, and these are unlikely because of the great care that both authors took to eliminate such artefacts. Thus the phase discrepancy remains a mystery, but should not be allowed to obscure the fact that the theory agrees very well with experiment as far as the amplitude response (fig. A.14) and the general shape of the response throughout the cycle (see especially fig. A.13(b)) are concerned.

A.5 Departures from boundary layer theory for a short hot-film

Whatever the value of the Péclet number, l^2 in (A.5), there are regions near the leading and trailing edges of the hot-film where longitudinal diffusion is important and hence the boundary layer approximation is invalid. Only if l^2 is sufficiently large will these regions have a negligible effect on the overall heat transfer, $\hat{Q}$, which in steady flow is given by (A.12a) according to boundary layer theory. It is the purpose of the rest of this appendix to analyse short hot-films in steady flow. In this section we seek to predict the value of l^2 at which (A.12a) ceases to be accurate and the corrections that should be made at smaller (but not too small) values; the analysis will be that of Springer & Pedley (1973) and of Springer (1974). In the next section we outline the recent theory of Ackerberg, Patel & Gupta (1978) for very short films ($l \ll 1$).

The analysis of this section is based on the assumption that the leading- and trailing-edge regions in which boundary layer theory is inaccurate are independent of each other. This is equivalent to the statement that there is a region in the middle of the hot-film, albeit very short, in which the temperature field is accurately described by boundary layer theory. Ling (1963) performed a numerical solution of the problem of a finite hot-film in a steady, uniform shear, and his results can be interpreted as suggesting that each 'non-boundary-layer' region has a length of about $1.2\,(\kappa/\hat{S})^{1/2}$. Thus boundary layer theory might be expected to be valid somewhere on the film if $l^2 \geqslant 6$. The results of this section suggest that in fact the trailing-edge region is somewhat longer than the leading-edge region, and that the value of 6 should be replaced by about 16.

The problem to be solved is made dimensionless by the scaling (A.6), and can be stated mathematically as follows: we seek the function $\theta(x, y)$ that satisfies the equation

$$y\theta_x = \theta_{xx} + \theta_{yy} \qquad (A.40)$$

and the boundary conditions

$$\text{on } y = 0, \quad \theta = 1 \quad \text{for } 0 \leqslant x \leqslant l, \quad \theta_y = 0 \text{ elsewhere,} \\ \text{as } (x^2 + y^2)^{1/2} \to \infty, \quad \theta \to 0. \qquad \left.\right\} \qquad (A.41a)$$

Results should be expressed in terms of the dimensionless heat transfer, either per unit length ($q(x)$ from (A.11)) or total (Q from (A.12)).

A.5.1 *The leading edge*

Here we suppose that 'far' downstream from $x = 0$ the temperature distribution is given by the boundary layer solution (A.9). This means that the film can be regarded as effectively semi-infinite, and the boundary conditions (A.41) should be replaced with

$$\text{on } y = 0, \quad \theta = e^{-ax} \text{ for } x \geq 0, \quad \theta_y = 0 \text{ for } x < 0,$$
$$\text{as } y \to \infty \text{ (all } x) \quad \text{and} \quad x \to -\infty \text{ (all } y) \quad \theta \to 0. \tag{A.41b}$$

Here a is a small positive quantity that is introduced to ensure existence of the Fourier transform of θ, but that will be allowed to tend to zero as soon as is convenient. We further assume that the last condition can be strengthened to

$$\theta = o(e^{bx}) \quad \text{as } x \to -\infty \tag{A.41c}$$

for all y and for some real number $b > 0$; the consistency of this assumption is confirmed *a posteriori*. The temperature field $\theta(x, y)$ is expected to be continuous and bounded everywhere. However, its gradient at the wall, $\theta_y(x, 0) = -q(x)$, must be discontinuous at $x = y = 0$. Now, as $(x^2 + y^2)^{1/2} \to 0$, (A.40) reduces to Laplace's equation, and the least singular solution turns out to be such that

$$|q(x)| = O(x^{-1/2}) \quad \text{as } x \to 0+. \tag{A.42}$$

The problem is solved by means of the Wiener–Hopf technique. We therefore begin by taking Fourier transforms with respect to x, and define

$$\tilde{\theta}(k, y) = \int_{-\infty}^{\infty} \theta(x, y)\, e^{-ikx}\, dx,$$

so that (A.40) transforms into

$$\tilde{\theta}_{yy} - ik(y - ik)\tilde{\theta} = 0. \tag{A.43}$$

The solution of this that satisfies the boundary condition at infinity is

$$\tilde{\theta} = H(k)Ai(s), \tag{A.44}$$

where
$$s = (0+ik)^{1/3}(y-ik)$$

and $H(k)$ is a function to be found, as long as $-\frac{1}{3}\pi < \arg s < \frac{1}{3}\pi$. This means that the k-plane should be cut along the positive imaginary axis, with $-\frac{3}{2}\pi < \arg k < \frac{1}{2}\pi$.

We now split $\tilde{\theta}(k, y)$ into two functions, analytic in upper and lower halves of the k-plane respectively, and denoted by subscripts $+$, $-$: we define

$$\tilde{\theta}_-(k, y) = \int_0^\infty \theta(x, y)\, e^{-ikx}\, dx,$$

$$\tilde{\theta}_+(k, y) = \int_{-\infty}^0 \theta(x, y)\, e^{-ikx}\, dx,$$

so that $\tilde{\theta} = \tilde{\theta}_+ + \tilde{\theta}_-$. From the boundary conditions on $y = 0$ and the fact that θ is bounded, we see that $\tilde{\theta}_+$ is analytic in the upper half-plane Im $k > -b$, and that $\tilde{\theta}_-$ is analytic in the lower half-plane Im $k < a$. From (A.44) we have

$$\left. \begin{aligned} \tilde{\theta}_+(k, 0) + \tilde{\theta}_-(k, 0) &= H(k)Ai(s_0), \\ \tilde{\theta}'_+(k, 0) + \tilde{\theta}'_-(k, 0) &= H(k)Ai'(s_0)(0+ik)^{1/3}, \end{aligned} \right\} \quad \text{(A.45)}$$

where primes on $\tilde{\theta}_\pm$ denote differentiation with respect to y, and

$$s_0 = -ik(0+ik)^{1/3}, \qquad \text{(A.46)}$$

the value of s at $y = 0$. Now, if we transform the wall conditions in (A.41b), we obtain

$$\tilde{\theta}_-(k, 0) = 1/i(k-ia), \qquad \tilde{\theta}'_+(k, 0) = 0,$$

so that on elimination of $H(k)$, (A.45) gives

$$(0+ik)^{-1/3}[Ai(s_0)/Ai'(s_0)]\tilde{\theta}'_-(k, 0) - \tilde{\theta}_+(k, 0) - 1/i(k-ia) = 0. \tag{A.47}$$

This equation is in a form to which the Wiener–Hopf technique can readily be applied in the k-plane. We note that $Ai(s_0)$ and $Ai'(s_0)$ both have zeros on the negative imaginary k-axis, as well as a branch point at $k = 0$. Thus if b is chosen to be less than the magnitude of the first zero of either $Ai(s_0)$ or $Ai'(s_0)$ every term in (A.47) is defined and non-zero everywhere in the strip $-b < \text{Im } k < 0$. We also note that $\tilde{q}(k) = -\tilde{\theta}'_-(k, 0)$ is the Fourier transform of the heat-transfer function $q(x)$.

In order to obtain an expression for $\tilde{q}(k)$, we must split the left-hand side of (A.47) into the sum of two functions, one analytic in Im $k < 0$ (a 'lower function') and one analytic in Im $k > -b$ (an 'upper function'). To this end we first split the factor multiplying $\tilde{q}(k)$ into the product of an upper function and a lower function. If we define

$$F(k) \equiv -(0 + ik)^{1/3} Ai'(s_0)/Ai(s_0), \qquad (A.48)$$

then upper and lower functions $K_+(k)$ and $K_-(k)$ with the property that

$$F(k) = K_+(k)K_-(k)$$

are given by

$$\log K_+ = \frac{1}{2\pi i} \int_{-\infty}^{\infty} \log[F(z)] \frac{dz}{z-k}, \qquad \text{Im } k > 0, \qquad (A.49a)$$

$$\log K_- = \frac{-1}{2\pi i} \int_{-\infty}^{\infty} \log[F(z)] \frac{dz}{z-k}, \qquad \text{Im } k < 0, \qquad (A.49b)$$

where the integral is in each case taken along the real axis, indented below the branch point at $z = 0$ (these definitions correspond to those used by Stewartson (1968) in a related problem). The relevant properties of $K_{\pm}$ will be derived below.

With $K_{\pm}$ determined, (A.47) can be rewritten

$$\tilde{q}(k)/K_-(k) - \tilde{\theta}_+(k, 0)K_+(k) - K_+(k)/i(k - ia) = 0.$$

The first term is a lower function and the second is an upper function, but the third is neither because of the pole at $k = ia$. However, this difficulty can be removed by writing

$$K_+(k)/i(k - ia) = [K_+(k) - K_+(ia)]/i(k - ia) + K_+(ia)/i(k - ia);$$

the first of these terms is now an upper function, say $R_+(k)$, and the second is a lower function, $R_-(k)$. Thus we have

$$\tilde{q}(k)/K_-(k) - R_-(k) = \tilde{\theta}_+(k, 0)K_+(k) + R_+(k), \qquad (A.50)$$

of which the terms on the left-hand side are analytic for Im $k < 0$, and those on the right-hand side are analytic for Im $k > -b$. By the principle of analytic continuation, therefore, there exists an entire function, $J(k)$, equal to either side of (A.50) wherever that side is defined. Furthermore, all the terms in (A.50) tend to zero as

$|k| \to \infty$ (since from (A.42) $|\tilde{q}(k)| \sim |k|^{-1/2}$, and we show below that $|K_{\pm}(k)| \sim |k|^{1/2}$, as $|k| \to \infty$), so $J(k)$ is identically zero by Liouville's theorem. Hence, letting $a \to 0$, we have

$$\tilde{q}(k) = K_-(k)R_-(k) = K_+(0)K_-(k)/ik,$$

and inversion yields

$$q(x) = \frac{K_+(0)}{2\pi i} \int_{-\infty}^{\infty} \frac{K_-(k)}{k} e^{ikx} \, dk. \qquad (A.51)$$

Properties of $K_{\pm}(k)$

Following Stewartson (1968), we first examine K_-, noting that the argument s_0 (see (A.46)) of the Airy functions occurring in $F(k)$ is real and positive on the straight lines $\arg k = \frac{1}{4}\pi$ and $\arg k = -\frac{5}{4}\pi$. We therefore deform the contour of integration in the Cauchy integral (A.49b) to lie along these straight lines, together with two arcs of large radius, $z = R\, e^{i\theta}$ with $\frac{1}{4}\pi > \theta > 0$ and $-\pi > \theta > -\frac{5}{4}\pi$, and a small indentation round the origin that does not contribute to the integral (see fig. A.15). The contribution to $\log K_-$ from the two circular arcs is

$$\tfrac{1}{4} \log R - \tfrac{1}{8}i\pi$$

for large R. When $k = r\, e^{i\pi/4}$, $F(k) = e^{i\pi/4}M(r)$, where

$$M(r) = -r^{1/3}Ai'(r^{4/3})/Ai(r^{4/3}),$$

while when $k = r\, e^{-5i\pi/4}$, $F(k) = e^{-i\pi/4}M(r)$. The integrals along the straight lines from $r = 0$ to $r = R$ can then be combined to give the following contribution to $\log K_-$:

$$\frac{-1}{2\pi i} \int_0^R \log[M(r)] \frac{\sqrt{2}k\, dr}{r^2 + rki\sqrt{2} - k^2} + \tfrac{1}{4}\log k - \tfrac{1}{4}\log R + \frac{i\pi}{8}.$$

Hence, writing $t = 0 + ik$, we obtain

$$\log K_- = \frac{t}{\pi\sqrt{2}} \int_0^{\infty} \frac{\log[M(r)]\, dr}{r^2 + rt\sqrt{2} + t^2} + \tfrac{1}{4}\log t - \tfrac{1}{8}i\pi, \qquad (A.52)$$

the first term of which is real when t is real, i.e. on the negative imaginary k-axis, and on either side of the branch cut along the positive imaginary k-axis. This means that, apart from factors of the form $e^{i\phi}$, the integral in (A.51) can be reduced to a real integral if

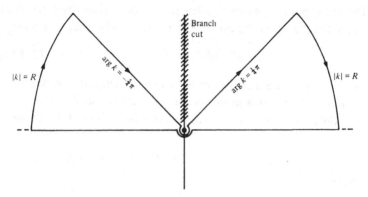

Fig. A.15. Sketch of the complex k-plane, showing the deformation of the
original contour of integration (the indented real axis) in (A.51).

the inversion contour along the real axis is deformed to run along
both sides of the branch cut. It is then in a form suitable for
numerical evaluation.

As a check on the numerical work, and to aid physical understanding of the results, it is important to derive asymptotic expan-
sions for $K_\pm$ at large and small values of $|k|$. We consider large $|k|$
first, noting that

$$M(r) = r[1 + 1/4r^2 + O(r^{-4})] \quad \text{as } r \to \infty.$$

The first term on the right-hand side of (A.52) can be rewritten as

$$\frac{t}{\pi\sqrt{2}} \int_0^\infty \frac{\log r \, dr}{r^2 + rt\sqrt{2} + t^2} + \frac{1}{\pi t\sqrt{2}} \int_0^\infty \log\left(\frac{M}{r}\right) dr + O(t^{-2} \log t)$$

(Springer, 1974), which is equal to

$$\tfrac{1}{4} \log t + \gamma_1/t + O(t^{-2} \log t),$$

where

$$\gamma_1 = \frac{1}{\pi\sqrt{2}} \int_0^\infty \log\left(\frac{M}{r}\right) dr = 0.1797.$$

Thus

$$\log K_- \sim \tfrac{1}{2} \log t - \tfrac{1}{8}i\pi + \gamma_1/t + O(t^{-2} \log t) \text{ as } |t| \to \infty,$$

or (in terms of k),

$$K_- \sim e^{-i\pi/8}(0 + ik)^{1/2}[1 + \gamma_1/ik + O(k^{-2} \log|k|)] \quad \text{as } |k| \to \infty.$$
$$\text{(A.53}a\text{)}$$

The corresponding expansion for K_+ can be derived from (A.53a) and a direct asymptotic expansion of $F(k)$, and the leading term is

$$K_+ \sim e^{i\pi/8}(0-ik)^{1/2}[1+O(k^{-1})]. \qquad (A.53b)$$

The results in (A.53a, b) confirm the stated asymptotic behaviour of the terms in (A.50), and justify the conclusion that $J(k) \equiv 0$.

In order to derive the expansions at small values of $|k|$, we note that

$$M(r) = pr^{1/3}[1+pr^{4/3}+O(r^{8/3})] \quad \text{as } r \to 0,$$

where

$$p = -Ai'(0)/Ai(0) = 3^{1/3}\Gamma(\tfrac{2}{3})/\Gamma(\tfrac{1}{3}) = 0.7290. \qquad (A.54)$$

Thus we may rewrite (A.52) in the form

$$\log K_- = \tfrac{1}{4}\log t - \tfrac{1}{8}i\pi + \frac{t}{\pi\sqrt{2}}\int_0^\infty \frac{\log(pr^{1/3})\,dr}{r^2+rt\sqrt{2}+t^2}$$

$$+ \frac{t}{\pi\sqrt{2}}\left\{\int_1^\infty \frac{\log(M/pr^{1/3})}{r^2}\,dr \right.$$

$$+ \int_0^1 \frac{[\log(M/pr^{1/3})-pr^{4/3}]}{r^2}\,dr$$

$$\left. + p\int_0^1 \frac{r^{4/3}\,dr}{r^2+rt\sqrt{2}+t^2} + O(t)\right\}$$

$$= \tfrac{1}{3}\log t - \tfrac{1}{8}i\pi + \tfrac{1}{4}\log p + \beta_1 t + O(t^{4/3}),$$

where

$$\beta_1 = \frac{1}{\pi\sqrt{2}}\int_0^\infty \frac{\log(M/pr^{1/3})}{r^2}\,dr = 0.6856.$$

Hence

$$K_-(k) = p^{1/4}e^{-i\pi/8}(0+ik)^{1/3}[1+\beta_1 ik+O(k^{4/3})] \quad \text{as } |k| \to 0. \qquad (A.55a)$$

The leading term of the corresponding expansion for K_+, which is required for the evaluation of $q(x)$ from (A.51), is derived directly from (A.55a) and the definition of $F(k)$ and is given by

$$K_+(k) \sim e^{i\pi/8}p^{3/4}[1-\beta_1 ik+O(k^2)] \quad \text{as } |k| \to 0. \qquad (A.55b)$$

Results

The heat-transfer function $q(x)$ can be expanded asymptotically for large x by using the small-k expansion of $K_-(k)$. Thus, from (A.51), (A.55a) and (A.55b), we obtain

$$q(x) \sim \frac{p}{2\pi i} \int_{-\infty}^{\infty} (0+ik)^{1/3}[1 + \beta_1 ik + O(k^{4/3})]\, e^{ikx}\, \frac{dk}{k}$$

$$\sim [3^{5/6}\Gamma(\tfrac{2}{3})/2\pi x^{1/3}]\{1 - \beta_1/3x + O(x^{-4/3})\}. \tag{A.56}$$

The leading term of this can be seen to be identical to that obtained from Lévêque's (1928) boundary layer solution (see (A.11)), if the identity $\Gamma(\tfrac{1}{3})\Gamma(\tfrac{2}{3}) = 2\pi/\sqrt{3}$ is used. Springer & Pedley (1973), who developed the analysis ⌐f $K_\pm$ in terms of an infinite product, rather than the Cauchy integrals (A.49a and b), took the expansion (A.56) as far as terms of $O(x^{-11/3})$ in the curly brackets. However the numerical results show that this complicated expansion is actually less useful than its leading term. In fig. A.16, q is plotted against x on a log–log plot. The continuous curve was obtained by direct numerical integration of the real integral derived from (A.51), the broken curve represents the full expansion of Springer & Pedley, while the dot–dash curve represents the first term of that expansion, the boundary layer result (A.11). It can be seen that, as x is decreased from a very large value, q at first falls slightly below its asymptotic form before rising above it again for $x \leqslant 1.7$. The full asymptotic expansion faithfully follows this deviation from the leading term, until near $x = 2.5$ it becomes wildly inaccurate. However, the leading term itself is accurate to within 2% for all $x > 1.0$, confirming Ling's (1963) numerical conclusion.

At small x, the continuous curve asymptotically approaches the straight (dotted) line

$$q(x) = 0.445x^{-1/2}.$$

This too can be verified by using the large-k expansion of $K_-(k)$ in (A.51). Using (A.53a) and (A.53b) we obtain, as $x \to 0$,

$$q(x) \sim \frac{p^{3/4}}{2\pi} \int_{-\infty}^{\infty} \frac{e^{ikx}\, dk}{(0+ik)^{1/2}} = \frac{p^{3/4}}{(\pi x)^{1/2}} = 0.4451x^{-1/2}. \tag{A.57}$$

This expression, together with the leading term of (A.56), can be seen to give a very accurate representation of $q(x)$ for all x.

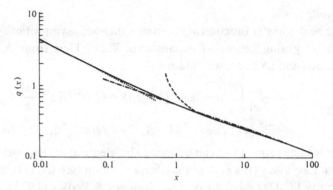

Fig. A.16. Dimensionless heat transfer $q(x)$ plotted against x (logarithmic scales). Continuous curve, exact solution; broken curve, seven-term asymptotic expansion for large x; dot–dash line, Lévêque's boundary layer solution (A.11); dotted line, leading term of the small-x expansion (A.57). (After Springer & Pedley, 1973.)

Finally, we should check that $\theta(x, y) = O(e^{bx})$ as $x \to -\infty$ for some $b > 0$. Assuming that it is adequate to carry out the check at $y = 0$, we note from (A.50) that

$$\tilde{\theta}_+(k, 0) = (1/ik)(K_+(0)/K_+(k) - 1).$$

This is regular at $k = 0$, but has an infinite number of poles on the negative imaginary axis (the zeros of $Ai'(s_0)$), the first of which is at $k = -ik_1 \approx -1.014i$. Completing the inversion contour in the lower half-plane for $x < 0$, therefore, shows that the behaviour as $x \to -\infty$ is given by

$$\theta(x, 0) = O(e^{(k_1 - \delta)x}),$$

where δ is an arbitrary small positive number. This confirms the assumed upstream condition.

A.5.2 The trailing edge

In this case, we shall suppose that the boundary layer solution (A.9) holds far upstream from the trailing edge $x = l$. If we define a new x-variable, $x^* = x - l$, and a new θ-variable

$$\theta^*(x^*, y) = \theta(x^*, y) - \theta_{\text{bl}}(x^*, y),$$

where

$$\theta_{bl}(x^*, y) = c_0 \int_\eta^\infty e^{-s^3} ds, \qquad \eta = y[9(x^* + l)]^{-1/3},$$

and $c_0 = 1/\Gamma(\tfrac{4}{3})$, then the governing equation remains (A.40) (with starred variables) and the boundary conditions become

$$\text{on } y = 0 \begin{cases} \theta^* = 0 \quad \text{for } x^* < 0, \\ \theta_y^* = -\theta_{bly} = c_0/[9(x^* + l)]^{1/3} \quad \text{for } x^* > 0; \end{cases} \quad \text{(A.58)}$$
$$\text{as } (x^{*2} + y^2)^{1/2} \to \infty, \, \theta^* \to 0.$$

As in § A.5.1, we assume that the supplementary condition can be applied, that $\theta^* = O(e^{bx^*})$ as $x^* \to -\infty$ for some $b > 0$; we also note that θ^* will be continuous at $(0, 0)$, but that $\theta_y^*(x^*, 0)$ will behave like $|x^*|^{-1/2}$ as $x^* \to 0-$.

The development of the Wiener–Hopf analysis is exactly the same as for the leading edge, and the equation corresponding to (A.47) is

$$\tilde{\theta}_-(k)F(k) + \tilde{\theta}'_+(k) + G_-(k) = 0, \qquad \text{(A.59)}$$

where $\tilde{\theta}$ is the Fourier transform of θ^*, $F(k)$ is again defined by (A.48), and

$$G_-(k) = \tilde{\theta}'_-(k, 0) = c_0 \int_0^\infty \frac{e^{-ikx^*} dx^*}{[9(x^* + l)]^{1/3}}. \qquad \text{(A.60)}$$

If $K_\pm(k)$ are defined as in (A.49), and if $G_-(k)/K_+(k)$ is expressed as the sum of an upper and a lower function, $H_+(k) + H_-(k)$, then (A.59) can be written

$$\tilde{\theta}_-(k, 0)K_-(k) + H_-(k) = -[\tilde{\theta}'_+(k, 0)/K_+(k) + H_+(k)]. \quad \text{(A.59a)}$$

The theory of analytic continuation and the asymptotic properties of the functions involved again show that each side of this equation is identically zero, so that $\tilde{\theta}_-(k, 0)$ and $\tilde{\theta}'_+(k, 0)$ are both, in principle, determined.

The function $\tilde{\theta}'_+(k, 0)$ is the Fourier transform of the heat-transfer function, $q(x)$, less its boundary layer component, and it might be thought that to calculate the heat transfer from the whole, finite hot-film it would be necessary to invert this function and integrate a composite of it and the function plotted in fig. A.16 over the whole length of the film. That, however, would be very laborious, and ignores the identity established in § A.2, by which the dimensionless heat transfer, Q, is shown to be proportional to the coefficient of the leading term, $x^{-2/3}$, in the large-x expansion of $\theta(x, 0)$ (see (A.15) and (A.16)). It is therefore more convenient to calculate $\tilde{\theta}_-(k, 0)$ and evaluate the leading terms in the asymptotic expansion of its inverse:

$$\theta^*(x^*, 0) = -\frac{1}{2\pi} \int_{-\infty}^{\infty} \frac{H_-(k)}{K_-(k)} e^{ikx^*} \, dk, \quad \text{for } x^* > 0. \quad (A.61)$$

Properties of $H_-(k)$
From (A.60) we deduce that

$$G_-(k) = p\left\{\frac{e^{ikl}}{(0+ik)^{2/3}} - l^{2/3} \sum_{n=0}^{\infty} \frac{(ikl)^n}{\Gamma(n+\frac{5}{3})}\right\}.$$

The second term of this is an entire function, and when divided by K_+ it will become an upper function. Thus only the first term, with the branch point at $k = 0$, contributes to H_-, which can be defined by the following Cauchy integral:

$$H_-(k) = \frac{-p}{2\pi i} \int_{-\infty}^{\infty} \frac{e^{izl} \, dz}{K_+(z)(0+iz)^{2/3}(z-k)}, \quad \text{Im } k < 0;$$

the integral is taken along the real axis, indented below the origin. If the contour is deformed to run along the two sides of the branch cut on the positive imaginary z-axis, this integral becomes

$$H_-(k) = \frac{p\sqrt{3}}{2\pi} \int_0^{\infty} \frac{e^{-rl} \, dr}{K_+(ir)r^{2/3}(r+ik)}, \quad (A.62)$$

where

$$K_+(ir) \sim e^{i\pi/8} p^{3/4}[1 + \beta_1 r + O(r^2)] \quad \text{as } r \to 0$$

from (A.55b). To obtain the small-k expansion of H_-, we follow Springer (1974) and split (A.62) into two parts:

$$H_- = \frac{p^{1/4}\sqrt{3} e^{-i\pi/8}}{2\pi} \left[\int_0^{\infty} \frac{(1-\beta_1 r) e^{-rl}}{r^{2/3}(r+ik)} \, dr + \int_0^{\infty} \frac{N(r) e^{-rl}}{r^{2/3}(r+ik)}\right] dr,$$

where

$$N(r) = p^{-3/4} e^{-i\pi/8} K_+(ir) - 1 - \beta_1 r = O(r^2) \quad \text{as } r \to 0.$$

The first integral has the small-k expansion

$$\frac{-2\pi l^{2/3}}{\sqrt{3}\,\Gamma(\frac{2}{3})}\left[\left(\frac{3}{2}+\frac{\beta_1}{l}\right)+\left(\frac{9}{10}+\frac{3\beta_1}{2l}\right)(ikl)+O(kl)^2\right]$$

$$+\frac{2\pi}{\sqrt{3}}\frac{e^{ikl}}{(0+ik)^{2/3}}[1+\beta_1 ik+O(k^2)],$$

while the second has the expansion

$$\delta_0(l)-ik\delta_1(l)+O(k^2),$$

where

$$\delta_n(l)=\int_0^\infty \frac{N(r)\,e^{-rl}}{r^{5/3+n}}\,dr;$$

both δ_0 and δ_1 exist because $N=O(r^2)$ as $r\to 0$. Thus, finally,

$$H_-(k)\sim p^{1/4}\,e^{-i\pi/8}\{g_0-ikg_1+O(k^2)$$

$$+[e^{ikl}/(0+ik)^{2/3}][1+\beta_1 ik+O(k^2)]\}\quad\text{as }|k|\to 0,\quad\text{(A.63)}$$

where

$$g_0(l)=\frac{\sqrt{3}\,\delta_0(l)}{2\pi}-\frac{l^{2/3}}{\Gamma(\frac{2}{3})}\left(\frac{3}{2}+\frac{\beta_1}{l}\right),\qquad\text{(A.64)}$$

$$g_1(l)=\frac{\sqrt{3}\,\delta_1(l)}{2\pi}+\frac{l^{5/3}}{\Gamma(\frac{2}{3})}\left(\frac{9}{10}+\frac{3\beta_1}{2l}\right).$$

Results
The large-x^* expansion for $\theta^*(x^*,0)$ is obtained by substituting (A.63) and (A.55a) into (A.61), and inverting term by term. The resulting series is

$$\theta^*(x^*,0)\sim -1-[g_0/\Gamma(\tfrac{1}{3})]x^{*-2/3}+O(x^{*-4/3}).$$

Hence, using (A.16), we deduce that the dimensionless heat transfer from the film, Q, is given by

$$Q=-pg_0(l).\qquad\text{(A.65)}$$

For large values of l, this yields the boundary layer result (A.12a), $Q=\frac{1}{2}3^{1/3}c_0 l^{2/3}$, but (A.64) shows that there is an error of $O(l^{-1/3})$. Springer (1974) computed Q, as given by (A.65), for various values of l. They are compared with the boundary layer values in table A.3

Table A.3. *Comparison of calculated heat transfer with the boundary layer predictions*

l	Q_{bl}	Q
1	0.80755	1.11688
2	1.28190	1.54488
3	1.67977	1.91632
4	2.03490	2.25340
5	2.36129	2.56631
6	2.66647	2.86085
7	2.95507	3.14075
8	3.23020	3.40856
9	3.49406	3.66615
10	3.74831	3.91493
11	3.99421	4.15599
12	4.23275	4.39022
13	4.46476	4.61834
14	4.69088	4.84093
15	4.91168	5.05851
16	5.12762	5.27143
17	5.33910	5.48023
18	5.54648	5.68507
19	5.75005	5.88627
20	5.95007	6.08409

and with experiment in fig. A.17. The table shows that the error in using the boundary layer approximation everywhere on the plate is about 2% when $l = 20$, about 8% when $l = 5$ and about 20% when $l = 2$, if the present results are accurate for such a small value.

To assess the accuracy of the present results, it is necessary to consider the heat-transfer function $-\theta_y^*(x^*, 0)$ for $x^* < 0$. Springer (1974) did not perform an exact numerical inversion of the function $\tilde{\theta}'_+(k, 0)$, obtained by setting the right-hand side of (A.59a) equal to zero. We can, however, use an argument similar to that at the end of § A.5.1, to suggest that θ^* tends to zero like $e^{l_1 x^*}$ as $x^* \to -\infty$, where l_1 is the first pole of $K_+(k)$, i.e. the first zero of $Ai(s_0)$; $l_1 \approx 1.891$. On the assumption that e^{-5} is negligibly small, then, we predict that the perturbation to the boundary layer heat-transfer function will be negligible for $x^* \leqslant 2.64$. Together with the results for the leading edge, this suggests that the boundary layer solution is valid somewhere on the hot-film, and so the results of table A.3 are accurate, if $l \geqslant 4$ or the Péclet number $l^2 \geqslant 16$.

Springer (1974) did calculate the first few terms of an expansion of $-\theta_y^*(x^*, 0)$ in powers of x^*, for $x^* < 0$. They show that in the neighbourhood of the trailing edge, as well as in that of the leading edge, the heat-transfer function $q(x)$ is increased above its boundary layer value. This is to be expected because of the influence of longitudinal diffusion, so that (A.40) increasingly resembles Laplace's equation as either edge is approached. The consequence is that the heat transfer becomes less and less dependent on the flow-rate as l is decreased (cf. fig. A.17). Such a result is also likely to be true in unsteady flow, and has the implication that the shorter a hot-film is, the less responsive its heat transfer will be to fluctuations in the local wall shear. This suggests that the amplitude of the heat-transfer fluctuation would be less than for a longer film, if the behaviour is quasi-steady. However, the shorter a hot-film is, the more likely it is to behave quasi-steadily according to boundary layer theory (§ A.3), so it is not obvious how the shortness of the hot-film could account for the observed phase lag between heat transfer and fluid velocity.

A.6 Steady heat transfer from a very short hot-film

To complete the picture, we now outline Ackerberg et al.'s (1978) analysis of the case of very small Péclet number, $l \ll 1$. In this case boundary layer theory cannot be accurate anywhere over the film, and the first approximation to the temperature field near the film might be expected to be a purely diffusive solution, with the influence of the shear flow being more important far away. A matched asymptotic expansion, such as that developed by Proudman & Pearson (1957) for slow flow past a circular cylinder, is clearly called for. We shall limit ourselves to presenting merely the leading terms of the inner and outer expansions (as it were the Stokes and Oseen approximations), since they are enough to compute the heat transfer with satisfactory accuracy.

We follow Ackerberg *et al.* in choosing a new origin at the mid-point of the film and non-dimensionalising $\hat{x}$ and $\hat{y}$ with respect to $\frac{1}{2}\hat{l}$ so that $(x', y') = (2/\hat{l})(\hat{x}, \hat{y})$. The governing equation for the dimensionless temperature field $\theta(x', y')$ is

$$\varepsilon y' \theta_{x'} = \theta_{x'x'} + \theta_{y'y'}, \qquad (A.66)$$

where $\varepsilon = \frac{1}{4}l^2$. The boundary conditions are:

$$\text{on } y' = 0 \begin{cases} \theta = 1 & \text{for } |x'| < 1, & (A.67a) \\ \theta_{y'} = 0 & \text{for } |x'| > 1; & (A.67b) \end{cases}$$

$$\text{as } (x'^2 + y'^2) \to \infty, \qquad \theta \to 0. \qquad (A.67c)$$

First-order inner solution. The coordinates x' and y' are clearly suitable inner variables, and if we set $\varepsilon = 0$ in (A.66), we see that any function of the form

$$\theta = 1 + g(\varepsilon)\theta_0(x', y'), \qquad (A.68)$$

where $g(\varepsilon)$ is arbitrary, $\nabla^2 \theta_0 = 0$ and $\theta_0(x', 0) = 0$ for $|x'| < 1$, satisfies both the equation and the first boundary condition (A.67a). However, no such solution can be found that satisfies the other boundary conditions, and the last of them, (A.67c), must be abandoned. There is then a unique solution that both satisfies (A.67b) and does not lead to non-integrable singularities in $\theta_{y'}(x', 0)$ at the ends of the strip (cf. the constraint (A.42)). This is

$$\theta_0(x', y') = Re \log [z + (z^2 - 1)^{1/2}], \qquad (A.69)$$

where $z = x' + iy'$, and $(z^2 - 1)^{1/2}$ is analytic in the z-plane that has been cut along the strip $y' = 0$, $|x'| \leq 1$. The principal value of the logarithm is to be taken. For matching purposes we shall need the expansion of (A.69) as $r' = |z| \to \infty$; it is

$$\theta_0 \sim \log 2r' + O(r'^{-2}) \quad \text{as } r' \to \infty. \qquad (A.70)$$

First-order outer solution. At large distances the hot-film will resemble a point source of heat on the wall, and advection will balance diffusion in determining the temperature field. Appropriate outer variables are then

$$(x, y) = \varepsilon^{1/2}(x', y'),$$

which are exactly the same as the (x, y) of (A.6) and of the last section. In terms of these, (A.66) is the full equation (A.40), while the boundary condition on the wall becomes

$$\theta_y(x, 0) = -Q'(\varepsilon)\delta(x)$$

for some constant $Q'(\varepsilon)$, which is to be determined by matching, and which we expect to be equal to the dimensionless heat transfer from the film. The solution of this problem, obtained as usual by means of Fourier transforms, is

$$\theta = -\frac{Q'}{2\pi} \int_{-\infty}^{\infty} \frac{Ai[(0+ik)^{1/3}(y-ik)]}{(0+ik)^{1/3}Ai'(s_0)} e^{ikx} \, dk, \qquad (A.71)$$

where s_0 is defined by (A.46).

For positive x, the contour of integration can be deformed to pass along both sides of the branch cut on the positive imaginary k-axis, and (A.71) becomes

$$\theta(x, y) = \frac{Q'}{\pi} \text{Im} \int_0^{\infty} \frac{Ai[e^{i\pi/3}t^{1/3}(y+t)]}{Ai'[e^{i\pi/3}t^{4/3}]} \frac{e^{-i\pi/3} e^{-tx}}{t^{1/3}} \, dt. \qquad (A.72)$$

We may note that for large positive values of x, and $y = 0$, the Airy functions may be expanded in powers of t, to give

$$\theta(x, 0) = \frac{Q'\sqrt{3}}{2\pi p} \int_0^{\infty} \frac{e^{-tx}}{t^{1/3}} [1 + O(t^{8/3})] \, dt,$$

where p is given by (A.54). Hence

$$\theta(x, 0) \sim Q'\sqrt{3}\Gamma(\tfrac{2}{3})/2\pi p x^{2/3},$$

and comparison with (A.16) shows that $Q' = Q$, confirming the above expectation.

Matching. The determination of Q requires that the solution (A.71) be expanded for small values of $r = (x^2 + y^2)^{1/2}$ and matched to (A.68) and (A.70). To this end we rewrite (A.72) as

$$\frac{\pi}{Q} \theta(x, y) = \text{Im} \left\{ \int_0^1 \frac{e^{-i\pi/3} Ai(s)}{t^{1/3} Ai'(s_0)} e^{-tx} \, dt \right.$$

$$+ \int_1^{\infty} \left[\frac{e^{-i\pi/3} Ai(s)}{t^{1/3} Ai'(s_0)} - \frac{i}{t} e^{-ity} \left(1 + \frac{i}{4t^2}\right) \right] e^{-tx} \, dt$$

$$\left. + \int_1^{\infty} \frac{i}{t} \left(1 + \frac{i}{4t^2}\right) e^{-t(x+iy)} \, dt \right\},$$

where $s = e^{i\pi/3} t^{1/3}(y + t)$, $s_0 = e^{i\pi/3} t^{4/3}$. This re-arrangement means that all integrals except the last are convergent at $x = y = 0$, and can be expanded in powers of x and y; the last integral is the sum of exponential integrals. We deduce that, as $r \to 0$,

$$\theta(x, y) = -(Q/\pi)[\log r + a_0 + O(r)], \qquad (A.73)$$

where

$$a_0 = \gamma + \frac{3}{4}\left\{ \int_0^1 \frac{Ai(s)}{Ai'(s)} \frac{ds}{s^{1/2}} + \int_1^\infty \left[\frac{Ai(s)}{Ai'(s)} \cdot \frac{1}{s^{1/2}} + \frac{1}{s} \right] ds \right\}$$

$$= -1.0559$$

(Ackerberg et al, 1978), and γ is Euler's constant. To match the inner solution with this we rewrite it in terms of r by setting $r' = \varepsilon^{-1/2} r$, so that (A.68), (A.70) and (A.73) give

$$-(Q/\pi)(\log r + a_0) = 1 + g(\varepsilon) \log(2\varepsilon^{-1/2} r).$$

This requires both that $g(\varepsilon) = -Q/\pi$ (for the log r terms to agree) and that $g(\varepsilon) = -[\log (2\varepsilon^{-1/2}) - a_0]^{-1}$ (for the constant terms to agree). These expressions provide the leading term of the small-ε expansion for Q; rewritten in terms of l, the square root of the Péclet number, it is

$$Q = \pi[\log (4/l) - a_0]^{-1}. \qquad (A.74)$$

By taking further terms in the inner and outer expansions for θ, Ackerberg et al. showed that the next correction to Q is $O[\varepsilon g^2(\varepsilon)]$. However their numerical results suggest that the leading term alone is a better approximation than the two-term expansion for $l \geqslant 1$, while the two are virtually indistinguishable for $l \leqslant 1$, so there is no need to go further here.

Results. In addition to their theory, Ackerberg et al. (1978) performed a very careful experiment to measure the electrochemical mass transfer from effectively two-dimensional electrodes, in known shear flows, at small values of l. Their results are shown in fig. A.17, on a log–log plot. The experimental points agree very well with (A.74) for $l < 2$, and with the results of table A.3 (Springer, 1974) for $l > 6$; the Lévêque boundary layer solution (A.12a) can also be seen to be accurate for $l > 20$, as indicated by Springer's

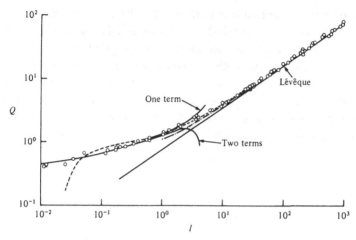

Fig. A.17. Dimensionless heat transfer Q plotted against l, the square root of the Péclet number (logarithmic scales). Continuous curves, present theory (one- and two-term expansions) and Lévêque's boundary layer solution; dot–dash curve, Springer (1974); broken curve, Newman's (1973) results. The circles are experimental points. (After Ackerberg et al., 1978.)

work. The broken curve, which agrees well with the experiments for all $l > 0.5$, was constructed from a numerical solution of Newman (1973). It is clear that the behaviour of two-dimensional hot-films in steady shear flow has been accurately analysed for virtually all values of l, and is an area of fluid mechanics in which theory and experiment agree to a satisfying extent.

It cannot, however, be said that the unsteady fluid mechanics of hot-film anemometers is completely understood. The approximate boundary layer analysis of § A.4 should clearly be improved in order both to explain the troublesome phase difference between experiment and theory (fig. A.13) and to establish it on a more 'rational' basis. Then unsteady analysis should be performed for shorter hot-films; Springer (1973) began to extend his work to the case of fluctuating film temperature in steady flow, but did not complete the solution. Perhaps an unsteady version of the theory of this section would also be possible. Finally, there are three-dimensional effects, which must be important for the blood velocity probes. Again, the analysis of this section may perhaps be extended to cover very small films of finite width. Alternatively, edge effects for longer films of finite width can be analysed if we suppose the film to be long enough in the flow direction for longitudinal diffusion (in the x-direction) to be negligible over most of it, while near the edge $z = 0$ lateral diffusion (in the z-direction) is not negligible. However, we suppose the film to be long enough in the z-direction for the boundary layer solution to be applicable as $z \to +\infty$. The problem is therefore to solve:

$$y\theta_x = \theta_{yy} + \theta_{zz}$$

with

$$\text{on } y = 0 \begin{cases} \theta = 1 & \text{for } z > 0, \\ \theta_y = 0 & \text{for } z < 0; \end{cases}$$

$$\text{as } y \to \infty, \ \theta \to 0.$$

The expected validity of boundary layer theory as $z \to \infty$ suggests we rewrite the equation in terms of similarity variables,

$$\eta = y(9x)^{-1/3}, \qquad \zeta = z(9x)^{-1/3},$$

to obtain

$$\theta_{\eta\eta} + 3\eta^2\theta_\eta + \theta_{\zeta\zeta} = 0.$$

The solution will be sought by taking Fourier transforms in the ζ-direction, and using the Wiener–Hopf technique (cf. § A.5), but before significant progress can be made it will be necessary to investigate the analytical properties of the transformed equation

$$\tilde{\theta}_{\eta\eta} + 3\eta^2\tilde{\theta}_\eta - k^2\tilde{\theta} = 0,$$

where

$$\tilde{\theta}(k, \eta) = \int_{-\infty}^{\infty} e^{-ik\zeta}\theta(\zeta, \eta) \, d\zeta.$$

This equation is not of any standard form.

REFERENCES

Abramowitz, M. & Stegun, I. A. (1965) *Handbook of Mathematical Functions*. National Bureau of Standards, Washington.

Ackerberg, R. C., Patel, R. D. & Gupta, S. K. (1978) The heat/mass transfer to a finite strip at small Péclet numbers. *J. Fluid Mech.* **86**, 49–65.

Ackerberg, R. C. & Phillips, J. H. (1972) The unsteady laminar boundary layer on a semi-infinite flat plate due to small fluctuations in the magnitude of the free-stream velocity. *J. Fluid Mech.* **51**, 137–57.

Adler, M. (1934) Strömung in gekrümmten Rohren. *Z. Angew. Math. Mech.* **14**, 257–75.

Agrawal, Y. C. (1975) Laser velocimeter study of entrance flows in curved pipes. PhD thesis, University of California, Berkeley.

Agrawal, Y., Talbot, L. & Gong, K. (1978) Laser anemometer study of flow development in curved circular pipes. *J. Fluid Mech.* **85**, 497–518.

Anliker, M. (1972) Toward a nontraumatic study of the circulatory system. In *Biomechanics: Its Foundations and Objectives*, ed. Y. C. Fung, N. Perrone & M. Anliker, chapter 15. Prentice-Hall, Englewood Cliffs, New Jersey.

Anliker, M. & Dorfman, M. (1970) Theoretical model studies of wave transmission in semicircular canal ducts. *Ing. Arch.* **39**, 390–406.

Anliker, M., Histand, M. B. & Ogden, E. (1968*a*) Dispersion and attenuation of small artificial pressure waves in the canine aorta. *Circ. Res.* **23**, 539–51.

Anliker, M. & Maxwell, J. A. (1966) Dispersion of waves in blood vessels. In *Biomechanics Symposium*, ed. Y. C. Fung, American Society of Mechanical Engineers, New York.

Anliker, M., Moritz, W. E. & Ogden, E. (1968*b*) Transmission characteristics of axial waves in blood vessels. *J. Biomech.* **1**, 235–46.

Anliker, M. & Raman, K. R. (1966) Korotkoff sounds at diastole – a phenomenon of dynamic instability of fluid-filled shells. *Int. J. Solids Struct.* **2**, 467–91.

Anliker, M., Rockwell, R. L. & Ogden, E. (1971) Nonlinear analysis of flow pulses and shock waves in arteries. I & II. *Z. Angew. Math. Phys.* **22**, 217–46 & 563–81.

Anliker, M., Wells, M. K. & Ogden, E. (1969) The transmission characteristics of large and small pressure waves in the abdominal vena cava. *IEEE Trans. Bio-Med. Eng.* **BME-16**, 262–73.

ASME (1966) *Biomedical Fluid Mechanics Symposium*. American Society of Mechanical Engineers, New York.

Atabek, H. B. (1968) Wave propagation through a viscous liquid contained in a tethered, initially stressed, orthotropic elastic tube. *Biophys. J.* **8**, 626–49.

Atabek, H. B. & Chang, C. C. (1961) Oscillatory flow near the entry of a circular tube. *Z. Angew. Math. Phys.* **12**, 185–201.

Atabek, H. B., Ling, S. C. & Patel, D. J. (1975) Analysis of coronary flow fields in thoracotomized dogs. *Circ. Res.* **37**, 752–61.

Attinger, E. Ò. (ed.) (1964) *Pulsatile Blood Flow*. McGraw-Hill, New York.

Attinger, E. O. (1969) Wall properties of veins. *IEEE Trans. Bio-Med. Eng.* **BME-16**, 253–61.

Barua, S. N. (1963) On secondary flow in stationary curved pipes. *Q. J. Mech. Appl. Math.* **16**, 61–77.

Batchelor, G. K. (1956) On steady laminar flow with closed streamlines at large Reynolds number. *J. Fluid Mech.* **1**, 177–90.

Bellhouse, B. J. (1969) Velocity and pressure distributions in the aortic valve. *J. Fluid Mech.* **37**, 587–600.

Bellhouse, B. J. (1972) The fluid mechanics of heart valves. In *Cardiovascular Fluid Dynamics*, ed. D. H. Bergel, chapter 8. Academic Press, London & New York.

Bellhouse, B. J. & Schultz, D. L. (1967) The determination of fluctuating velocity in air with heated thin film gauges. *J. Fluid Mech.* **29**, 289–95.

Bellhouse, B. J. & Talbot, L. (1969) The fluid mechanics of the aortic valve. *J. Fluid Mech.* **35**, 721–35.

Benedict, J. V., Walker, L. B. & Harris, E. H. (1968) Stress–strain characteristics and tensile strength of unembalmed human tendon. *J. Biomech.* **1**, 53–63.

Bergel, D. H. (1961*a*) The static elastic properties of the arterial wall. *J. Physiol.* **156**, 445–57.

Bergel, D. H. (1961*b*) The dynamic elastic properties of the arterial wall. *J. Physiol.* **156**, 458–69.

Bergel, D. H. (ed.) (1972*a*) *Cardiovascular Fluid Dynamics*, 2 vols. Academic Press, London & New York.

Bergel, D. H. (1972*b*) The properties of blood vessels. In *Biomechanics: Its Foundations and Objectives*, ed. Y. C. Fung, N. Perrone & M. Anliker, chapter 5. Prentice-Hall, Englewood Cliffs, New Jersey.

Bergel, D. H. (1972*c*) The measurement of lengths and dimensions. In *Cardiovascular Fluid Dynamics*, ed. D. H. Bergel, chapter 4. Academic Press, London & New York.

Bergel, D. H. & Schultz, D. L. (1971) Arterial elasticity and fluid dynamics. *Prog. Biophys. Mol. Biol.* **22**, 1–35.

Berger, C., Calvet, P., & Jacquemin, C. (1972) Structure d'écoulements de gaz dans des systèmes tubulaires bifurques. Report of DER en mesures. Office nationale d'études et de recherches aérospatiales – Centre d'études et de recherches de Toulouse.

Bevir, M. K. (1970) The theory of induced voltage electromagnetic flowmeters. *J. Fluid Mech.* **43**, 577–90.

Blennerhassett, P. (1976) Secondary motion and diffusion in unsteady flow in a curved pipe. PhD thesis, Imperial College, London.

Blinks, J. R. & Jewell, B. R. (1972) The meaning and measurement of myocardial contractility. In *Cardiovascular Fluid Dynamics*, ed. D. H. Bergel, chapter 7. Academic Press, London & New York.

Bodoia, J. R. & Osterle, J. F. (1961) Finite difference analysis of plane Poiseuille and Couette flow developments. *Appl. Sci. Res.* **A10**, 265–76.

Brech, R. & Bellhouse, B. J. (1973) Flow in branching vessels. *Cardiovasc. Res.* **7**, 593–600.

Brecher, G. A. (1952) Mechanism of venous flow under different degrees of aspiration. *Am. J. Physiol.* **169**, 423–33.

Brighton, P. W. M. (1977) Boundary layer and stratified flows past obstacles. PhD thesis, Cambridge University.

Brown, S. N. & Stewartson, K. (1973) On the propagation of disturbances in a laminar boundary layer. I. *Proc. Camb. Philos. Soc.* **73**, 493–503.

Brutsaert, D. L. & Sonnenblick, E. H. (1969) Force–velocity–length–time relations of the contractile elements in heart muscle of the cat. *Circ. Res.* **24**, 137–49.

Carew, T. E. (1971) Mechano-chemical response of canine aortic endothelium to elevated shear stress in vitro. PhD thesis, Catholic University of America, Washington DC.

Carlsson, E. (1969) Experimental studies of ventricular mechanics in dogs using the tantalum-labelled heart. *Fed. Proc.* **28**, 1324–9.

Caro, C. G., Fitz-Gerald, J. M. & Schroter, R. C. (1971) Atheroma and arterial wall shear: observation, correlation and proposal of a shear dependent mass transfer mechanism for atherogenesis. *Proc. R. Soc. Lond.* **B177**, 109–59.

Caro, C. G., Foley, T. H. & Sudlow, M. F. (1970) Forearm vasodilation following release of venous congestion. *J. Physiol.* **207**, 257–69

Caro, C. G. & Nerem, R. M. (1973) Transport of ^{14}C-4-cholesterol between serum and wall in perfused dog common carotid artery. *Circ. Res.* **32**, 187–205.

Caro, C. G., Pedley, T. J., Schroter, R. C. & Seed, W. A. (1978) *The Mechanics of the Circulation.* Oxford University Press.

Caro, C. G., Pedley, T. J. & Seed, W. A. (1974) Mechanics of the circulation. In *Cardiovascular Physiology*, ed. A. C. Guyton, *MTP International Review of Science, Physiology*, ser. 1, vol. 1. Butterworths, London.

Carrier, G. F. & Di Prima, R. C. (1956) On the unsteady motion of a viscous fluid past a semi-infinite flat plate. *J. Math. & Phys.* **35**, 359–83.

Carton, T. W., Dainauskas, J. & Clark, J. W. (1962) Elastic properties of single elastic fibres. *J. Appl. Physiol.* **17**, 547–51.

Charm, S. E. & Kurland, G. S. (1972) Blood rheology. In *Cardiovascular Fluid Dynamics*, ed. D. H. Bergel, chapter 15. Academic Press, London & New York.

Clark, C. (1974) Thin film gauges for fluctuating velocity measurements in blood. *J. Phys., Ser. E, Sci. Instrum.* **7**, 548–56.

Clark, C. & Schultz, D. L. (1973) Velocity distribution in aortic flow. *Cardiovasc. Res.* **7**, 601–13.

Clément, J., van de Woestijne, K. P. & Pardaens, J. (1973) A general theory of respiratory mechanics applied to forced expiration. *Respir. Physiol.* **19**, 60–79.

Cokelet, G. R. (1972) The rheology of human blood. In *Biomechanics: Its Foundations and Objectives*, ed. Y. C. Fung, N. Perrone & M. Anliker, chapter 4. Prentice-Hall, Englewood Cliffs, New Jersey.

Cole, J. D. (1968) *Perturbation Methods in Applied Mathematics*. Blaisdell, Waltham, Massachusetts.

Collins, W. M. & Dennis, S. C. R. (1975) The steady motion of a viscous fluid in a curved tube. *Q. J. Mech. Appl. Math.* **28**, 133–56.

Collins, R., Flaud, P., Geiger, D., Kivity, Y. & Oddou, C. (1976) Propagation of shock-like waves in long visco-elastic tubes. *Biomechanika, Sofia.*

Collins, R. & Tedgui A. (1979) The role of axial tension in the opening and closing characteristics of fluid-filled collapsible tubes. Paper presented at Euromech 118, Zuoz, Switzerland.

Conrad, W. A. (1969) Pressure–flow relationships in collapsible tubes. *IEEE Trans. Bio-Med. Eng.* **BME-16**, 284–95.

Conrad, W. A., Cohen, M. L. & McQueen, D. M. (1978) Note on the oscillations of collapsible tubes. *Med. Biol. Eng. Comput.* **16**, 211–14.

Coppel, W. A. (1960) On a differential equation of boundary-layer theory. *Philos. Trans. R. Soc. Lond.* **A253**, 101–36.

Cornhill, J. F. & Roach, M. R. (1976) A quantitative study of the localization of atherosclerotic lesions in the rabbit aorta. *Atherosclerosis* **23**, 489–501.

Cumming, G., Henderson, R., Horsfield, K. & Singhal, S. (1969) The functional morphology of the pulmonary circulation. In *The Pulmonary Circulation and Interstitial Space*, ed. A. P. Fishman & H. H. Hecht. University of Chicago Press.

Davis, S. H (1976) The stability of time-periodic flows. *Annu. Rev. Fluid Mech.* **8**, 57–74.

Dawson, S. V. & Elliott, E. A. (1977) Wave-speed limitation on expiratory flow – a unifying concept. *J. Appl. Physiol., Respir. Environ. Exercise Physiol.* **43**, 498–515.

Dean, W. R. (1928) The stream-line motion of fluid in a curved pipe. *Philos. Mag.* ser. 7, **5**, 673–95.

Dennis, S. C. R. (1972) The motion of a viscous fluid past an impulsively started semi-infinite flat plate. *J. Inst. Math. Appl.* **10**, 105–17.

Douglass, R., (1973) Flow in a human lung model at high Reynolds numbers. MS thesis, Duke University, North Carolina.

Dressler, R. F. (1949) Mathematical solution of the problem of roll waves in inclined open channels. *Commun. Pure Appl. Math.* **2**, 149–94.

Duck, P. W. (1979) Viscous flow through unsteady symmetric channels. *J. Fluid Mech.* (in press).

Eagles, P. M. & Weissman, M. (1975) On the stability of slowly-varying flow: the divergent channel. *J. Fluid Mech.* **69**, 241–62.

Ettinger, S. J. & Suter, P. F. (1970) *Canine Cardiology.* W. B. Saunders, Philadelphia.

Farthing, S. P. (1977) Flow in the thoracic aorta and its relation to atherogenesis. PhD thesis, Cambridge University.

Flaherty, J. E., Keller, J. B. & Rubinow, S. I. (1972a) Post-buckling behaviour of elastic tubes and rings with opposite sides in contact. *SIAM J. Appl. Math.* **23**, 446–55.

Flaherty, J. T., Pierce, J. E., Ferrans, V. J., Patel, D. J., Tucker, W. K. & Fry, D. L. (1972b) Endothelial nuclear patterns in the canine arterial tree with particular reference to hemodynamic events. *Circ. Res.* **30**, 23–33.

Flügge, W. (1973) *Stresses in Shells.* Springer-Verlag, Berlin, Heidelberg & New York.

Fry, D. L. (1968) Acute vascular endothelial changes associated with increased blood velocity gradients. *Circ. Res.* **22**, 165–97.

Fry, D. L. (1973) Responses of the arterial wall to certain physical factors. In *Atherogenesis: Initiating Factors*, ed. R. Porter & J. Knight. Associated Scientific Publishers, Amsterdam.

Fung, Y. C. (ed.) (1966) *Biomechanics Symposium.* American Society of Mechanical Engineers, New York.

Fung, Y. C. (1970) Mathematical representation of the mechanical properties of the heart muscle. *J. Biomech.* **3**, 381–404.

Fung, Y. C., Perrone, N. & Anliker, M. (eds.) (1972) *Biomechanics: Its Foundations and Objectives.* Prentice-Hall, Englewood Cliffs, New Jersey.

Gaster, M. (1974) On the effects of boundary layer growth on flow stability. *J. Fluid Mech.* **66**, 465–80.

Goldstein, S. (1938) *Modern Developments in Fluid Dynamics.* Clarendon Press, Oxford.

Gonzalez, F. (1974) The origin of Korotkoff sounds and their role in sphygmomanometry. PhD Thesis, University of Florida, Gainesville.

Gow, B. S. (1972) The influence of vascular smooth muscle on the viscoelastic properties of blood vessels. In *Cardiovascular Fluid Dynamics*, ed. D. H. Bergel, chapter 12. Academic Press, London & New York.

Gradshteyn, I. S. & Ryzhik, I. M. (1965) *Table of Integrals, Series and Products*. Academic Press, London & New York.

Griffiths, D. J. (1971*a, b, c*) Hydrodynamics of male micturition. I, II & III. *Med. Biol. Eng.* **9**, 581–8, 589–96 & 597–602.

Griffiths, D. J. (1975*a, b, c*) Negative resistance effects in flow through collapsible tubes: I. Relaxation oscillations; II. Two-dimensional theory of flow near an elastic constriction; III. Two-dimensional treatment of the elastic properties of elastic constriction. *Med. Biol. Eng.* **13**, 785–90, 791–6 & 797–802.

Grosch, C. E. & Salwen, H. (1968) The stability of steady and time-dependent plane Poiseuille flow. *J. Fluid Mech.* **34**, 177–205.

Hall, M. G. (1969) The boundary layer over an impulsively started flat plate. *Proc. R. Soc. Lond.* **A310**, 401–14.

Hall, P. (1974) Unsteady viscous flow in a pipe of slowly varying cross-section. *J. Fluid Mech.* **64**, 209–26.

Hall, P. & Parker, K. (1976) The stability of the decaying flow in a suddenly blocked channel. *J. Fluid Mech.* **75**, 305–14.

Harper, J. F. (1963) On boundary layers in two-dimensional flow with vorticity. *J. Fluid Mech.* **17**, 141–53.

Hill, A. V. (1938) Heat of shortening and dynamic constants of muscle. *Proc. R. Soc. Lond.* **B126**, 136–95.

Histand, M. B. & Anliker, M. (1973) Influence of flow and pressure on wave propagation in the canine aorta. *Circ. Res.* **32**, 524–9.

Holt, J. P. (1969) Flow through collapsible tubes and through *in situ* veins. *IEEE Trans. Bio-Med. Eng.* **BME-16**, 274–83.

Horlock, J. H. & Lakshminarayana, B. (1973) Secondary flows: theory, experiment and application in turbomachinery aerodynamics. *Annu. Rev. Fluid Mech.* **5**, 247–80.

Horsfield, K., Dart, G., Olson, D. E., Filley, G. F. & Cumming, G. (1971) Models of the human bronchial tree. *J. Appl. Physiol.* **31**, 207–17.

Hultgren, H. N. (1962) Venous pistol shot sounds. *Am. J. Cardiol.* **10**, 667–72.

Hunter, P. (1975) Finite element analysis of cardiac muscle mechanics. PhD thesis, Oxford University.

Ito, H. (1969) Laminar flow in curved pipes. *Z. Angew. Math. Mech.* **49**, 653–63.

Jackson, P. S. (1973) The flow round obstacles in boundary layers. PhD thesis, University of Cambridge.

Jaffrin, M. Y. & Hennessey, T. V. (1972) Pressure distribution in a model of the central airways for sinusoidal flow. *Bull. Physio-Path. Resp.* **8**, 375–90.

Jones, E., Anliker, M. & Chang, I-Dee (1971) Effects of viscosity and constraints on the dispersion and dissipation of waves in large blood vessels. I & II. *Biophys. J.* **11**, 1085–120 & 1121–34.

Katz, A. I., Chen, Y. & Moreno, A. H. (1969) Flow through a collapsible tube. *Biophys. J.* **9**, 1261–79.

Kivity, Y. & Collins, R. (1974) Steady state fluid flow in viscoelastic tubes. Application to blood flow in human arteries. *Arch. Mech., Warsaw* **26**, 921–31.

Knowlton, F. P. & Starling, E. H. (1912) The influence of variations in temperature and blood-pressure on the performance of the isolated mammalian heart. *J. Physiol.* **44**, 206–19.

Krueger, J. W. & Pollack, G. H. (1975) Myocardial sarcomere dynamics during isometric contraction. *J. Physiol.* **251**, 627–43.

Kuchar, N. R. & Ostrach, S. (1966) Flows in the entrance regions of circular elastic tubes. In *Biomedical Fluid Dynamics Symposium*, pp. 45–69. American Society of Mechanical Engineers, New York.

Lambert, R. K. & Wilson, T. A. (1972) Flow limitation in a collapsible tube. *J. Appl. Physiol.* **33**, 150–3.

Lambert, R. K. & Wilson,T. A. (1973) A model for the elastic properties of the lung and their effect on expiratory flow. *J. Appl. Physiol.* **34**, 34–48.

Learoyd, B. M. & Taylor, M. G. (1966) Alterations with age in the viscoelastic properties of human arterial walls. *Circ. Res.* **18**, 278–92.

Lee, J. S. & Fung, Y. C. (1970) Flow in locally constricted tubes at low Reynolds numbers. *Trans. ASME, Ser E*, **37**, 9–16.

Lévêque, M. A. (1928) Transmission de Chaleur par convection. *Ann. Mines*, **13**, 201–362.

Lewis, J. A. & Carrier, G. F. (1949) Some remarks on the flat plate boundary layer. *Q. Appl. Math.* **7**, 228–34.

Libby, P. A. & Fox, H. (1963) Some perturbation solutions in laminar boundary-layer theory. I. The momentum equation. *J. Fluid Mech.* **17**, 433–49.

Liepmann, H. W. & Skinner, G. T. (1954) Shearing-stress measurements by use of a heated element. *NACA Technical Note*, no. 3268.

Lighthill, M. J. (1954) The response of laminar skin friction and heat transfer to fluctuations in the stream velocity. *Proc. R. Soc. Lond.* **A224**, 1–23.

Lighthill, M. J. (1958) *Fourier Analysis and Generalised Functions.* Cambridge University Press.

Lighthill, M. J. (1975) *Mathematical Biofluiddynamics.* Society for Industrial and Applied Mathematics, Philadelphia.

Lighthill, M. J. (1978) *Waves in Fluids.* Cambridge University Press.

Lin, C. C. (1956) Motion in the boundary layer with a rapidly oscillating external flow. *Proceedings of the 11th International Congress on Theoretical and Applied Mechanics, Brussels*, vol. 4, p. 155.

Ling, S. C. (1963) Heat transfer from a small isothermal spanwise strip on an insulated boundary. *Trans. ASME, Ser. C, J. Heat Transfer.* **85**, 230–6.

Ling, S. C. & Atabek, H. B. (1972) Nonlinear analysis of pulsatile flow in arteries. *J. Fluid Mech.* **55**, 493–511.

Ling, S. C., Atabek, H. B., Fry, D. L., Patel, D. J. & Janicki, J. S. (1968) Application of heated film velocity and shear probes to hemodynamic studies. *Circ. Res.* **23**, 789–801.

Ling, S. C., Atabek, H. B., Letzing, W. G. & Patel, D. J. (1973) Nonlinear analysis of aortic flow in living dogs. *Circ. Res.* **33**, 198–212.

Love, A. E. H. (1927) *Treatise on the Mathematical Theory of Elasticity.* Reprinted by Dover Publications, New York.

Lusza, G. (1974) *X-Ray Anatomy of the Vascular System.* Butterworths, London.

Lutz, R. J., Cannon, J. N., Bischoff, K. B. & Dedrick, R. L. (1977) Wall shear stress distribution in a model canine artery during steady flow. *Circ. Res.* **41**, 391–9.

Lyne, W. H. (1971) Unsteady viscous flow in a curved pipe. *J. Fluid Mech.* **45**, 13–31.

McConalogue, D. J. & Srivastava, R. S. (1968) Motion of fluid in a curved tube. *Proc. R. Soc. Lond.* **A307**, 37–53

McCutcheon, E. P. & Rushmer, R. F. (1967) Korotkoff sounds: an experimental critique. *Circ. Res.* **20**, 149–69.

McDonald, D. A. (1960, 1974) *Blood Flow in Arteries,* 2 edns. Arnold, London.

McDonald, D. A. & Gessner, U. (1968) Wave attenuation in visco-elastic arteries. In *Hemorheology,* ed. A. L. Copley, pp. 113–25. Pergamon Press, Oxford.

Maloney, J. E., Rooholamini, S. A. & Wexler, L. (1970) Pressure-diameter relations of small blood vessels in isolated dog lung. *Microvasc. Res.* **2**, 1–12.

Manton, M. J. (1971) Low Reynolds number flow in slowly varying axisymmetric tubes. *J. Fluid Mech.* **49**, 451–9.

Mills, C. J. (1972) Measurement of pulsatile flow and flow velocity. In *Cardiovascular Fluid Dynamics,* ed. D. H. Bergel, chapter 3. Academic Press, London & New York.

Mills, C. J., Gabe, I. T., Gault, J. H., Mason, D. T., Ross, J., Braunwald, E. & Shillingford, J. P. (1970) Pressure-flow relationships and vascular impedance in man. *Cardiovasc. Res.* **4**, 405–17.

Mills, C. J. & Shillingford, J. P. (1967) A catheter-tip electro-magnetic velocity probe and its evaluation. *Cardiovasc. Res.* **1**, 263–73.

Milnor, W. R. (1972) Pulmonary hemodynamics. In *Cardiovascular Fluid Dynamics,* ed. D. H. Bergel, chapter 18. Academic Press, London & New York.

Milnor, W. R. (1975) Arterial impedance as ventricular afterload. *Circ. Res.* **36**, 565–70.

Minton, P. & Selvalingam, S. (1970) Flow in an oscillating pipe. In *The Measurement of Pulsating Flow*. Institute of Measurement and Control Symposium, University of Surrey.

Moore, D. W. (1963) The boundary layer on a spherical gas bubble. *J. Fluid Mech.* **16**, 161–76.

Moore, F. K. (1951) Unsteady laminar boundary layer flow. *NACA Technical Note*, no. 2471.

Moore, F. K. (1957) Aerodynamic effects of boundary layer unsteadiness. *Proceedings of the 6th Anglo-American Conference, Royal Aeronautical Society, Folkestone*, Royal Aeronautical Society, London, pp. 439–76.

Moreno, A. H., Katz, A. I., Gold, L. D. & Reddy, R. V. (1970) Mechanics of distension of dog veins and other very thin-walled tubular structures. *Circ. Res.* **27**, 1069–80.

Murata, S., Miyake, Y. & Inaba, T. (1976) Laminar flow in a curved pipe with varying curvature. *J. Fluid Mech.* **73**, 735–52.

Nerem, R. M., Mosberg, A. T. & Schwerin, W. D. (1976) Transendothelial transport of ^{131}I-albumin. *Biorheology* **13**, 71–7.

Nerem, R. M., Rumberger, J. A., Gross, D. R., Hamlin, R. L. & Geiger, G. L. (1974a) Hot-film anemometer velocity measurements of arterial blood flow in horses. *Circ. Res.* **34**, 193–203.

Nerem, R. M., Rumberger, J. A., Gross, D. R., Hamlin, R. L. & Geiger, G. L. (1974b) Hot-film measurements of coronary blood flow in horses. In *Fluid Dynamic Aspects of Arterial Disease, Proceedings of a Specialists' Meeting*, Columbus, Ohio.

Nerem, R. M. & Seed, W. A. (1972) An *in vivo* study of the nature of aortic flow disturbances. *Cardiovasc. Res.* **6**, 1–14.

Nerem, R. M., Seed, W. A. & Wood, N. B. (1972) An experimental study of the velocity distribution and transition to turbulence in the aorta. *J. Fluid Mech.* **52**, 137–60.

Newman, D. L. & Bowden, L. N. R. (1973) Effect of reflection from an unmatched junction on the abdominal aortic impedance. *Cardiovasc. Res.* **7**, 827–33.

Newman, D. L., Gosling, R. G., Bowden, N. L. R. & King, D. H. (1973) Pressure amplitude increase on unmatching the aortic-iliac junction of the dog. *Cardiovasc. Res.* **7**, 6–13.

Newman, H. A. I. & Zilversmit, D. B. (1962) Quantitative aspects of cholesterol flux in rabbit atheromatous lesions. *J. Biol. Chem.* **237**, 2078–84.

Newman, J. (1973) The fundamental principles of current distribution and mass transport in electrochemical cells. In *Electroanalytical Chemistry*, ed. A. J. Bard, vol. 6, p. 187. Marcel Dekker, New York.

Noble, M. I. M. (1968) The contribution of blood momentum to left ventricular ejection in the dog. *Circ. Res.* **23**, 663–70.

Noble, M. I. M., Trenchard, D. & Guz, A. (1966) Left ventricular ejection in conscious dogs. I. Measurement and significance of the maximum acceleration of blood from the left ventricle. *Circ. Res.* **19**, 139–47.

Obremski, H. J., Morkovin, M. V. & Landahl, M. (1969) Portfolio of the stability characteristics of incompressible boundary layers. *AGAR-Dograph*, no. 134.

Olsen, J. H. & Shapiro, A. H. (1967) Large amplitude unsteady flow in liquid filled elastic tubes. *J. Fluid Mech.* **29**, 513–38.

Olson, D. E. (1971) Fluid mechanics relevant to respiration – flow within curved or elliptical tubes and bifurcating systems. PhD thesis, Imperial College, London.

Olson, R. M. (1968) Aortic blood pressure and velocity as a function of time and position. *J. Appl. Physiol.* **24**, 563–9.

O'Rourke, M. F. and Taylor, M. G. (1966) Vascular impedance of the femoral bed. *Circ. Res.* **18**, 126–39.

O'Rourke, M. F. & Taylor, M. G. (1967) Input impedance of the systemic circulation. *Circ. Res.* **20**, 365–80.

Ostrach, S. (1964) Laminar flows with body forces. In *Theory of Laminar Flows*, ed. F. K. Moore, *High Speed Aerodynamics and Jet Propulsion*, vol. IV. Oxford University Press.

Pacome, J-J. (1975) Structures d'écoulement et pertes de charges calculées dans le modèle d'arbre bronchique de Weibel. Doctoral thesis, Paul Sabatier University, Toulouse.

Parker, K. H. (1977) Instability in arterial blood flow. In *Cardiovascular Flow Dynamics and Measurement*, ed. N. H. S. Hwang & N. A. Normann, pp 633–63. University Park Press, Baltimore, Maryland.

Patel, D. J., de Freitas, F. M. & Fry, D. L. (1963*a*) Hydraulic input impedance to aorta and pulmonary artery in dogs. *J. Appl. Physiol.* **18**, 134–40.

Patel, D. J., de Freitas, F. M., Greenfield, J. C. & Fry, D. L. (1963*b*) Relationship of radius to pressure along the aorta in living dogs. *J. Appl. Physiol.* **18**, 1111–17.

Patel, D. J. & Fry, D. L. (1966) Longitudinal tethering of arteries in dogs. *Circ. Res.* **19**, 1011–21.

Patel, D. J., Janicki, J. S., Vaishnav, R. N. & Young, J. T. (1973) Dynamic anisotropic viscoelastic properties of the aorta in living dogs. *Circ. Res.* **32**, 93–107.

Patel, D. J. & Vaishnav, R. N. (1972) The rheology of large blood vessels. In *Cardiovascular Fluid Dynamics*, ed. D. H. Bergel, chapter 11. Academic Press, London & New York.

Pedley, T. J. (1972*a*) On the forced heat transfer from a hot film embedded in the wall in two-dimensional unsteady flow. *J. Fluid Mech.* **55**, 329–57.

Pedley, T. J. (1972*b*) Two-dimensional boundary layers in a free stream which oscillates without reversing. *J. Fluid Mech.* **55**, 359–83.

Pedley, T. J. (1976a) Viscous boundary layers in reversing flow. *J. Fluid Mech.* **74**, 59–79.

Pedley, T. J. (1976b) Heat transfer from a hot film in reversing shear flow. *J. Fluid Mech.* **78**, 513–34.

Pedley, T. J. (1977) Pulmonary fluid dynamics. *Annu. Rev. Fluid Mech.* **9**, 229–74.

Pedley, T. J. (1978) The fluid mechanics of circulatory systems. In *Comparative Physiology – Water, Ions and Fluid Mechanics*, ed. K. Schmidt-Nielsen, L. Bolis & S. H. P. Maddrell. Cambridge University Press.

Pedley, T. J., Schroter, R. C. & Sudlow, M. F. (1970a) Energy losses and pressure drop in models of human airways. *Respir. Physiol.* **9**, 371–86.

Pedley, T. J., Schroter, R. C. & Sudlow, M. F. (1970b) The prediction of pressure drop and variation of resistance within the human bronchial airways. *Respir. Physiol.* **9**, 387–405.

Pedley, T. J., Schroter, R. C. & Sudlow, M. F. (1971) Flow and pressure drop in systems of repeatedly branching tubes. *J. Fluid Mech.* **46**, 365–83.

Pedley, T. J., Schroter, R. C. & Sudlow, M. F. (1977) Gas flow and mixing in the airways. In *Bio-engineering Aspects of the Lung*, ed. J. B. West, chapter 3. Marcel Dekker, New York.

Pedley, T. J. & Seed, W. A. (1977) The fluid mechanics of left ventricular ejection. In *Cardiovascular and Pulmonary Dynamics*, ed. M. Y. Jaffrin, pp. 311–19. *Proceedings of Euromech 92.* Éditions INSERM. Institut national de la santé et de la recherche medicale, Paris.

Peronneau, P., Deloche, A., Bui-Mong-Hung & Hinglais, J. (1969) Débitmétrie ultrasonore: développements et applications experimentales. *Eur. Surg. Res.* **1**, 147–56.

Peronneau, P., Leger, F., Hinglais, J., Pellet, M., & Schwartz, P. Y. (1970) Vélocimètre sanguin à effet Doppler à émission ultrasonore pulsée. *Onde électr.*, **50**, 369–89.

Porter, B. (1967) *Stability Criteria for Linear Dynamical Systems.* Oliver & Boyd, Edinburgh.

Porter, R. & Knight, J. (eds.) (1973) *Atherogenesis: Initiating Factors*, CIBA Foundation Symposium. Associated Scientific Publishers, Amsterdam.

Proudman, I. & Pearson, J. R. A. (1957) Expansions at small Reynolds numbers for the flow past a sphere and a circular cylinder. *J. Fluid Mech.* **2**, 237–62.

Raines, J. K., Jaffrin, M. Y. & Shapiro, A. H. (1974) A computer simulation of arterial dynamics in the human leg. *J. Biomech.* **7**, 77–91.

Reiss, L. P. & Hanratty, T. J. (1962) Measurement of instantaneous rates of mass transfer to a small sink on a wall. *AIChEJ.* **8**, 245–7.

Reyn, J. W. (1974) On the mechanism of self-excited oscillations in the flow through collapsible tubes. *Delft Prog. Rep.* **F1**, 51–67.

Riley, N. (1965) Oscillating viscous flows. *Mathematika* **12**, 161–75.

Riley, N. & Dennis, S. C. R. (1976) Flow in a curved pipe at high Dean numbers. Paper presented to the Workshop on Viscous Interaction and Boundary-layer Separation, Columbus, Ohio.

Rittgers, S. E., Karayannacos, P. E., Barrera, J. G., Talukder, N., Nerem, R. M. & Vasko, J. S. (1976) Acute and chronic studies of velocity profiles in arterial vein grafts. In *Proceedings of the 29th American Conference on Engineering in Medicine and Biology, Boston.* Alliance for Engineering in Medicine and Biology, Chevy Chase, Maryland.

Roach, M. R. (1972) Poststenotic dilatation in arteries. In *Cardiovascular Fluid Dynamics,* ed. D. H. Bergel, chapter 13. Academic Press, London & New York.

Roach, M. R. & Burton, A. C. (1957) The reason for the shape of the distensibility curves of arteries. *Can. J. Biochem. Physiol.* **35**, 181–90.

Rosenhead, L. (ed.) (1963) *Laminar Boundary Layers.* Clarendon Press, Oxford.

Rowe, M. (1970) Measurements and computations of flow in pipe bends. *J. Fluid Mech.* **43**, 771–83.

Rubinow, S. I. & Keller, J. B. (1972) Flow of a viscous fluid through an elastic tube with applications to blood flow. *J. Theor. Biol.* **35**, 299–313.

Rumberger, J. A. (1976) A non-linear mathematical model of coronary blood flow. PhD thesis, Ohio State University.

Rumberger, J. A. & Nerem, R. M. (1977) A method-of-characteristics calculation of coronary blood flow. *J. Fluid Mech.* **82**, 429–48.

Sarpkaya, T. (1966) Experimental determination of the critical Reynolds number for pulsating Poiseuille flow. *Trans. ASME, Ser. D, J. Basic Eng.* **88**, 589–98.

Schaaf, B. W. & Abbrecht, P. H. (1972) Digital computer simulation of human systemic arterial pulse wave transmission: a nonlinear model. *J. Biomech.* **5**, 345–64.

Scherer, P. W. (1972) A model for high Reynolds number flow in a human bronchial bifurcation. *J. Biomech.* **5**, 223–9.

Scherer, P. W., Kamm, R. D. & Shapiro, A. H. (1975a) External pneumatic compression for the prevention of deep venous thrombosis. *Proceedings of the San Diego Biomedical Symposium,* vol. 14. Academic Press, London & New York.

Scherer, P. W., Shendalman, L. H., Greene, N. M. & Bouhuys, A., (1975b) Measurement of axial diffusivities in a model of the bronchial airways. *J. Appl. Physiol.* **38**, 719–23.

Schlichting, H. (1968) *Boundary Layer Theory,* 6th edn. McGraw-Hill, New York.

Schoenberg, M. (1968) Pulse wave propagation in elastic tubes having longitudinal changes in area and stiffness. *Biophys. J.* **8**, 991–1008.

Schoendorfer, D. W. & Shapiro, A. H. (1977) The collapsible tube as a prosthetic vocal source. *Proceedings of the San Diego Biomedical Symposium,* vol. 16, 349–56. Academic Press, London & New York.

Schreck, R. M. & Mockros, L. F. (1970) Fluid dynamics in the upper pulmonary airways. In *AIAA 3rd Fluid and Plasma Dynamics Conference, Los Angeles*. American Institute of Aeronautics and Astronautics, New York.

Schroter, R. C. & Sudlow, M. F. (1969) Flow patterns in models of the human bronchial airways. *Respir. Physiol.* **7**, 341–55.

Schultz, D. L. (1972) Pressure and flow in large arteries. In *Cardiovascular Fluid Dynamics*, ed. D. H. Bergel, chapter 9. Academic Press, London & New York.

Schultz, D. L., Tunstall-Pedoe, D. L., Lee, G. de J., Gunning, A. J. & Bellhouse, B. J. (1969) Velocity distribution and transition in the arterial system. In *Circulatory and Respiratory Mass Transport*, ed. G. E. W. Wolstenholme & J. Knight, *CIBA Symposium*. J. & A. Churchill, Edinburgh.

Schwartz, C. J., Bell, F. P., Somer, J. B. & Gerrity, R. (1974) Focal and regional differences in aortic permeability to macromolecules. In *Fluid Dynamic Aspects of Arterial Disease, Proceedings of a Specialists' Meeting*, Columbus, Ohio.

Seed, W. A. & Wood, N. B. (1970a) Development and evaluation of a hot-film velocity probe for cardiovascular studies. *Cardiovasc. Res.* **4**, 253–63.

Seed, W. A. and Wood, N. B. (1970b) Use of a hot film velocity probe for cardiovascular studies. *J. Phys., Ser. E., Sci. Instrum.* **3**, 377–84.

Seed, W. A. & Wood, N. B. (1971) Velocity patterns in the aorta. *Cardiovasc. Res.* **5**, 319–30.

Seeley, B. D. & Young, D. F. (1976) Effect of geometry on pressure losses across models of arterial stenoses. *J. Biomech.* **9**, 439–48.

Seminara, G. & Hall, P. (1976) Centrifugal instability of a Stokes layer: linear theory. *Proc. R. Soc. Lond.* **A350**, 299–316.

Seminara, G. & Hall, P. (1977) Centrifugal instability of a Stokes layer: non-linear theory. *Proc. R. Soc. Lond.* **A354**, 119–26.

Seymour, B. R. (1975) Unsteady flow in flexible tubes: a modulated simple wave. *Int. J. Eng. Sci.* **13**, 579–94.

Shapiro, A. H. (1977) Steady flow in collapsible tubes. *Trans. ASME, Ser. K., J. Biomech. Eng.* **99**, 126–47.

Shayo, L. K. & Ellen, C. H. (1974) The stability of finite length circular cross-section pipes conveying inviscid fluid. *J. Sound Vib.* **37**, 535–45.

Singh, M. P. (1974) Entry flow in a curved pipe. *J. Fluid Mech.* **65**, 517–39.

Skalak, R. (1972) Synthesis of a complete circulation. In *Cardiovascular Fluid Dynamics*, ed. D. H. Bergel, chapter 19. Academic Press, London & New York.

Skalak, R. & Stathis, T. (1966) A porous tapered elastic tube model of a vascular bed. In *Biomechanics Symposium*, ed. Y. C. Fung. American Society of Mechanical Engineers, New York.

Smith, F. T. (1974) Laminar flow over a small hump on a flat plate. *J. Fluid Mech.* **57**, 803–24.

Smith, F. T. (1975) Pulsatile flow in curved pipes. *J. Fluid Mech.* **71**, 15–42.

Smith, F. T. (1976a) Steady motion within a curved pipe. *Proc. R. Soc. Lond.* **A347**, 345–70.

Smith, F. T. (1976b) Fluid flow into a curved pipe. *Proc. R. Soc. Lond.* **A351**, 71–87.

Smith, F. T. (1976c) Pipeflows distorted by non-symmetric indentation or branching. *Mathematika* **23**, 62–83.

Smith, F. T. (1976d, e) Flow through constricted or dilated pipes and channels. I & II. *Q. J. Mech. Appl. Math.* **29**, 343–64 and 365–83.

Smith, F. T. (1977a) Steady motion through a branching tube. *Proc. R. Soc. Lond.* **A355**, 167–87.

Smith, F. T. (1977b) Upstream interactions in channel flows. *J. Fluid Mech.* **79**, 631–55.

Smith, F. T., Sykes, R. I. & Brighton, P. W. M. (1977) A two-dimensional boundary layer encountering a three-dimensional hump. *J. Fluid Mech.* **83**, 163–76.

Smith, K. A., Colton, C. K. & Freedman, R. W. (1974) Shear stress measurements at bifurcations. In *Fluid Dynamic Aspects of Arterial Disease, Proceedings of a Specialists' Meeting*, Columbus, Ohio.

Sobey, I. J. (1976a) Inviscid secondary flow in a tube of slowly varying ellipticity. *J. Fluid Mech.* **73**, 621–39.

Sobey, I. J. (1976b) Bio-fluid dynamics of bifurcations. PhD thesis, University of Cambridge.

Sobey, I. J. (1977) Laminar boundary-layer flow past a two-dimensional slot. *J. Fluid Mech.* **83**, 33–47.

Sokolnikoff, I. S. (1956) *Mathematical Theory of Elasticity*. McGraw-Hill, New York.

Springer, S. G. (1973) The solution of heat transfer problems by the Wiener–Hopf technique. PhD thesis, Imperial College, London.

Springer, S. G. (1974) The solution of heat-transfer problems by the Wiener–Hopf technique. II. Trailing edge of a hot film. *Proc. R. Soc. Lond.* **A337**, 395–412.

Springer, S. G. & Pedley, T. J. (1973) The solution of heat-transfer problems by the Wiener–Hopf technique. I. Leading edge of a hot film. *Proc. R. Soc. Lond.* **A333**, 347–62.

Stahl, W. R. (1967) Scaling of respiratory variables in mammals. *J. Appl. Physiol.* **22**, 453–60.

Steinfeld, L., Alexander, H. & Cohen, M. L. (1974) Updating sphygmomanometry. *Am. J. Cardiology* **33**, 107–10.

Stewartson, K. (1957) On asymptotic expansions in the theory of boundary layers. *J. Math. & Phys.* **36**, 173–91.

Stewartson, K. (1958) On rotating laminar boundary layers. In *Proceedings of the IUTAM Symposium on Boundary Layer Research*, ed. H. Görtler, p. 59. Springer-Verlag, Berlin, Heidelberg & New York.

Stewartson, K. (1968) On the flow near the trailing edge of a flat plate. *Proc. R. Soc. Lond.* **A306**, 275–90.

Stewartson, K. (1973) On the impulsive motion of a flat plate in a viscous fluid. II. *Q. J. Mech. Appl. Math.* **26**, 143–52.

Stewartson, K. (1974) Multistructured boundary layers on flat plates and related bodies. *Adv. Appl. Mech.* **14**, 145–239.

Streeter, D. D., Vaishnav, R. N., Patel, D. J., Spotnitz, H. M., Ross, J. & Sonnenblick, E. H. (1970) Stress distribution in the canine left ventricle during diastole and systole. *Biophys. J.* **10**, 345–63.

Streeter, V. L., Keitzer, W. F. & Bohr, F. F. (1963) Pulsatile pressure and flow through distensible vessels. *Circ. Res.* **13**, 3–20.

Talukder, N. (1975) An investigation on the flow characteristics in arterial branchings. *Trans. ASME* **75-APMB-4**.

Talukder, N. & Nerem, R. M. (1978) Flow characteristics in vascular graft models. *Digest of the First International Conference on Mechanics in Medicine and Biology, Aachen*, pp. VII 281–4.

Taylor, D. E. M. & Wade, J. D. (1970) The pattern of flow around the atrioventricular valves during diastolic ventricular filling. *J. Physiol.* **207**, 71–2.

Taylor, G. I. (1929) The criterion for turbulence in curved pipes. *Proc. R. Soc. Lond.* **A124**, 243–9.

Taylor, M. G. (1965) Wave travel in a non-uniform transmission line, in relation to pulses in arteries. *Phys. Med. Biol.* **10**, 539–50.

Taylor, M. G. (1966) The input impedance of an assembly of randomly branching elastic tubes. *Biophys. J.* **6**, 29–51.

Ur, A. & Gordon, M. (1970) Origin of Korotkoff sounds. *Am. J. Physiol.* **218**, 524–9.

Van Dyke, M. (1970) Entry flow in a channel. *J. Fluid Mech.* **44**, 813–23.

Van Dyke, M. (1975) *Perturbation Methods in Fluid Mechanics*, 2nd edn. Parabolic Press, Stanford, California.

Van Dyke, M. (1978) Extended Stokes series: laminar flow through a loosely coiled pipe. *J. Fluid Mech.* **86**, 129–45.

van Steenhofen, A. A. & van Dongen, M. E. H. (1979) Model studies of the closing behaviour of the aortic valve. *J. Fluid Mech.* **90**, 21–32.

Weibel, E. R. (1963) *Morphometry of the human lung*. Springer-Verlag, Berlin, Heidelberg & New York.

Weinbaum, S. & Caro, C. G. (1976) A macromolecule transport model for the arterial wall and endothelium based on the ultrastructural specialization observed in electron microscopic studies. *J. Fluid Mech.* **74**, 611–40.

Weinbaum, S. & Parker, K. H. (1975) The laminar decay of suddenly blocked channel and pipe flows. *J. Fluid Mech.* **69**, 729–52.

Wells, M. K., Winter, D. C., Nelson, A. W. & McCarthy, T. C. (1974) Hemodynamic patterns in coronary arteries. In *Fluid Dynamic Aspects of Arterial Disease, Proceedings of a Specialists' Meeting*, Columbus, Ohio.

West, J. B., Glazier, J. B., Hughes, J. M. B. & Maloney, J. E. (1969) Pulmonary capillary flow, diffusion ventilation and gas exchange. In *Circulatory and Respiratory Mass Transport*, ed. G. E. W. Wolstenholme & J. Knight, *CIBA Symposium*. J. & A. Churchill, Edinburgh.

Wexler, L., Bergel, D. H., Gabe, I. T., Makin, G. S. & Mills, C. J. (1968) Velocity of blood flow in normal human venae cavae. *Circ. Res.* **23**, 349–59.

White, C. M. (1929) Streamline flow through curved pipes. *Proc. R. Soc. Lond.* **A123**, 645–63.

Whitham, G. B. (1974) *Linear and Non-Linear Waves*. Wiley, New York.

Whitmore, R. L. (1968) *Rheology of the Circulation*. Pergamon Press, Oxford.

Wiederhielm, C. A. (1972) The interstitial space. In *Biomechanics: Its Foundations and Objectives*, ed. Y. C. Fung, N. Perrone & M. Anliker, chapter 11. Prentice-Hall, Englewood Cliffs, New Jersey.

Wiener, F., Morkin, E., Skalak, R. & Fishman, A. P. (1966) Wave propagation in the pulmonary circulation. *Circ. Res.* **19**, 834–50.

Wild, R., Pedley, T. J. & Riley, D. S. (1977) Viscous flow in collapsible tubes of slowly-varying elliptical cross-section. *J. Fluid Mech.* **81**, 273–94.

Wilson, S. D. R. (1971) Entry flow in a channel. II. *J. Fluid Mech.* **46**, 787–99.

Womersley, J. R. (1955) Method for the calculation of velocity, rate of flow and viscous drag in arteries when the pressure gradient is known. *J. Physiol.* **127**, 553–63.

Womersley, J. R. (1957) The mathematical analysis of the arterial circulation in a state of oscillatory motion. *Wright Air Development Centre, Technical Report* WADC-TR 56-614.

Yao, L-S. & Berger, S. A. (1975) Entry flow in a curved pipe. *J. Fluid Mech.* **67**, 177–96.

Young, D. F. & Tsai, F. Y. (1973a, b) Flow characteristics in models of arterial stenosis. I. Steady flow. II. Unsteady flow. *J. Biomech.* **6**, 395–410 & 547–60.

Young, T. (1809) On the functions of the heart and arteries. *Philos. Trans. R. Soc. Lond.* **99**, 1–31.

Zeller, H., Talukder, N. & Lorenz, J. (1970) Model studies of pulsating flow in arterial branches and wave propagation in blood vessels. In *Fluid Dynamics of Blood Circulation and Respiratory Flow*, *AGARD Conference Proceedings*, no. 65.

Zweifach, B. W. (1974) Quantitative studies of microcirculatory structure and function. I & II. *Circ. Res.* **34**, 843–57 & 858–66.

INDEX

acceleration, aortic flow, 66
admittance
  characteristic, 100, 102, 110
  effective, 110
  input: aorta, 112-3; pulmonary
    circulation, 114
afterload, ventricular, 62-3
airways
  collapse, 302
  lung, 235-6
Airy functions, 227, 273, 283, 405-7,
  419
alveoli, 1, 32
anemometer
  electromagnetic, 36, 47, 131
  hot-film, 38-9, 47, 131-6, 263, 369-
    422; frequency response, 133, 380;
    unsteady calibration experiments
    371, 396, 401; yawed, 402; see also
    hot-film
  hot-wire 218, 239
  laser-Doppler, 218
  pulsed wire, 239
  ultrasonic Doppler, 131
aorta
  area, 13
  curvature, 9, 46
  Dean number, 183
  diameter, 1, 11
  distensibility, 20, 24
  elastic properties, 99
  entry flow, 44-5, 60, 121
  frequency parameter, 11
  geometry, 4, 9-12
  input impedance, 63
  instability of flow, 48-51, 291-300
  length, 11
  pressure, 34-5

aorta (cont.)
  Reynolds number, 11; steady stream-
    ing, 183
  taper, 9-12
  turbulence, 48-51, 291-2
  velocity, 11, 34
  velocity profiles, 42-6, 60
  velocity waveform, 39-40, 137
  viscoelasticity, 99
  wall shear, 215-7
  wall shear waveform, 156-8
  wall thickness, 11
  wave speed, 11
  Young's modulus, 11, 20
area ratio, bifurcation, 236, 260
arteries, systemic
  anatomy, 9-22
  elastic properties, 13-22
arteriole, 1
  mechanical properties, 11
artery
  backflow, 36, 40, 45, 122
  carotid, 10; mechanical properties,
    11
  collapse, 56-7, 311
  coronary, 8-10, geometry, 12, 14;
    oscillations, 41-2, 124; pulse wave,
    124, 128-30; velocity profiles, 47;
    velocity waveform, 40-1; wall
    shear, 128-30
  elastic properties, 120
  elasticity, orthotropic, 88
  external pressure, 116
  femoral, 10; mechanical properties,
    11; input admittance, 113
  flow rate, 99-100
  geometry, 1
  iliac, 9-10